MAXWELL'S EQUATIONS AND THE PRINCIPLES OF ELECTROMAGNETISM

MAXWELL'S EQUATIONS AND THE PRINCIPLES OF ELECTROMAGNETISM

RICHARD FITZPATRICK, PH.D.
University of Texas at Austin

INFINITY SCIENCE PRESS LLC
Hingham, Massachusetts
New Delhi

Publisher: David Pallai

INFINITY SCIENCE PRESS LLC
11 Leavitt Street
Hingham, MA 02043
Tel. 877-266-5796 (toll free)
Fax 781-740-1677
info@infinitysciencepress.com
www.infinitysciencepress.com

This book is printed on acid-free paper.

Richard Fitzpatrick. *Maxwell's Equations and the Principles of Electromagnetism.*
ISBN: 978-1-934015-20-9

Library of Congress Cataloging-in-Publication Data

Fitzpatrick, Richard.
Maxwell's equations and the principles of electromagnetism / Richard Fitzpatrick.
p. cm.
Includes bibliographical references and index.
ISBN-13: 978-1-934015-20-9 (hardcover with cd-rom : alk. paper)
1. Maxwell equations. 2. Electromagnetic theory. I. Title.
QC670.F545 2008
530.14′1–dc22

2007050220

Printed in the United States of America
08 09 10 5 4 3 2 1

Our titles are available for adoption, license or bulk purchase by institutions, corporations, etc. For additional information, please contact the Customer Service Dept. at 877-266-5796 (toll free).

Requests for replacement of a defective CD-ROM must be accompanied by the original disc, your mailing address, telephone number, date of purchase and purchase price. Please state the nature of the problem, and send the information to Infinity Science Press, 11 Leavitt Street, Hingham, MA 02043.

The sole obligation of Infinity Science Press to the purchaser is to replace the disc, based on defective materials or faulty workmanship, but not based on the operation or functionality of the product.

For Faith

CONTENTS

Chapter 1 INTRODUCTION

The main topic of this book is *Maxwell's Equations*. These are a set of *eight*, scalar, first-order partial differential equations which constitute a *complete* description of classical electric and magnetic phenomena. To be more exact, Maxwell's equations constitute a complete description of the classical behavior of electric and magnetic *fields*.

Electric and magnetic fields were first introduced into electromagnetic theory merely as mathematical constructs designed to facilitate the calculation of the forces exerted between electric charges and between current carrying wires. However, physicists soon came to realize that the physical existence of these fields is key to making Classical Electromagnetism consistent with Einstein's Special Theory of Relativity. In fact, Classical Electromagnetism was the first example of a so-called *field theory* to be discovered in Physics. Other, subsequently discovered, field theories include General Relativity, Quantum Electrodynamics, and Quantum Chromodynamics.

At any given point in space, an electric or magnetic field possesses two properties—a *magnitude* and a *direction*. In general, these properties vary (continuously) from point to point. It is conventional to represent such a field in terms of its components measured with respect to some conveniently chosen set of Cartesian axes (*i.e.*, the standard x-, y-, and z-axes). Of course, the orientation of these axes is *arbitrary*. In other words, different observers may well choose differently aligned coordinate axes to describe the same field. Consequently, the same electric and magnetic fields may have different components according to different observers. It can be seen that any description of electric and magnetic fields is going to depend on two seperate things. Firstly, the nature of the fields themselves, and, secondly, the arbitrary choice of the coordinate axes with respect to which these fields are measured. Likewise, Maxwell's equations—the equations which describe the behavior of electric and magnetic fields—depend on two separate things. Firstly, the fundamental laws of Physics which govern the behavior of electric and magnetic fields, and, secondly, the arbitrary choice of coordinate axes. It would be helpful to be able to easily distinguish those elements of Maxwell's equations which depend on Physics from those which only depend on coordinates. In fact, this goal can be achieved by employing a branch of mathematics called *vector field theory*. This formalism enables Maxwell's equations to be written in a manner which is *completely independent* of the choice of coordinate axes. As an added bonus, Maxwell's equations look a lot simpler when written in a coordinate-free fashion. Indeed, instead of *eight* first-order partial differential equations, there are only *four* such equations within the context of vector field theory.

Electric and magnetic fields are useful and interesting because they interact *strongly* with ordinary matter. Hence, the primary application of Maxwell's equations is the study of this interaction. In order to facilitate this study, materials are generally divided into three broad classes: *conductors*, *dielectrics*, and *magnetic materials*. Conductors contain free charges which drift in response to an applied electric field. Dielectrics are made up of atoms and molecules which develop electric dipole moments in the presence of an applied electric field. Finally, magnetic materials are made up of atoms and molecules which develop magnetic dipole moments in response to an applied magnetic field. Generally speaking, the interaction of electric and magnetic fields with these three classes of materials is usually investigated in two limits. Firstly, the *low-frequency limit*, which is appropriate to the study of the electric and magnetic fields found in conventional electrical circuits. Secondly, the *high-frequency limit*, which is appropriate to the study of the electric and magnetic fields which occur in electromagnetic waves. In the low-frequency limit, the interaction of a conducting body with electric and magnetic fields is conveniently parameterized in terms of its *resistance*, its *capacitance*, and its *inductance*. Resistance measures the resistance of the body to the passage of electric currents. Capacitance measures its capacity to store charge. Finally, inductance measures the magnetic field generated by the body when a current flows through it. Conventional electric circuits are can be represented as networks of pure resistors, capacitors, and inductors.

This book commences in Chapter 1 with a review of vector field theory. In Chapters 2 and 3, vector field theory is employed to transform the familiar laws of electromagnetism (*i.e.*, Coulomb's law, Ampère's law, Faraday's law, *etc.*) into Maxwell's equations. The general properties of these equations and their solutions are then discussed. In particular, it is explained why it is necessary to use fields, rather than forces alone, to fully describe electric and magnetic phenomena. It is also demonstrated that Maxwell's equations are soluble, and that their solutions are unique. In Chapters 4 to 6, Maxwell's equations are used to investigate the interaction of low-frequency electric and magnetic fields with conducting, dielectric, and magnetic media. The related concepts of resistance, capacitance, and inductance are also examined. The interaction of high-frequency radiation fields with various different types of media is discussed in Chapter 8. In particular, the emission, absorption, scattering, reflection, and refraction of electromagnetic waves is investigated in detail. Chapter 7 contains a demonstration that Maxwell's equations conserve both energy and momentum. Finally, in Chapter 9 it is shown that Maxwell's equations are fully consistent with Einstein's Special Theory of Relativity, and can, moreover, be written in a manifestly Lorentz invariant manner. The relativistic form of Maxwell's equations is then used to examine radiation by accelerating charges.

This book is primarily intended to accompany a single-semester upper-division Classical Electromagnetism course for physics majors. It assumes a knowledge of elementary physics, advanced calculus, partial differential equations, vector algebra, vector calculus, and complex analysis.

Much of the material appearing in this book was gleaned from the excellent references listed in Appendix D. Furthermore, the contents of Chapter 2 are partly based on my recollection of a series of lectures given by Dr. Stephen Gull at the University of Cambridge.

Chapter 2

VECTORS AND VECTOR FIELDS

2.1 INTRODUCTION

This chapter outlines those aspects of vector algebra, vector calculus, and vector field theory which are required to derive and understand Maxwell's equations.

2.2 VECTOR ALGEBRA

Physical quantities are (predominately) represented in Mathematics by two distinct classes of objects. Some quantities, denoted *scalars*, are represented by *real numbers*. Others, denoted *vectors*, are represented by directed line elements in space: *e.g.*, $\overrightarrow{\rm PQ}$—see Figure 2.1. Note that line elements (and, therefore, vectors) are *movable*, and do not carry intrinsic position information (*i.e.*, in Figure 2.2, $\overrightarrow{\rm PS}$ and $\overrightarrow{\rm QR}$ are considered to be the *same* vector). In fact, vectors just possess a magnitude and a direction, whereas scalars possess a magnitude but no direction. By convention, vector quantities are denoted by boldfaced characters (*e.g.*, **a**) in typeset documents. Vector addition can be represented using a parallelogram: $\overrightarrow{\rm PR} = \overrightarrow{\rm PQ} + \overrightarrow{\rm QR}$—see Figure 2.2. Suppose that $\mathbf{a} \equiv \overrightarrow{\rm PQ} \equiv \overrightarrow{\rm SR}$, $\mathbf{b} \equiv \overrightarrow{\rm QR} \equiv \overrightarrow{\rm PS}$, and $\mathbf{c} \equiv \overrightarrow{\rm PR}$. It is clear, from Figure 2.2, that vector

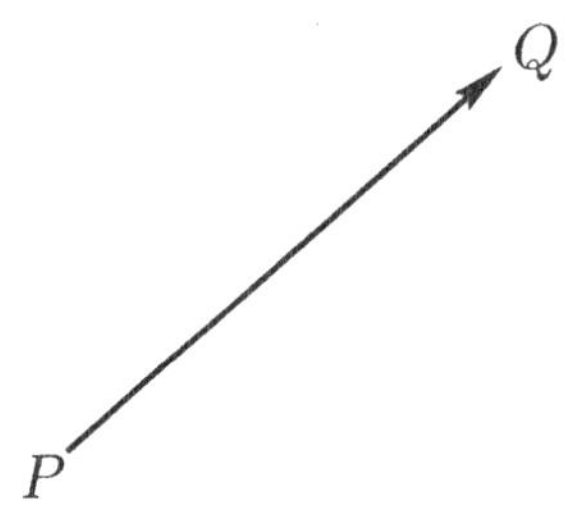

Figure 2.1: *A directed line element.*

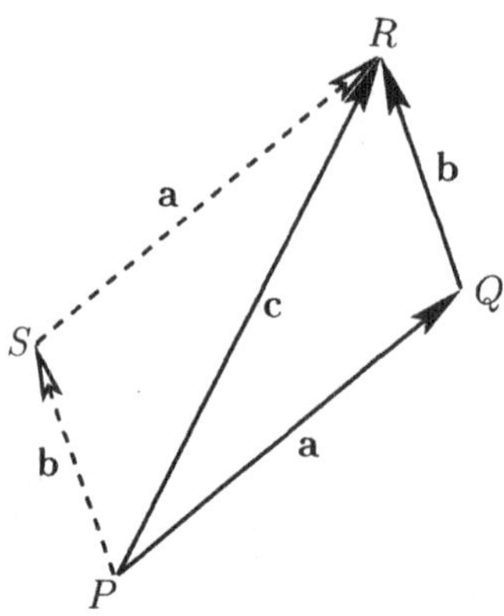

Figure 2.2: *Vector addition.*

addition is *commutative*: *i.e.*, $\mathbf{a} + \mathbf{b} = \mathbf{b} + \mathbf{a}$. It can also be shown that the *associative* law holds: *i.e.*, $\mathbf{a} + (\mathbf{b} + \mathbf{c}) = (\mathbf{a} + \mathbf{b}) + \mathbf{c}$.

There are two approaches to vector analysis. The *geometric* approach is based on line elements in space. The *coordinate* approach assumes that space is defined in terms of Cartesian coordinates, and uses these to characterize vectors. In Physics, we generally adopt the second approach, because it is far more convenient than the first.

In the coordinate approach, a vector is denoted as the row matrix of its components (*i.e.*, perpendicular projections) along each of three mutually perpendicular Cartesian axes (the x-, y-, and z-axes, say):

$$\mathbf{a} \equiv (a_x,\ a_y,\ a_z). \tag{2.1}$$

If $\mathbf{a} \equiv (a_x, a_y, a_z)$ and $\mathbf{b} \equiv (b_x, b_y, b_z)$ then vector addition is defined

$$\mathbf{a} + \mathbf{b} \equiv (a_x + b_x,\ a_y + b_y,\ a_z + b_z). \tag{2.2}$$

If $\mathbf{a}$ is a vector and n is a scalar then the product of a scalar and a vector is defined

$$n\,\mathbf{a} \equiv (n\,a_x,\ n\,a_y,\ n\,a_z). \tag{2.3}$$

Note that $n\,\mathbf{a}$ is interpreted as a vector which is parallel (or antiparallel if $n < 0$) to $\mathbf{a}$, and of length $|n|$ times that of $\mathbf{a}$. It is clear that vector algebra is *distributive* with respect to scalar multiplication: *i.e.*, $n\,(\mathbf{a} + \mathbf{b}) = n\,\mathbf{a} + n\,\mathbf{b}$. It is also easily demonstrated that $(n + m)\,\mathbf{a} = n\,\mathbf{a} + m\,\mathbf{a}$, and $n\,(m\,\mathbf{a}) = (n\,m)\,\mathbf{a}$, where m is a second scalar.

Unit vectors can be defined in the x-, y-, and z-directions as $\mathbf{e}_x \equiv (1, 0, 0)$, $\mathbf{e}_y \equiv (0, 1, 0)$, and $\mathbf{e}_z \equiv (0, 0, 1)$. Any vector can be written in

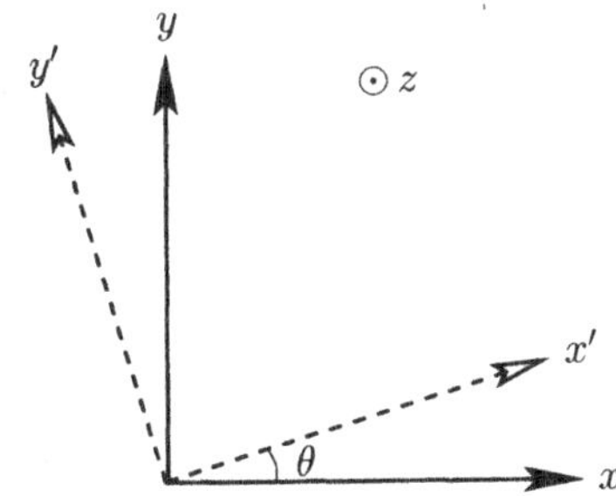

Figure 2.3: *Rotation of the basis about the z-axis.*

terms of these unit vectors:

$$\mathbf{a} = a_x \, \mathbf{e}_x + a_y \, \mathbf{e}_y + a_z \, \mathbf{e}_z. \tag{2.4}$$

In mathematical terminology, three vectors used in this manner form a *basis* of the vector space. If the three vectors are mutually perpendicular then they are termed *orthogonal basis vectors*. However, any set of three non-coplanar vectors can be used as basis vectors.

Examples of vectors in Physics are displacements from an origin,

$$\mathbf{r} = (x, \, y, \, z), \tag{2.5}$$

and velocities,

$$\mathbf{v} = \frac{d\mathbf{r}}{dt} = \lim_{\delta t \to 0} \frac{\mathbf{r}(t + \delta t) - \mathbf{r}(t)}{\delta t}. \tag{2.6}$$

Suppose that we transform to a new orthogonal basis, the x'-, y'-, and z'-axes, which are related to the x-, y-, and z-axes via a rotation through an angle θ around the z-axis—see Figure 2.3. In the new basis, the coordinates of the general displacement $\mathbf{r}$ from the origin are (x', y', z'). These coordinates are related to the previous coordinates via the transformation:

$$x' = x \cos\theta + y \sin\theta, \tag{2.7}$$

$$y' = -x \sin\theta + y \cos\theta, \tag{2.8}$$

$$z' = z. \tag{2.9}$$

We do not need to change our notation for the displacement in the new basis. It is still denoted $\mathbf{r}$. The reason for this is that the magnitude and direction of $\mathbf{r}$ are *independent* of the choice of basis vectors. The coordinates of **r** *do* depend on the choice of basis vectors. However, they must depend in a very specific manner [*i.e.*, Equations (2.7)–(2.9)] which preserves the magnitude and direction of **r**.

Since any vector can be represented as a displacement from an origin (this is just a special case of a directed line element), it follows that the components of a general vector **a** must transform in an analogous manner to Equations (2.7)–(2.9). Thus,

$$a_{x'} = a_x \cos\theta + a_y \sin\theta, \tag{2.10}$$

$$a_{y'} = -a_x \sin\theta + a_y \cos\theta, \tag{2.11}$$

$$a_{z'} = a_z, \tag{2.12}$$

with analogous transformation rules for rotation about the y- and z-axes. In the coordinate approach, Equations (2.10)–(2.12) constitute the *definition* of a vector. The three quantities (a_x, a_y, a_z) are the components of a vector provided that they transform under rotation like Equations (2.10)–(2.12). Conversely, (a_x, a_y, a_z) *cannot* be the components of a vector if they do not transform like Equations (2.10)–(2.12). Scalar quantities are *invariant* under transformation. Thus, the individual components of a vector (a_x, say) are real numbers, but they are *not* scalars. Displacement vectors, and all vectors derived from displacements, automatically satisfy Equations (2.10)–(2.12). There are, however, other physical quantities which have both magnitude and direction, but which are not obviously related to displacements. We need to check carefully to see whether these quantities are vectors.

2.3 VECTOR AREAS

Suppose that we have planar surface of scalar area S. We can define a vector area **S** whose magnitude is S, and whose direction is perpendicular to the plane, in the sense determined by the right-hand grip rule on the rim, assuming that a direction of circulation around the rim is specified—see Figure 2.4. This quantity clearly possesses both magnitude and direction. But is it a true vector? We know that if the normal to the surface makes an angle α_x with the x-axis then the area seen looking along the x-direction is $S \cos\alpha_x$. This is the x-component of **S**. Similarly, if the normal makes an angle α_y with the y-axis then the area seen looking along the y-direction is $S \cos\alpha_y$. This is the y-component of **S**. If we limit ourselves to a surface whose normal is perpendicular to the z-direction then $\alpha_x = \pi/2 - \alpha_y = \alpha$. It follows that $\mathbf{S} = S\,(\cos\alpha, \sin\alpha, 0)$.

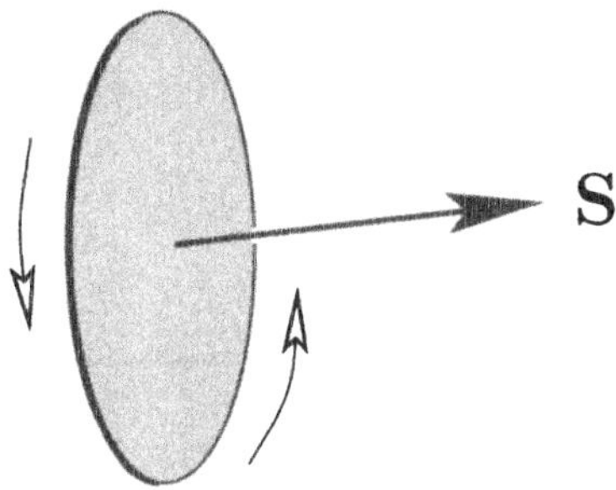

Figure 2.4: *A vector area.*

If we rotate the basis about the z-axis by θ degrees, which is equivalent to rotating the normal to the surface about the z-axis by −θ degrees, then

$$S_{x'} = S\cos(\alpha - \theta) = S\cos\alpha\cos\theta + S\sin\alpha\sin\theta = S_x\cos\theta + S_y\sin\theta, \tag{2.13}$$

which is the correct transformation rule for the x-component of a vector. The other components transform correctly as well. This proves that a vector area is a true vector.

According to the vector addition theorem, the projected area of two plane surfaces, joined together at a line, looking along the x-direction (say) is the x-component of the resultant of the vector areas of the two surfaces. Likewise, for many joined-up plane areas, the projected area in the x-direction, which is the same as the projected area of the *rim* in the x-direction, is the x-component of the resultant of all the vector areas:

$$\mathbf{S} = \sum_i \mathbf{S}_i. \tag{2.14}$$

If we approach a limit, by letting the number of plane facets increase, and their areas reduce, then we obtain a continuous surface denoted by the resultant vector area

$$\mathbf{S} = \sum_i \delta\mathbf{S}_i. \tag{2.15}$$

It is clear that the projected area of the rim in the x-direction is just S_x. Note that the rim of the surface determines the vector area rather than the nature of the surface. So, two different surfaces sharing the same rim both possess the *same* vector area.

In conclusion, a loop (not all in one plane) has a vector area **S** which is the resultant of the vector areas of any surface ending on the loop. The components of $\mathbf{S}$ are the projected areas of the loop in the directions of the basis vectors. As a corollary, a closed surface has $\mathbf{S} = \mathbf{0}$, since it does not possess a rim.

2.4 THE SCALAR PRODUCT

A scalar quantity is invariant under all possible rotational transformations. The individual components of a vector are not scalars because they change under transformation. Can we form a scalar out of some combination of the components of one, or more, vectors? Suppose that we were to define the "percent" product,

$$\mathbf{a} \,\%\, \mathbf{b} = a_x\, b_z + a_y\, b_x + a_z\, b_y = \text{scalar number}, \tag{2.16}$$

for general vectors $\mathbf{a}$ and $\mathbf{b}$. Is $\mathbf{a} \,\%\, \mathbf{b}$ invariant under transformation, as must be the case if it is a scalar number? Let us consider an example. Suppose that $\mathbf{a} = (0, 1, 0)$ and $\mathbf{b} = (1, 0, 0)$. It is easily seen that $\mathbf{a} \,\%\, \mathbf{b} = 1$. Let us now rotate the basis through 45° about the z-axis. In the new basis, $\mathbf{a} = (1/\sqrt{2}, 1/\sqrt{2}, 0)$ and $\mathbf{b} = (1/\sqrt{2}, -1/\sqrt{2}, 0)$, giving $\mathbf{a} \,\%\, \mathbf{b} = 1/2$. Clearly, $\mathbf{a} \,\%\, \mathbf{b}$ is *not* invariant under rotational transformation, so the above definition is a bad one.

Consider, now, the *dot product* or *scalar product*:

$$\mathbf{a} \cdot \mathbf{b} = a_x\, b_x + a_y\, b_y + a_z\, b_z = \text{scalar number}. \tag{2.17}$$

Let us rotate the basis though θ degrees about the z-axis. According to Equations (2.10)–(2.12), in the new basis $\mathbf{a} \cdot \mathbf{b}$ takes the form

$$\begin{aligned}\mathbf{a} \cdot \mathbf{b} &= (a_x \cos\theta + a_y \sin\theta)\,(b_x \cos\theta + b_y \sin\theta) \\ &\quad + (-a_x \sin\theta + a_y \cos\theta)\,(-b_x \sin\theta + b_y \cos\theta) + a_z\, b_z \\ &= a_x\, b_x + a_y\, b_y + a_z\, b_z.\end{aligned} \tag{2.18}$$

Thus, $\mathbf{a} \cdot \mathbf{b}$ *is* invariant under rotation about the z-axis. It can easily be shown that it is also invariant under rotation about the x- and y-axes. Clearly, $\mathbf{a} \cdot \mathbf{b}$ is a true scalar, so the above definition is a good one. Incidentally, $\mathbf{a} \cdot \mathbf{b}$ is the *only* simple combination of the components of two vectors which transforms like a scalar. It is easily shown that the dot product is commutative and distributive: *i.e.*,

$$\begin{aligned}\mathbf{a} \cdot \mathbf{b} &= \mathbf{b} \cdot \mathbf{a}, \\ \mathbf{a} \cdot (\mathbf{b} + \mathbf{c}) &= \mathbf{a} \cdot \mathbf{b} + \mathbf{a} \cdot \mathbf{c}.\end{aligned} \tag{2.19}$$

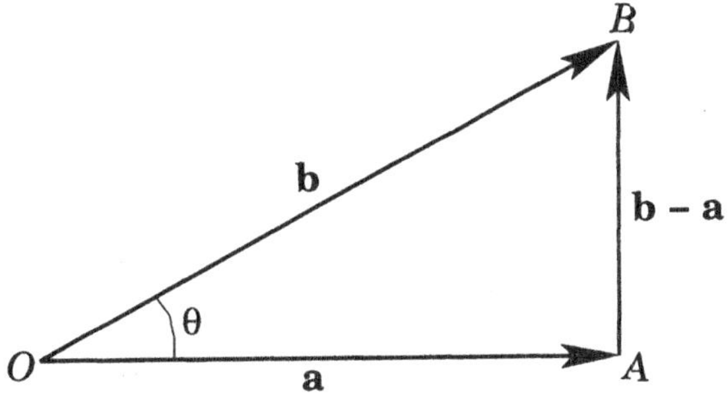

Figure 2.5: *A vector triangle.*

The associative property is meaningless for the dot product, because we cannot have $(\mathbf{a} \cdot \mathbf{b}) \cdot \mathbf{c}$, since $\mathbf{a} \cdot \mathbf{b}$ is scalar.

We have shown that the dot product $\mathbf{a} \cdot \mathbf{b}$ is coordinate independent. But what is the physical significance of this? Consider the special case where $\mathbf{a} = \mathbf{b}$. Clearly,

$$\mathbf{a} \cdot \mathbf{b} = a_x^2 + a_y^2 + a_z^2 = \text{Length } (OP)^2, \tag{2.20}$$

if $\mathbf{a}$ is the position vector of P relative to the origin O. So, the invariance of $\mathbf{a} \cdot \mathbf{a}$ is equivalent to the invariance of the length, or magnitude, of vector $\mathbf{a}$ under transformation. The length of vector $\mathbf{a}$ is usually denoted $|\mathbf{a}|$ ("the modulus of a") or sometimes just a, so

$$\mathbf{a} \cdot \mathbf{a} = |\mathbf{a}|^2 = a^2. \tag{2.21}$$

Let us now investigate the general case. The length squared of AB in the vector triangle shown in Figure 2.5 is

$$(\mathbf{b} - \mathbf{a}) \cdot (\mathbf{b} - \mathbf{a}) = |\mathbf{a}|^2 + |\mathbf{b}|^2 - 2\,\mathbf{a} \cdot \mathbf{b}. \tag{2.22}$$

However, according to the "cosine rule" of trigonometry,

$$(AB)^2 = (OA)^2 + (OB)^2 - 2\,(OA)\,(OB)\,\cos\theta, \tag{2.23}$$

where (AB) denotes the length of side AB. It follows that

$$\mathbf{a} \cdot \mathbf{b} = |\mathbf{a}|\,|\mathbf{b}|\,\cos\theta. \tag{2.24}$$

Clearly, the invariance of $\mathbf{a} \cdot \mathbf{b}$ under transformation is equivalent to the invariance of the angle subtended between the two vectors. Note that if

$\mathbf{a} \cdot \mathbf{b} = 0$ then either $|\mathbf{a}| = 0$, $|\mathbf{b}| = 0$, or the vectors **a** and **b** are perpendicular. The angle subtended between two vectors can easily be obtained from the dot product:

$$\cos\theta = \frac{\mathbf{a} \cdot \mathbf{b}}{|\mathbf{a}|\,|\mathbf{b}|}. \tag{2.25}$$

The work W performed by a constant force **F** moving an object through a displacement **r** is the product of the magnitude of **F** times the displacement in the direction of **F**. If the angle subtended between **F** and **r** is θ then

$$W = |\mathbf{F}|\,(|\mathbf{r}|\,\cos\theta) = \mathbf{F} \cdot \mathbf{r}. \tag{2.26}$$

The rate of flow of liquid of constant velocity **v** through a loop of vector area **S** is the product of the magnitude of the area times the component of the velocity perpendicular to the loop. Thus,

$$\text{Rate of flow} = \mathbf{v} \cdot \mathbf{S}. \tag{2.27}$$

2.5 THE VECTOR PRODUCT

We have discovered how to construct a scalar from the components of two general vectors **a** and **b**. Can we also construct a vector which is not just a linear combination of **a** and **b**? Consider the following definition:

$$\mathbf{a} * \mathbf{b} = (a_x\, b_x,\ a_y\, b_y,\ a_z\, b_z). \tag{2.28}$$

Is $\mathbf{a} * \mathbf{b}$ a proper vector? Suppose that $\mathbf{a} = (0,\ 1,\ 0)$, $\mathbf{b} = (1,\ 0,\ 0)$. Clearly, $\mathbf{a} * \mathbf{b} = \mathbf{0}$. However, if we rotate the basis through 45° about the z-axis then $\mathbf{a} = (1/\sqrt{2},\ 1/\sqrt{2},\ 0)$, $\mathbf{b} = (1/\sqrt{2},\ -1/\sqrt{2},\ 0)$, and $\mathbf{a} * \mathbf{b} = (1/2,\ -1/2,\ 0)$. Thus, $\mathbf{a} * \mathbf{b}$ does not transform like a vector, because its magnitude depends on the choice of axes. So, the above definition is a bad one.

Consider, now, the *cross product* or *vector product*:

$$\mathbf{a} \times \mathbf{b} = (a_y\, b_z - a_z\, b_y,\ a_z\, b_x - a_x\, b_z,\ a_x\, b_y - a_y\, b_x) = \mathbf{c}. \tag{2.29}$$

Does this rather unlikely combination transform like a vector? Let us try rotating the basis through θ degrees about the z-axis using

Equations (2.10)–(2.12). In the new basis,

$$\begin{aligned} c_{x'} &= (-a_x \sin\theta + a_y \cos\theta)\, b_z - a_z\,(-b_x \sin\theta + b_y \cos\theta) \\ &= (a_y\, b_z - a_z\, b_y) \cos\theta + (a_z\, b_x - a_x\, b_z) \sin\theta \\ &= c_x \cos\theta + c_y \sin\theta. \end{aligned} \tag{2.30}$$

Thus, the x-component of $\mathbf{a} \times \mathbf{b}$ transforms correctly. It can easily be shown that the other components transform correctly as well, and that all components also transform correctly under rotation about the y- and z-axes. Thus, $\mathbf{a} \times \mathbf{b}$ is a proper vector. Incidentally, $\mathbf{a} \times \mathbf{b}$ is the *only* simple combination of the components of two vectors that transforms like a vector (which is non-coplanar with $\mathbf{a}$ and $\mathbf{b}$). The cross product is *anticommutative*,

$$\mathbf{a} \times \mathbf{b} = -\mathbf{b} \times \mathbf{a}, \tag{2.31}$$

distributive,

$$\mathbf{a} \times (\mathbf{b} + \mathbf{c}) = \mathbf{a} \times \mathbf{b} + \mathbf{a} \times \mathbf{c}, \tag{2.32}$$

but is *not* associative,

$$\mathbf{a} \times (\mathbf{b} \times \mathbf{c}) \neq (\mathbf{a} \times \mathbf{b}) \times \mathbf{c}. \tag{2.33}$$

Note that $\mathbf{a} \times \mathbf{b}$ can be written in the convenient, and easy-to-remember, determinant form

$$\mathbf{a} \times \mathbf{b} = \begin{vmatrix} \mathbf{e}_x & \mathbf{e}_y & \mathbf{e}_z \\ a_x & a_y & a_z \\ b_x & b_y & b_z \end{vmatrix}. \tag{2.34}$$

The cross product transforms like a vector, which means that it must have a well-defined direction and magnitude. We can show that $\mathbf{a} \times \mathbf{b}$ is *perpendicular* to both $\mathbf{a}$ and $\mathbf{b}$. Consider $\mathbf{a} \cdot \mathbf{a} \times \mathbf{b}$. If this is zero then the cross product must be perpendicular to $\mathbf{a}$. Now

$$\begin{aligned} \mathbf{a} \cdot \mathbf{a} \times \mathbf{b} &= a_x\,(a_y\, b_z - a_z\, b_y) + a_y\,(a_z\, b_x - a_x\, b_z) + a_z\,(a_x\, b_y - a_y\, b_x) \\ &= 0. \end{aligned} \tag{2.35}$$

Therefore, $\mathbf{a} \times \mathbf{b}$ is perpendicular to $\mathbf{a}$. Likewise, it can be demonstrated that $\mathbf{a} \times \mathbf{b}$ is perpendicular to $\mathbf{b}$. The vectors $\mathbf{a}$, $\mathbf{b}$, and $\mathbf{a} \times \mathbf{b}$ form a *right-handed* set, like the unit vectors $\mathbf{e}_x$, $\mathbf{e}_y$, and $\mathbf{e}_z$. In fact, $\mathbf{e}_x \times \mathbf{e}_y = \mathbf{e}_z$. This defines a unique direction for $\mathbf{a} \times \mathbf{b}$, which is obtained from the right-hand rule—see Figure 2.6.

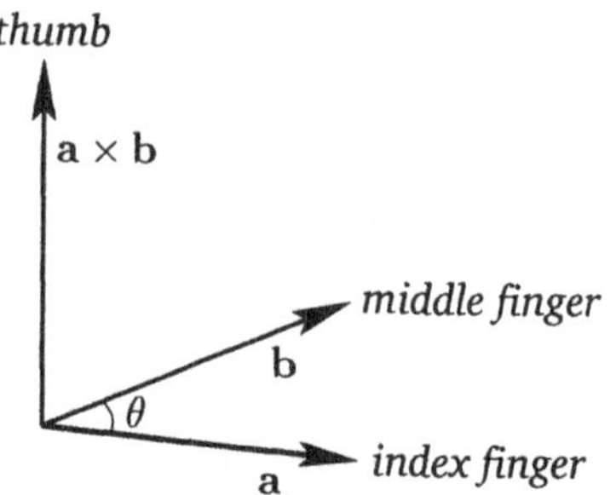

Figure 2.6: *The right-hand rule for cross products. Here, θ is less that 180°.*

Let us now evaluate the magnitude of $\mathbf{a} \times \mathbf{b}$. We have

$$\begin{aligned}(\mathbf{a} \times \mathbf{b})^2 &= (a_y\, b_z - a_z\, b_y)^2 + (a_z\, b_x - a_x\, b_z)^2 + (a_x\, b_y - a_y\, b_x)^2 \\ &= (a_x^{\,2} + a_y^{\,2} + a_z^{\,2})\,(b_x^{\,2} + b_y^{\,2} + b_z^{\,2}) - (a_x\, b_x + a_y\, b_y + a_z\, b_z)^2 \\ &= |\mathbf{a}|^2\, |\mathbf{b}|^2 - (\mathbf{a} \cdot \mathbf{b})^2 \\ &= |\mathbf{a}|^2\, |\mathbf{b}|^2 - |\mathbf{a}|^2\, |\mathbf{b}|^2 \cos^2\theta = |\mathbf{a}|^2\, |\mathbf{b}|^2 \sin^2\theta. \end{aligned} \tag{2.36}$$

Thus,

$$|\mathbf{a} \times \mathbf{b}| = |\mathbf{a}|\, |\mathbf{b}| \sin\theta, \tag{2.37}$$

where θ is the angle subtended between $\mathbf{a}$ and $\mathbf{b}$. Clearly, $\mathbf{a} \times \mathbf{a} = \mathbf{0}$ for any vector, since θ is always zero in this case. Also, if $\mathbf{a} \times \mathbf{b} = \mathbf{0}$ then either $|\mathbf{a}| = 0$, $|\mathbf{b}| = 0$, or $\mathbf{b}$ is parallel (or antiparallel) to $\mathbf{a}$.

Consider the parallelogram defined by vectors $\mathbf{a}$ and $\mathbf{b}$—see Figure 2.7. The scalar area is $a\, b \sin\theta$. The vector area has the magnitude of the scalar area, and is normal to the plane of the parallelogram, which means that it is perpendicular to both $\mathbf{a}$ and $\mathbf{b}$. Clearly, the vector

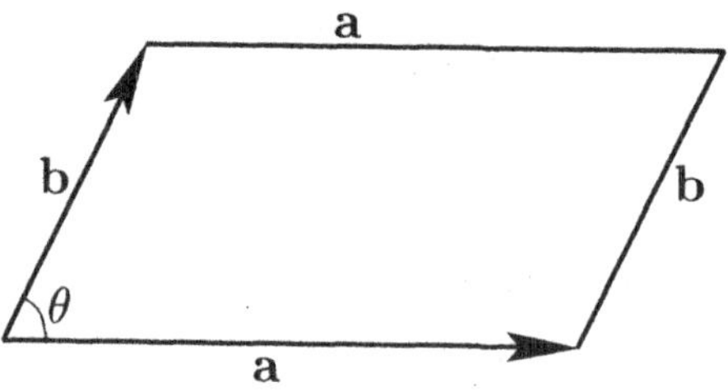

Figure 2.7: *A vector parallelogram.*

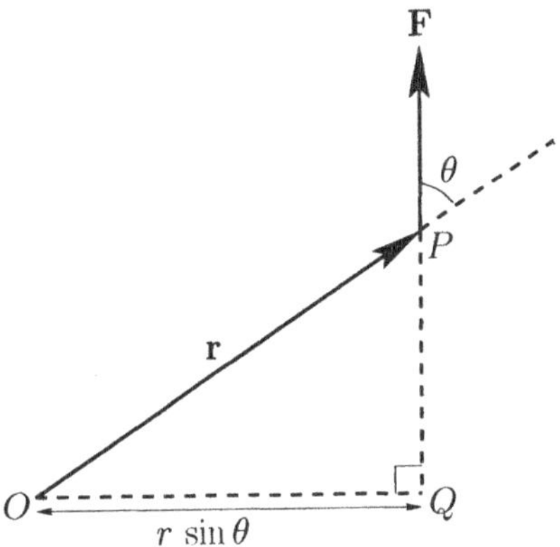

Figure 2.8: *A torque.*

area is given by

$$\mathbf{S} = \mathbf{a} \times \mathbf{b}, \tag{2.38}$$

with the sense obtained from the right-hand grip rule by rotating **a** on to **b**.

Suppose that a force **F** is applied at position **r**—see Figure 2.8. The torque about the origin O is the product of the magnitude of the force and the length of the lever arm OQ. Thus, the magnitude of the torque is $|\mathbf{F}|\,|\mathbf{r}|\,\sin\theta$. The direction of the torque is conventionally the direction of the axis through O about which the force tries to rotate objects, in the sense determined by the right-hand grip rule. It follows that the vector torque is given by

$$\boldsymbol{\tau} = \mathbf{r} \times \mathbf{F}. \tag{2.39}$$

2.6 ROTATION

Let us try to define a rotation vector $\boldsymbol{\theta}$ whose magnitude is the angle of the rotation, θ, and whose direction is the axis of the rotation, in the sense determined by the right-hand grip rule. Unfortunately, this is not a good vector. The problem is that the addition of rotations is not commutative, whereas vector addition is commuative. Figure 2.9 shows the effect of applying two successive 90° rotations, one about the x-axis, and the other about the z-axis, to a standard six-sided die. In the left-hand case, the z-rotation is applied before the x-rotation, and *vice versa* in the right-hand case. It can be seen that the die ends up in two completely

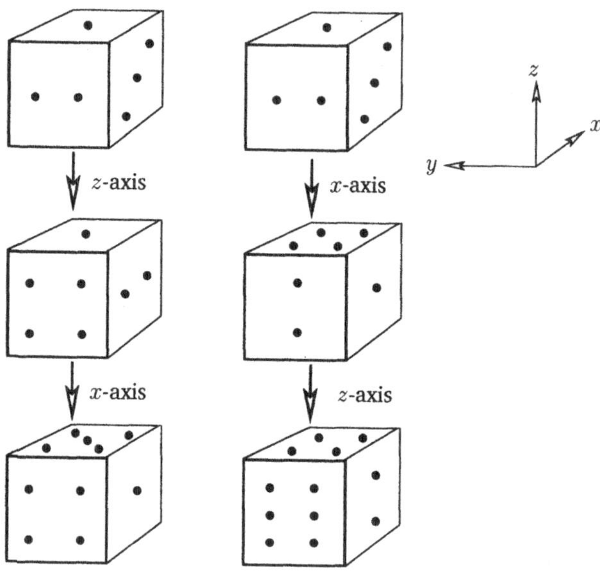

Figure 2.9: *Effect of successive rotations about perpendicular axes on a six-sided die.*

different states. Clearly, the z-rotation plus the x-rotation does not equal the x-rotation plus the z-rotation. This non-commuting algebra cannot be represented by vectors. So, although rotations have a well-defined magnitude and direction, they are *not* vector quantities.

But, this is not quite the end of the story. Suppose that we take a general vector **a** and rotate it about the z-axis by a *small* angle $\delta\theta_z$. This is equivalent to rotating the basis about the z-axis by $-\delta\theta_z$. According to Equations (2.10)–(2.12), we have

$$\mathbf{a}' \simeq \mathbf{a} + \delta\theta_z\,\mathbf{e}_z \times \mathbf{a}, \tag{2.40}$$

where use has been made of the small angle approximations $\sin\theta \simeq \theta$ and $\cos\theta \simeq 1$. The above equation can easily be generalized to allow small rotations about the x- and y-axes by $\delta\theta_x$ and $\delta\theta_y$, respectively. We find that

$$\mathbf{a}' \simeq \mathbf{a} + \delta\boldsymbol{\theta} \times \mathbf{a}, \tag{2.41}$$

where

$$\delta\boldsymbol{\theta} = \delta\theta_x\,\mathbf{e}_x + \delta\theta_y\,\mathbf{e}_y + \delta\theta_z\,\mathbf{e}_z. \tag{2.42}$$

Clearly, we can define a rotation vector $\delta\boldsymbol{\theta}$, but it only works for *small* angle rotations (*i.e.*, sufficiently small that the small angle approximations of sine and cosine are good). According to the above equation, a small z-rotation plus a small x-rotation is (approximately) equal to the two rotations applied in the opposite order. The fact that infinitesimal rotation is a vector implies that angular velocity,

$$\boldsymbol{\omega} = \lim_{\delta t \to 0} \frac{\delta \boldsymbol{\theta}}{\delta t}, \tag{2.43}$$

must be a vector as well. Also, if $\mathbf{a}'$ is interpreted as $\mathbf{a}(t + \delta t)$ in Equation (2.41) then it is clear that the equation of motion of a vector precessing about the origin with angular velocity $\boldsymbol{\omega}$ is

$$\frac{d\mathbf{a}}{dt} = \boldsymbol{\omega} \times \mathbf{a}. \tag{2.44}$$

2.7 THE SCALAR TRIPLE PRODUCT

Consider three vectors **a**, **b**, and **c**. The scalar triple product is defined $\mathbf{a} \cdot \mathbf{b} \times \mathbf{c}$. Now, $\mathbf{b} \times \mathbf{c}$ is the vector area of the parallelogram defined by **b** and **c**. So, $\mathbf{a} \cdot \mathbf{b} \times \mathbf{c}$ is the scalar area of this parallelogram times the component of **a** in the direction of its normal. It follows that $\mathbf{a} \cdot \mathbf{b} \times \mathbf{c}$ is the *volume* of the parallelepiped defined by vectors **a**, **b**, and **c**—see Figure 2.10. This volume is independent of how the triple product is formed from **a**, **b**, and **c**, except that

$$\mathbf{a} \cdot \mathbf{b} \times \mathbf{c} = -\mathbf{a} \cdot \mathbf{c} \times \mathbf{b}. \tag{2.45}$$

So, the "volume" is positive if **a**, **b**, and **c** form a right-handed set (*i.e.*, if **a** lies above the plane of **b** and **c**, in the sense determined from the

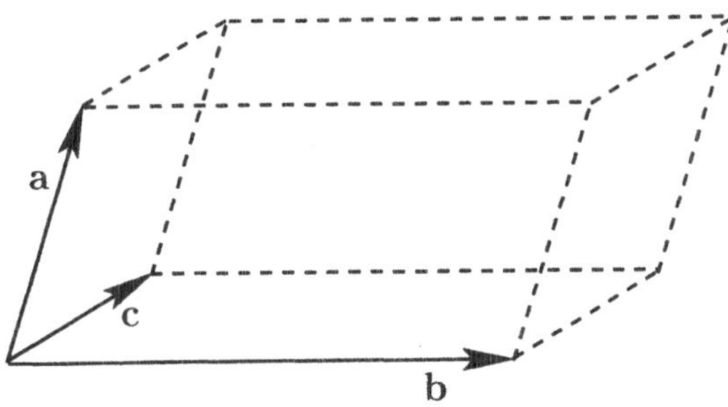

Figure 2.10: *A vector parallelepiped.*

right-hand grip rule by rotating **b** onto **c**) and negative if they form a left-handed set. The triple product is unchanged if the dot and cross product operators are interchanged,

$$\mathbf{a} \cdot \mathbf{b} \times \mathbf{c} = \mathbf{a} \times \mathbf{b} \cdot \mathbf{c}. \tag{2.46}$$

The triple product is also invariant under any cyclic permutation of **a**, **b**, and **c**,

$$\mathbf{a} \cdot \mathbf{b} \times \mathbf{c} = \mathbf{b} \cdot \mathbf{c} \times \mathbf{a} = \mathbf{c} \cdot \mathbf{a} \times \mathbf{b}, \tag{2.47}$$

but any anticyclic permutation causes it to change sign,

$$\mathbf{a} \cdot \mathbf{b} \times \mathbf{c} = -\mathbf{b} \cdot \mathbf{a} \times \mathbf{c}. \tag{2.48}$$

The scalar triple product is zero if any two of **a**, **b**, and **c** are parallel, or if **a**, **b**, and **c** are coplanar.

If **a**, **b**, and **c** are non-coplanar, then any vector **r** can be written in terms of them:

$$\mathbf{r} = \alpha\,\mathbf{a} + \beta\,\mathbf{b} + \gamma\,\mathbf{c}. \tag{2.49}$$

Forming the dot product of this equation with $\mathbf{b} \times \mathbf{c}$, we then obtain

$$\mathbf{r} \cdot \mathbf{b} \times \mathbf{c} = \alpha\,\mathbf{a} \cdot \mathbf{b} \times \mathbf{c}, \tag{2.50}$$

so

$$\alpha = \frac{\mathbf{r} \cdot \mathbf{b} \times \mathbf{c}}{\mathbf{a} \cdot \mathbf{b} \times \mathbf{c}}. \tag{2.51}$$

Analogous expressions can be written for β and γ. The parameters α, β, and γ are uniquely determined provided $\mathbf{a} \cdot \mathbf{b} \times \mathbf{c} \neq 0$: *i.e.*, provided that the three basis vectors are not coplanar.

2.8 THE VECTOR TRIPLE PRODUCT

For three vectors **a**, **b**, and **c**, the vector triple product is defined $\mathbf{a} \times (\mathbf{b} \times \mathbf{c})$. The brackets are important because $\mathbf{a} \times (\mathbf{b} \times \mathbf{c}) \neq (\mathbf{a} \times \mathbf{b}) \times \mathbf{c}$. In fact, it can be demonstrated that

$$\mathbf{a} \times (\mathbf{b} \times \mathbf{c}) \equiv (\mathbf{a} \cdot \mathbf{c})\,\mathbf{b} - (\mathbf{a} \cdot \mathbf{b})\,\mathbf{c} \tag{2.52}$$

and

$$(\mathbf{a} \times \mathbf{b}) \times \mathbf{c} \equiv (\mathbf{a} \cdot \mathbf{c})\,\mathbf{b} - (\mathbf{b} \cdot \mathbf{c})\,\mathbf{a}. \tag{2.53}$$

Let us try to prove the first of the above theorems. The left-hand side and the right-hand side are both proper vectors, so if we can prove this result in one particular coordinate system then it must be true in general. Let us take convenient axes such that the x-axis lies along $\mathbf{b}$, and $\mathbf{c}$ lies in the x-y plane. It follows that $\mathbf{b} = (b_x, 0, 0)$, $\mathbf{c} = (c_x, c_y, 0)$, and $\mathbf{a} = (a_x, a_y, a_z)$. The vector $\mathbf{b} \times \mathbf{c}$ is directed along the z-axis: $\mathbf{b} \times \mathbf{c} = (0, 0, b_x\, c_y)$. It follows that $\mathbf{a} \times (\mathbf{b} \times \mathbf{c})$ lies in the x-y plane: $\mathbf{a} \times (\mathbf{b} \times \mathbf{c}) = (a_y\, b_x\, c_y, -a_x\, b_x\, c_y, 0)$. This is the left-hand side of Equation (2.52) in our convenient axes. To evaluate the right-hand side, we need $\mathbf{a} \cdot \mathbf{c} = a_x\, c_x + a_y\, c_y$ and $\mathbf{a} \cdot \mathbf{b} = a_x\, b_x$. It follows that the right-hand side is

$$\begin{aligned} \text{RHS} &= ([a_x\, c_x + a_y\, c_y]\, b_x, 0, 0) - (a_x\, b_x\, c_x, a_x\, b_x\, c_y, 0) \\ &= (a_y\, c_y\, b_x, -a_x\, b_x\, c_y, 0) = \text{LHS}, \end{aligned} \tag{2.54}$$

which proves the theorem.

2.9 VECTOR CALCULUS

Suppose that vector $\mathbf{a}$ varies with time, so that $\mathbf{a} = \mathbf{a}(t)$. The time derivative of the vector is defined

$$\frac{d\mathbf{a}}{dt} = \lim_{\delta t \to 0} \left[\frac{\mathbf{a}(t + \delta t) - \mathbf{a}(t)}{\delta t} \right]. \tag{2.55}$$

When written out in component form this becomes

$$\frac{d\mathbf{a}}{dt} = \left(\frac{da_x}{dt}, \frac{da_y}{dt}, \frac{da_z}{dt} \right). \tag{2.56}$$

Suppose that $\mathbf{a}$ is, in fact, the product of a scalar $\phi(t)$ and another vector $\mathbf{b}(t)$. What now is the time derivative of $\mathbf{a}$? We have

$$\frac{da_x}{dt} = \frac{d}{dt}(\phi\, b_x) = \frac{d\phi}{dt}\, b_x + \phi\, \frac{db_x}{dt}, \tag{2.57}$$

which implies that

$$\frac{d\mathbf{a}}{dt} = \frac{d\phi}{dt}\, \mathbf{b} + \phi\, \frac{d\mathbf{b}}{dt}. \tag{2.58}$$

Moreover, it is easily demonstrated that

$$\frac{d}{dt}(\mathbf{a} \cdot \mathbf{b}) = \frac{d\mathbf{a}}{dt} \cdot \mathbf{b} + \mathbf{a} \cdot \frac{d\mathbf{b}}{dt}, \tag{2.59}$$

and

$$\frac{d}{dt}(\mathbf{a}\times\mathbf{b}) = \frac{d\mathbf{a}}{dt}\times\mathbf{b}+\mathbf{a}\times\frac{d\mathbf{b}}{dt}. \tag{2.60}$$

Hence, it can be seen that the laws of vector differentiation are analogous to those in conventional calculus.

2.10 LINE INTEGRALS

Consider a two-dimensional function $f(x, y)$ which is defined for all x and y. What is meant by the integral of f along a given curve from P to Q in the x-y plane? We first draw out f as a function of length l along the path—see Figure 2.11. The integral is then simply given by

$$\int_P^Q f(x, y)\, dl = \text{Area under the curve.} \tag{2.61}$$

As an example of this, consider the integral of $f(x, y) = x\,y^2$ between P and Q along the two routes indicated in Figure 2.12. Along route 1 we have $x = y$, so $dl = \sqrt{2}\,dx$. Thus,

$$\int_P^Q x\,y^2\, dl = \int_0^1 x^3\,\sqrt{2}\,dx = \frac{\sqrt{2}}{4}. \tag{2.62}$$

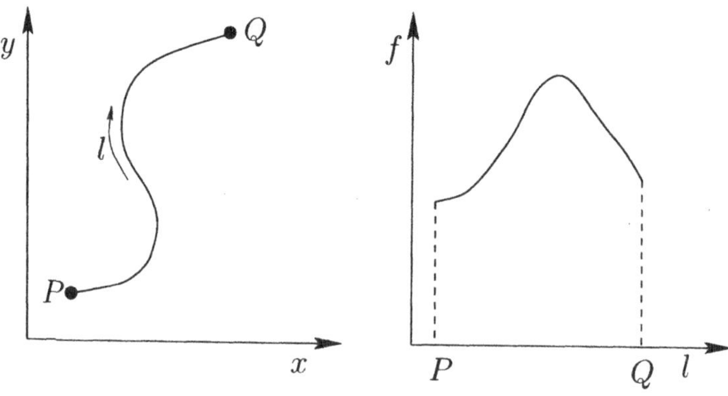

Figure 2.11: *A line integral.*

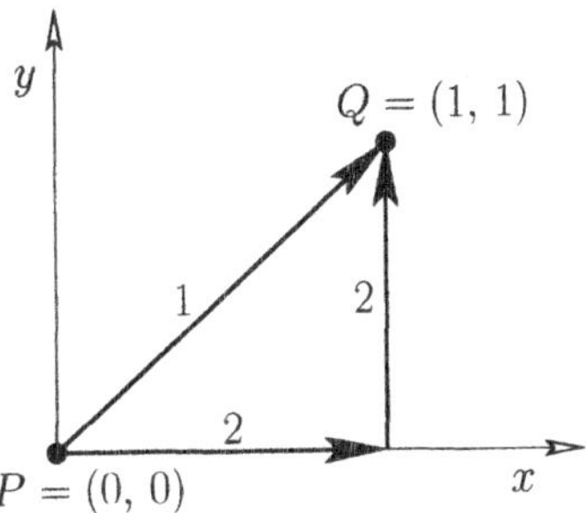

Figure 2.12: *An example line integral.*

The integration along route 2 gives

$$\int_P^Q x y^2\,dl = \int_0^1 x y^2\,dx\bigg|_{y=0} + \int_0^1 x y^2\,dy\bigg|_{x=1}$$

$$= 0 + \int_0^1 y^2\,dy = \frac{1}{3}. \tag{2.63}$$

Note that the integral depends on the route taken between the initial and final points.

The most common type of line integral is that in which the contributions from dx and dy are evaluated separately, rather that through the path length dl:

$$\int_P^Q [f(x,y)\,dx + g(x,y)\,dy]\,. \tag{2.64}$$

As an example of this, consider the integral

$$\int_P^Q \left[y\,dx + x^3\,dy\right] \tag{2.65}$$

along the two routes indicated in Figure 2.13. Along route 1 we have $x = y + 1$ and $dx = dy$, so

$$\int_P^Q = \int_0^1 \left[y\,dy + (y+1)^3\,dy\right] = \frac{17}{4}. \tag{2.66}$$

Along route 2,

$$\int_P^Q = \int_0^1 x^3\,dy\bigg|_{x=1} + \int_1^2 y\,dx\bigg|_{y=1} = \frac{7}{4}. \tag{2.67}$$

Again, the integral depends on the path of integration.

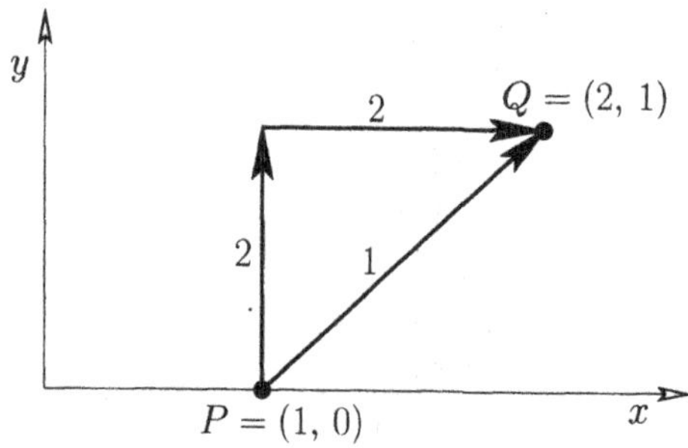

Figure 2.13: *An example line integral.*

Suppose that we have a line integral which does *not* depend on the path of integration. It follows that

$$\int_P^Q (f\,dx + g\,dy) = F(Q) - F(P) \tag{2.68}$$

for some function F. Given F(P) for one point P in the x-y plane, then

$$F(Q) = F(P) + \int_P^Q (f\,dx + g\,dy) \tag{2.69}$$

defines F(Q) for all other points in the plane. We can then draw a contour map of F(x, y). The line integral between points P and Q is simply the change in height in the contour map between these two points:

$$\int_P^Q (f\,dx + g\,dy) = \int_P^Q dF(x,y) = F(Q) - F(P). \tag{2.70}$$

Thus,

$$dF(x,y) = f(x,y)\,dx + g(x,y)\,dy. \tag{2.71}$$

For instance, if $F = x^3\,y$ then $dF = 3\,x^2\,y\,dx + x^3\,dy$ and

$$\int_P^Q \left(3\,x^2\,y\,dx + x^3\,dy\right) = \left[x^3\,y\right]_P^Q \tag{2.72}$$

is independent of the path of integration.

It is clear that there are two distinct types of line integral. Those which depend only on their endpoints and not on the path of integration, and those which depend both on their endpoints and the integration path. Later on, we shall learn how to distinguish between these two types.

2.11 VECTOR LINE INTEGRALS

A *vector field* is defined as a set of vectors associated with each point in space. For instance, the velocity $\mathbf{v}(\mathbf{r})$ in a moving liquid (*e.g.*, a whirlpool) constitutes a vector field. By analogy, a *scalar field* is a set of scalars associated with each point in space. An example of a scalar field is the temperature distribution $T(\mathbf{r})$ in a furnace.

Consider a general vector field $\mathbf{A}(\mathbf{r})$. Let $\mathbf{dl} = (dx, dy, dz)$ be the vector element of line length. Vector line integrals often arise as

$$\int_P^Q \mathbf{A} \cdot \mathbf{dl} = \int_P^Q (A_x\, dx + A_y\, dy + A_z\, dz). \tag{2.73}$$

For instance, if $\mathbf{A}$ is a force field then the line integral is the work done in going from P to Q.

As an example, consider the work done in a repulsive, inverse-square, central field, $\mathbf{F} = -\mathbf{r}/|r^3|$. The element of work done is $dW = \mathbf{F} \cdot \mathbf{dl}$. Take $P = (\infty, 0, 0)$ and $Q = (a, 0, 0)$. Route 1 is along the x-axis, so

$$W = \int_\infty^a \left(-\frac{1}{x^2}\right) dx = \left[\frac{1}{x}\right]_\infty^a = \frac{1}{a}. \tag{2.74}$$

The second route is, firstly, around a large circle (r = constant) to the point (a, ∞, 0), and then parallel to the y-axis—see Figure 2.14. In the first part, no work is done, since $\mathbf{F}$ is perpendicular to $\mathbf{dl}$. In the

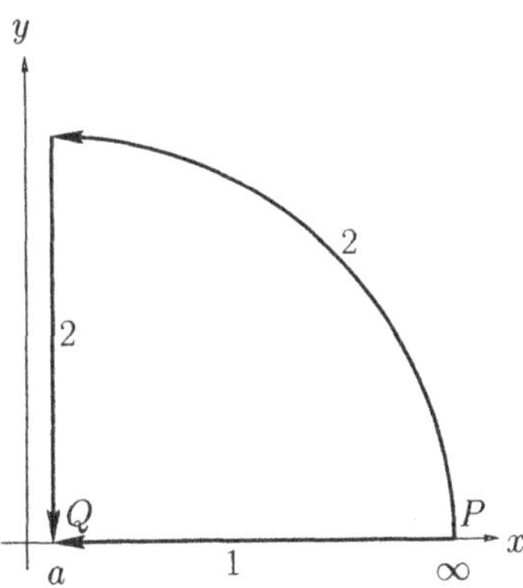

Figure 2.14: *An example vector line integral.*

second part,

$$W = \int_{\infty}^{0} \frac{-y\,dy}{(a^2 + y^2)^{3/2}} = \left[\frac{1}{(y^2 + a^2)^{1/2}}\right]_{\infty}^{0} = \frac{1}{a}. \tag{2.75}$$

In this case, the integral is independent of the path. However, not all vector line integrals are path independent.

2.12 SURFACE INTEGRALS

Let us take a surface S, which is not necessarily coplanar, and divide in up into (scalar) elements δS_i. Then

$$\iint_S f(x, y, z)\,dS = \lim_{\delta S_i \to 0} \sum_i f(x, y, z)\,\delta S_i \tag{2.76}$$

is a surface integral. For instance, the volume of water in a lake of depth $D(x, y)$ is

$$V = \iint D(x, y)\,dS. \tag{2.77}$$

To evaluate this integral we must split the calculation into two ordinary integrals. The volume in the strip shown in Figure 2.15 is

$$\left[\int_{x_1}^{x_2} D(x, y)\,dx\right] dy. \tag{2.78}$$

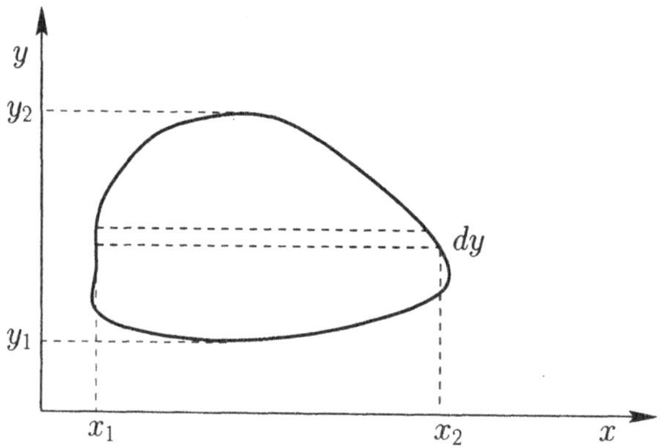

Figure 2.15: *Decomposition of a surface integral.*

Note that the limits x_1 and x_2 depend on y. The total volume is the sum over all strips:

$$V = \int_{y_1}^{y_2} dy \left[\int_{x_1(y)}^{x_2(y)} D(x, y)\, dx \right] \equiv \iint_S D(x, y)\, dx\, dy. \tag{2.79}$$

Of course, the integral can be evaluated by taking the strips the other way around:

$$V = \int_{x_1}^{x_2} dx \int_{y_1(x)}^{y_2(x)} D(x, y)\, dy. \tag{2.80}$$

Interchanging the order of integration is a very powerful and useful trick. But great care must be taken when evaluating the limits.

As an example, consider

$$\iint_S x\, y^2\, dx\, dy, \tag{2.81}$$

where S is shown in Figure 2.16. Suppose that we evaluate the x integral first:

$$dy \left(\int_0^{1-y} x\, y^2\, dx \right) = y^2\, dy \left[\frac{x^2}{2} \right]_0^{1-y} = \frac{y^2}{2} (1 - y)^2\, dy. \tag{2.82}$$

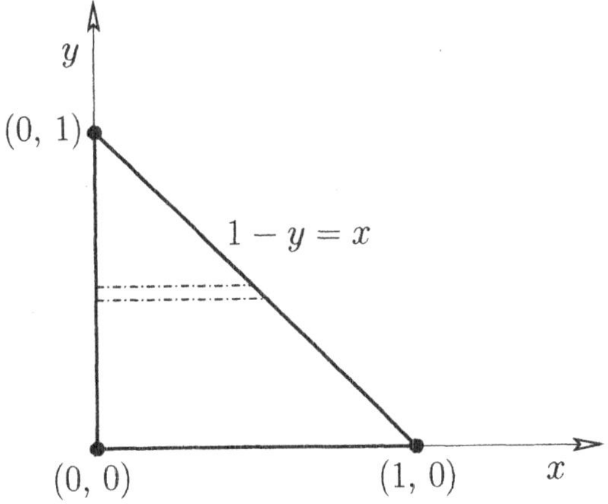

Figure 2.16: *An example surface integral.*

Let us now evaluate the y integral:

$$\int_0^1 \left(\frac{y^2}{2} - y^3 + \frac{y^4}{2}\right) dy = \frac{1}{60}. \tag{2.83}$$

We can also evaluate the integral by interchanging the order of integration:

$$\int_0^1 x\,dx \int_0^{1-x} y^2\,dy = \int_0^1 \frac{x}{3}(1-x)^3\,dx = \frac{1}{60}. \tag{2.84}$$

In some cases, a surface integral is just the product of two separate integrals. For instance,

$$\iint_S x^2 y\,dx\,dy \tag{2.85}$$

where S is a unit square. This integral can be written

$$\int_0^1 dx \int_0^1 x^2 y\,dy = \left(\int_0^1 x^2\,dx\right)\left(\int_0^1 y\,dy\right) = \frac{1}{3}\frac{1}{2} = \frac{1}{6}, \tag{2.86}$$

since the limits are both independent of the other variable.

2.13 VECTOR SURFACE INTEGRALS

Surface integrals often occur during vector analysis. For instance, the rate of flow of a liquid of velocity $\mathbf{v}$ through an infinitesimal surface of vector area $d\mathbf{S}$ is $\mathbf{v} \cdot d\mathbf{S}$. The net rate of flow through a surface $\mathbf{S}$ made up of lots of infinitesimal surfaces is

$$\iint_S \mathbf{v} \cdot d\mathbf{S} = \lim_{dS \to 0} \left[\sum v \cos\theta\, dS\right], \tag{2.87}$$

where θ is the angle subtended between the normal to the surface and the flow velocity.

Analogously to line integrals, most surface integrals depend both on the surface and the rim. But some (very important) integrals depend only on the rim, and not on the nature of the surface which spans it. As an example of this, consider incompressible fluid flow between two surfaces S_1 and S_2 which end on the same rim—see Figure 2.21. The volume between the surfaces is constant, so what goes in must come out, and

$$\iint_{S_1} \mathbf{v} \cdot d\mathbf{S} = \iint_{S_2} \mathbf{v} \cdot d\mathbf{S}. \tag{2.88}$$

It follows that

$$\iint \mathbf{v} \cdot d\mathbf{S} \tag{2.89}$$

depends only on the rim, and not on the form of surfaces S_1 and S_2.

2.14 VOLUME INTEGRALS

A volume integral takes the form

$$\iiint_V f(x, y, z)\, dV, \tag{2.90}$$

where V is some volume, and $dV = dx\, dy\, dz$ is a small volume element. The volume element is sometimes written $d^3\mathbf{r}$, or even $d\tau$. As an example of a volume integral, let us evaluate the center of gravity of a solid hemisphere of radius a (centered on the origin). The height of the center of gravity is given by

$$\bar{z} = \iiint z\, dV \Big/ \iiint dV. \tag{2.91}$$

The bottom integral is simply the volume of the hemisphere, which is $2\pi a^3/3$. The top integral is most easily evaluated in spherical polar coordinates, for which $z = r \cos\theta$ and $dV = r^2 \sin\theta\, dr\, d\theta\, d\phi$—see Section 2.19. Thus,

$$\begin{aligned} \iiint z\, dV &= \int_0^a dr \int_0^{\pi/2} d\theta \int_0^{2\pi} d\phi\; r \cos\theta\; r^2 \sin\theta \\ &= \int_0^a r^3\, dr \int_0^{\pi/2} \sin\theta \cos\theta\, d\theta \int_0^{2\pi} d\phi = \frac{\pi a^4}{4}, \end{aligned} \tag{2.92}$$

giving

$$\bar{z} = \frac{\pi a^4}{4} \frac{3}{2\pi a^3} = \frac{3a}{8}. \tag{2.93}$$

2.15 GRADIENT

A one-dimensional function $f(x)$ has a gradient df/dx which is defined as the slope of the tangent to the curve at x. We wish to extend this idea to cover scalar fields in two and three dimensions.

Consider a two-dimensional scalar field $h(x, y)$, which is (say) the height of a hill. Let $\mathbf{dl} = (dx, dy)$ be an element of horizontal distance. Consider dh/dl, where dh is the change in height after moving an infinitesimal distance $\mathbf{dl}$. This quantity is somewhat like the one-dimensional gradient, except that dh depends on the *direction* of $\mathbf{dl}$, as well as its magnitude. In the immediate vicinity of some point P, the slope reduces to an inclined plane—see Figure 2.17. The largest value of dh/dl is straight up the slope. For any other direction

$$\frac{dh}{dl} = \left(\frac{dh}{dl}\right)_{max} \cos\theta. \tag{2.94}$$

Let us define a two-dimensional vector, **grad** h, called the *gradient* of h, whose magnitude is $(dh/dl)_{max}$, and whose direction is the direction up the steepest slope. Because of the $\cos\theta$ property, the component of **grad** h in any direction equals dh/dl for that direction. [The argument, here, is analogous to that used for vector areas in Section 2.3. See, in particular, Equation (2.13).]

The component of dh/dl in the x-direction can be obtained by plotting out the profile of h at constant y, and then finding the slope of the

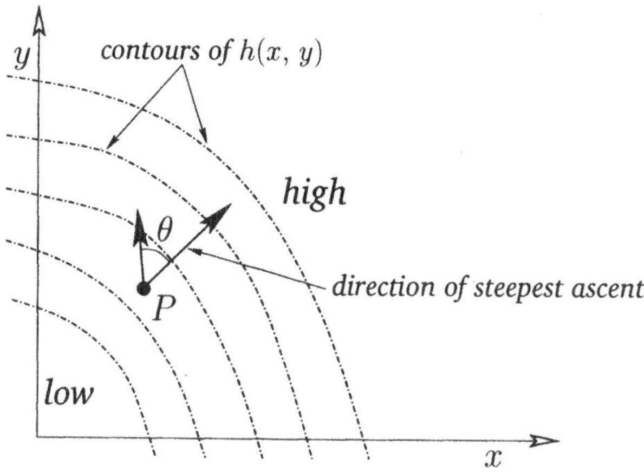

Figure 2.17: *A two-dimensional gradient.*

tangent to the curve at given x. This quantity is known as the *partial derivative* of h with respect to x at constant y, and is denoted $(\partial h/\partial x)_y$. Likewise, the gradient of the profile at constant x is written $(\partial h/\partial y)_x$. Note that the subscripts denoting constant-x and constant-y are usually omitted, unless there is any ambiguity. If follows that in component form

$$\mathbf{grad}\, h = \left(\frac{\partial h}{\partial x}, \frac{\partial h}{\partial y}\right). \tag{2.95}$$

Now, the equation of the tangent plane at $P = (x_0, y_0)$ is

$$h_T(x, y) = h(x_0, y_0) + \alpha\,(x - x_0) + \beta\,(y - y_0). \tag{2.96}$$

This has the same local gradients as $h(x, y)$, so

$$\alpha = \frac{\partial h}{\partial x}, \qquad \beta = \frac{\partial h}{\partial y}, \tag{2.97}$$

by differentiation of the above. For small $dx = x - x_0$ and $dy = y - y_0$, the function h is coincident with the tangent plane. We have

$$dh = \frac{\partial h}{\partial x}\,dx + \frac{\partial h}{\partial y}\,dy. \tag{2.98}$$

But, $\mathbf{grad}\, h = (\partial h/\partial x, \partial h/\partial y)$ and $\mathbf{dl} = (dx, dy)$, so

$$dh = \mathbf{grad}\, h \cdot \mathbf{dl}. \tag{2.99}$$

Incidentally, the above equation demonstrates that $\mathbf{grad}\, h$ is a proper vector, since the left-hand side is a scalar, and, according to the properties of the dot product, the right-hand side is also a scalar, provided that $\mathbf{dl}$ and $\mathbf{grad}\, h$ are both proper vectors ($\mathbf{dl}$ is an obvious vector, because it is directly derived from displacements).

Consider, now, a three-dimensional temperature distribution $T(x, y, z)$ in (say) a reaction vessel. Let us define $\mathbf{grad}\, T$, as before, as a vector whose magnitude is $(dT/dl)_{max}$, and whose direction is the direction of the maximum gradient. This vector is written in component form

$$\mathbf{grad}\, T = \left(\frac{\partial T}{\partial x}, \frac{\partial T}{\partial y}, \frac{\partial T}{\partial z}\right). \tag{2.100}$$

Here, $\partial T/\partial x \equiv (\partial T/\partial x)_{y,z}$ is the gradient of the one-dimensional temperature profile at constant y and z. The change in T in going from point

P to a neighboring point offset by $\mathbf{dl} = (dx, dy, dz)$ is

$$dT = \frac{\partial T}{\partial x}\,dx + \frac{\partial T}{\partial y}\,dy + \frac{\partial T}{\partial z}\,dz. \tag{2.101}$$

In vector form, this becomes

$$dT = \mathbf{grad}\,T \cdot \mathbf{dl}. \tag{2.102}$$

Suppose that $dT = 0$ for some $\mathbf{dl}$. It follows that

$$dT = \mathbf{grad}\,T \cdot \mathbf{dl} = 0. \tag{2.103}$$

So, $\mathbf{dl}$ is perpendicular to $\mathbf{grad}\,T$. Since $dT = 0$ along so-called "isotherms" (*i.e.*, contours of the temperature), we conclude that the isotherms (contours) are everywhere perpendicular to $\mathbf{grad}\,T$—see Figure 2.18. It is, of course, possible to integrate dT. The line integral from point P to point Q is written

$$\int_P^Q dT = \int_P^Q \mathbf{grad}\,T \cdot \mathbf{dl} = T(Q) - T(P). \tag{2.104}$$

This integral is clearly independent of the path taken between P and Q, so $\int_P^Q \mathbf{grad}\,T \cdot \mathbf{dl}$ must be path independent.

Consider a vector field $\mathbf{A}(\mathbf{r})$. In general, $\int_P^Q \mathbf{A} \cdot \mathbf{dl}$ depends on path, but for some special vector fields the integral is path independent. Such

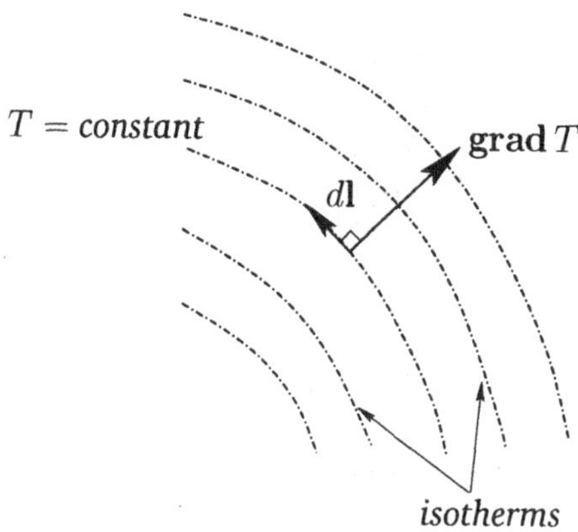

Figure 2.18: *Isotherms.*

fields are called *conservative* fields. It can be shown that if **A** is a conservative field then $\mathbf{A} = \mathbf{grad}\, V$ for some scalar field V. The proof of this is straightforward. Keeping P fixed, we have

$$\int_P^Q \mathbf{A} \cdot d\mathbf{l} = V(Q), \tag{2.105}$$

where V(Q) is a well-defined function, due to the path-independent nature of the line integral. Consider moving the position of the endpoint by an infinitesimal amount dx in the x-direction. We have

$$V(Q + dx) = V(Q) + \int_Q^{Q+dx} \mathbf{A} \cdot d\mathbf{l} = V(Q) + A_x\, dx. \tag{2.106}$$

Hence,

$$\frac{\partial V}{\partial x} = A_x, \tag{2.107}$$

with analogous relations for the other components of **A**. It follows that

$$\mathbf{A} = \mathbf{grad}\, V. \tag{2.108}$$

In Physics, the force due to gravity is a good example of a conservative field. If $\mathbf{A}(\mathbf{r})$ is a force field then $\int \mathbf{A} \cdot d\mathbf{l}$ is the work done in traversing some path. If **A** is conservative then

$$\oint \mathbf{A} \cdot d\mathbf{l} = 0, \tag{2.109}$$

where $\oint$ corresponds to the line integral around some closed loop. The fact that zero net work is done in going around a closed loop is equivalent to the conservation of energy (this is why conservative fields are called "conservative"). A good example of a non-conservative field is the force due to friction. Clearly, a frictional system loses energy in going around a closed cycle, so $\oint \mathbf{A} \cdot d\mathbf{l} \neq 0$.

It is useful to define the vector *operator*

$$\nabla \equiv \left(\frac{\partial}{\partial x}, \frac{\partial}{\partial y}, \frac{\partial}{\partial z}\right), \tag{2.110}$$

which is usually called the *grad* or *del* operator. This operator acts on everything to its right in an expression, until the end of the expression or a closing bracket is reached. For instance,

$$\mathbf{grad}\, f = \nabla f = \left(\frac{\partial f}{\partial x}, \frac{\partial f}{\partial y}, \frac{\partial f}{\partial z}\right). \tag{2.111}$$

For two scalar fields ϕ and ψ,

$$\mathbf{grad}\,(\phi\,\psi) = \phi\;\mathbf{grad}\,\psi + \psi\;\mathbf{grad}\,\phi \tag{2.112}$$

can be written more succinctly as

$$\nabla(\phi\,\psi) = \phi\,\nabla\psi + \psi\,\nabla\phi. \tag{2.113}$$

Suppose that we rotate the basis about the z-axis by θ degrees. By analogy with Equations (2.7)–(2.9), the old coordinates $(x,\ y,\ z)$ are related to the new ones $(x',\ y',\ z')$ via

$$x = x'\cos\theta - y'\sin\theta, \tag{2.114}$$

$$y = x'\sin\theta + y'\cos\theta, \tag{2.115}$$

$$z = z'. \tag{2.116}$$

Now,

$$\frac{\partial}{\partial x'} = \left(\frac{\partial x}{\partial x'}\right)_{y',z'}\frac{\partial}{\partial x} + \left(\frac{\partial y}{\partial x'}\right)_{y',z'}\frac{\partial}{\partial y} + \left(\frac{\partial z}{\partial x'}\right)_{y',z'}\frac{\partial}{\partial z}, \tag{2.117}$$

giving

$$\frac{\partial}{\partial x'} = \cos\theta\,\frac{\partial}{\partial x} + \sin\theta\,\frac{\partial}{\partial y}, \tag{2.118}$$

and

$$\nabla_{x'} = \cos\theta\,\nabla_x + \sin\theta\,\nabla_y. \tag{2.119}$$

It can be seen that the differential operator ∇ transforms like a proper vector, according to Equations (2.10)–(2.12). This is another proof that ∇f is a good vector.

2.16 DIVERGENCE

Let us start with a vector field $\mathbf{A}(\mathbf{r})$. Consider $\oint_S \mathbf{A}\cdot d\mathbf{S}$ over some closed surface S, where $d\mathbf{S}$ denotes an *outward*-pointing surface element. This surface integral is usually called the *flux* of $\mathbf{A}$ out of S. If $\mathbf{A}$ is the velocity of some fluid then $\oint_S \mathbf{A}\cdot d\mathbf{S}$ is the rate of fluid flow out of S.

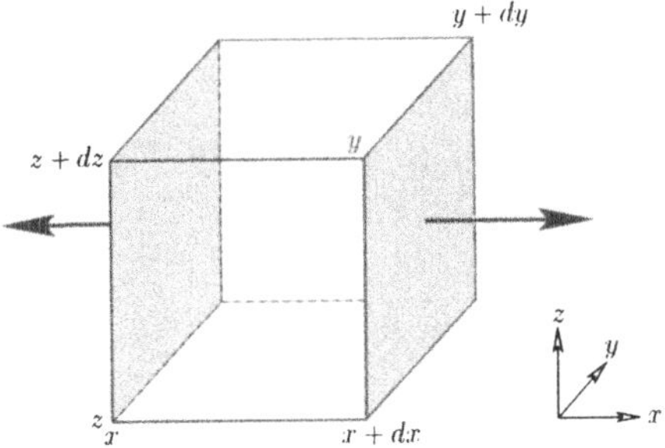

Figure 2.19: *Flux of a vector field out of a small box.*

If **A** is constant in space then it is easily demonstrated that the net flux out of S is zero,

$$\oint \mathbf{A} \cdot d\mathbf{S} = \mathbf{A} \cdot \oint d\mathbf{S} = \mathbf{A} \cdot \mathbf{S} = 0, \tag{2.120}$$

since the vector area **S** of a closed surface is zero.

Suppose, now, that **A** is not uniform in space. Consider a very small rectangular volume over which **A** hardly varies. The contribution to $\oint \mathbf{A} \cdot d\mathbf{S}$ from the two faces normal to the x-axis is

$$A_x(x + dx)\, dy\, dz - A_x(x)\, dy\, dz = \frac{\partial A_x}{\partial x}\, dx\, dy\, dz = \frac{\partial A_x}{\partial x}\, dV, \tag{2.121}$$

where $dV = dx\, dy\, dz$ is the volume element—see Figure 2.19. There are analogous contributions from the sides normal to the y- and z-axes, so the total of all the contributions is

$$\oint \mathbf{A} \cdot d\mathbf{S} = \left(\frac{\partial A_x}{\partial x} + \frac{\partial A_y}{\partial y} + \frac{\partial A_z}{\partial z} \right) dV. \tag{2.122}$$

The *divergence* of a vector field is defined

$$\operatorname{div} \mathbf{A} = \nabla \cdot \mathbf{A} = \frac{\partial A_x}{\partial x} + \frac{\partial A_y}{\partial y} + \frac{\partial A_z}{\partial z}. \tag{2.123}$$

Divergence is a good scalar (*i.e.*, it is coordinate independent), since it is the dot product of the vector operator ∇ with **A**. The formal definition of $\nabla \cdot \mathbf{A}$ is

$$\nabla \cdot \mathbf{A} = \lim_{dV \to 0} \frac{\oint \mathbf{A} \cdot d\mathbf{S}}{dV}. \tag{2.124}$$

This definition is independent of the shape of the infinitesimal volume element.

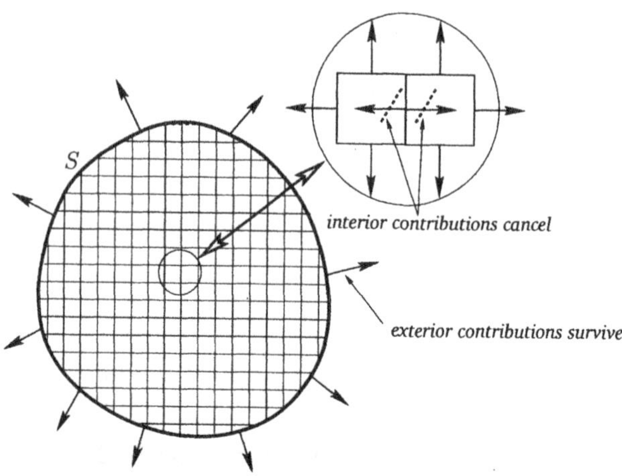

Figure 2.20: *The divergence theorem.*

One of the most important results in vector field theory is the so-called *divergence theorem* or *Gauss' theorem*. This states that for any volume V surrounded by a closed surface S,

$$\oint_S \mathbf{A} \cdot \mathbf{dS} = \int_V \nabla \cdot \mathbf{A}\ dV, \tag{2.125}$$

where $\mathbf{dS}$ is an outward-pointing volume element. The proof is very straightforward. We divide up the volume into lots of very small cubes, and sum $\int \mathbf{A} \cdot \mathbf{dS}$ over all of the surfaces. The contributions from the interior surfaces cancel out, leaving just the contribution from the outer surface—see Figure 2.20. We can use Equation (2.122) for each cube individually. This tells us that the summation is equivalent to $\int \nabla \cdot \mathbf{A}\ dV$ over the whole volume. Thus, the integral of $\mathbf{A} \cdot \mathbf{dS}$ over the outer surface is equal to the integral of $\nabla \cdot \mathbf{A}$ over the whole volume, which proves the divergence theorem.

Now, for a vector field with $\nabla \cdot \mathbf{A} = 0$,

$$\oint_S \mathbf{A} \cdot \mathbf{dS} = 0 \tag{2.126}$$

for any closed surface S. So, for two surfaces, S_1 and S_2, on the same rim,

$$\int_{S_1} \mathbf{A} \cdot \mathbf{dS} = \int_{S_2} \mathbf{A} \cdot \mathbf{dS} \tag{2.127}$$

—see Figure 2.21. (Note that the direction of the surface elements on S_1 has been reversed relative to those on the closed surface. Hence, the

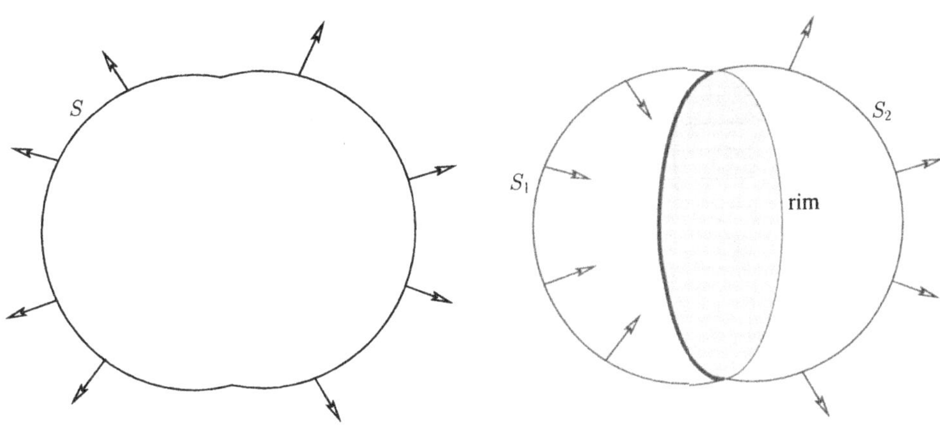

Figure 2.21: *Two surfaces spanning the same rim (right), and the equivalent closed surface (left).*

sign of the associated surface integral is also reversed.) Thus, if $\nabla \cdot \mathbf{A} = 0$ then the surface integral depends on the rim but not the nature of the surface which spans it. On the other hand, if $\nabla \cdot \mathbf{A} \neq 0$ then the integral depends on both the rim and the surface.

Consider an incompressible fluid whose velocity field is $\mathbf{v}$. It is clear that $\oint \mathbf{v} \cdot d\mathbf{S} = 0$ for any closed surface, since what flows into the surface must flow out again. Thus, according to the divergence theorem, $\int \nabla \cdot \mathbf{v}\, dV = 0$ for any volume. The only way in which this is possible is if $\nabla \cdot \mathbf{v}$ is everywhere zero. Thus, the velocity components of an incompressible fluid satisfy the following differential relation:

$$\frac{\partial v_x}{\partial x} + \frac{\partial v_y}{\partial y} + \frac{\partial v_z}{\partial z} = 0. \tag{2.128}$$

Consider, now, a compressible fluid of density ρ and velocity $\mathbf{v}$. The surface integral $\oint_S \rho\, \mathbf{v} \cdot d\mathbf{S}$ is the net rate of mass flow out of the closed surface S. This must be equal to the rate of decrease of mass inside the volume V enclosed by S, which is written $-(\partial/\partial t)(\int_V \rho\, dV)$. Thus,

$$\oint_S \rho\, \mathbf{v} \cdot d\mathbf{S} = -\frac{\partial}{\partial t}\left(\int_V \rho\, dV\right) \tag{2.129}$$

for any volume. It follows from the divergence theorem that

$$\nabla \cdot (\rho\, \mathbf{v}) = -\frac{\partial \rho}{\partial t}. \tag{2.130}$$

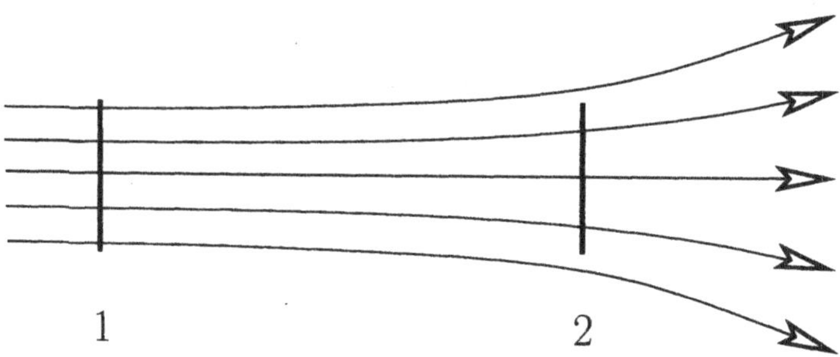

Figure 2.22: *Divergent lines of force.*

This is called the *equation of continuity* of the fluid, since it ensures that fluid is neither created nor destroyed as it flows from place to place. If ρ is constant then the equation of continuity reduces to the previous incompressible result, $\nabla \cdot \mathbf{v} = 0$.

It is sometimes helpful to represent a vector field $\mathbf{A}$ by *lines of force* or *field lines*. The direction of a line of force at any point is the same as the local direction of $\mathbf{A}$. The density of lines (*i.e.*, the number of lines crossing a unit surface perpendicular to $\mathbf{A}$) is equal to $|\mathbf{A}|$. For instance, in Figure 2.22, $|\mathbf{A}|$ is larger at point 1 than at point 2. The number of lines crossing a surface element $d\mathbf{S}$ is $\mathbf{A} \cdot d\mathbf{S}$. So, the net number of lines leaving a closed surface is

$$\oint_S \mathbf{A} \cdot d\mathbf{S} = \int_V \nabla \cdot \mathbf{A} \, dV. \tag{2.131}$$

If $\nabla \cdot \mathbf{A} = 0$ then there is no net flux of lines out of any surface. Such a field is called a *solenoidal* vector field. The simplest example of a solenoidal vector field is one in which the lines of force all form *closed loops*.

2.17 THE LAPLACIAN

So far we have encountered

$$\nabla \phi = \left(\frac{\partial \phi}{\partial x}, \frac{\partial \phi}{\partial y}, \frac{\partial \phi}{\partial z} \right), \tag{2.132}$$

which is a vector field formed from a scalar field, and

$$\nabla \cdot \mathbf{A} = \frac{\partial A_x}{\partial x} + \frac{\partial A_y}{\partial y} + \frac{\partial A_z}{\partial z}, \tag{2.133}$$

which is a scalar field formed from a vector field. There are two ways in which we can combine gradient and divergence. We can either form the vector field $\nabla(\nabla \cdot \mathbf{A})$ or the scalar field $\nabla \cdot (\nabla \phi)$. The former is not particularly interesting, but the scalar field $\nabla \cdot (\nabla \phi)$ turns up in a great many problems in Physics, and is, therefore, worthy of discussion.

Let us introduce the heat-flow vector $\mathbf{h}$, which is the rate of flow of heat energy per unit area across a surface perpendicular to the direction of $\mathbf{h}$. In many substances, heat flows directly down the temperature gradient, so that we can write

$$\mathbf{h} = -\kappa\, \nabla T, \tag{2.134}$$

where κ is the thermal conductivity. The net rate of heat flow $\oint_S \mathbf{h} \cdot d\mathbf{S}$ out of some closed surface S must be equal to the rate of decrease of heat energy in the volume V enclosed by S. Thus, we have

$$\oint_S \mathbf{h} \cdot d\mathbf{S} = -\frac{\partial}{\partial t}\left(\int c\, T\, dV\right), \tag{2.135}$$

where c is the specific heat. It follows from the divergence theorem that

$$\nabla \cdot \mathbf{h} = -c\,\frac{\partial T}{\partial t}. \tag{2.136}$$

Taking the divergence of both sides of Equation (2.134), and making use of Equation (2.136), we obtain

$$\nabla \cdot (\kappa\, \nabla T) = c\,\frac{\partial T}{\partial t}. \tag{2.137}$$

If κ is constant then the above equation can be written

$$\nabla \cdot (\nabla T) = \frac{c}{\kappa}\frac{\partial T}{\partial t}. \tag{2.138}$$

The scalar field $\nabla \cdot (\nabla T)$ takes the form

$$\begin{aligned}\nabla \cdot (\nabla T) &= \frac{\partial}{\partial x}\left(\frac{\partial T}{\partial x}\right) + \frac{\partial}{\partial y}\left(\frac{\partial T}{\partial y}\right) + \frac{\partial}{\partial z}\left(\frac{\partial T}{\partial z}\right) \\ &= \frac{\partial^2 T}{\partial x^2} + \frac{\partial^2 T}{\partial y^2} + \frac{\partial^2 T}{\partial z^2} \equiv \nabla^2 T.\end{aligned} \tag{2.139}$$

Here, the scalar differential operator

$$\nabla^2 \equiv \frac{\partial^2}{\partial x^2} + \frac{\partial^2}{\partial y^2} + \frac{\partial^2}{\partial z^2} \tag{2.140}$$

is called the *Laplacian*. The Laplacian is a good scalar operator (*i.e.*, it is coordinate independent) because it is formed from a combination of divergence (another good scalar operator) and gradient (a good vector operator).

What is the physical significance of the Laplacian? In one dimension, $\nabla^2 T$ reduces to $\partial^2 T/\partial x^2$. Now, $\partial^2 T/\partial x^2$ is positive if $T(x)$ is concave (from above) and negative if it is convex. So, if T is less than the average of T in its surroundings then $\nabla^2 T$ is positive, and *vice versa*.

In two dimensions,

$$\nabla^2 T = \frac{\partial^2 T}{\partial x^2} + \frac{\partial^2 T}{\partial y^2}. \tag{2.141}$$

Consider a local minimum of the temperature. At the minimum, the slope of T increases in all directions, so $\nabla^2 T$ is positive. Likewise, $\nabla^2 T$ is negative at a local maximum. Consider, now, a steep-sided valley in T. Suppose that the bottom of the valley runs parallel to the x-axis. At the bottom of the valley $\partial^2 T/\partial y^2$ is large and positive, whereas $\partial^2 T/\partial x^2$ is small and may even be negative. Thus, $\nabla^2 T$ is positive, and this is associated with T being less than the average local value.

Let us now return to the heat conduction problem:

$$\nabla^2 T = \frac{c}{\kappa}\frac{\partial T}{\partial t}. \tag{2.142}$$

It is clear that if $\nabla^2 T$ is positive then T is locally less than the average value, so $\partial T/\partial t > 0$: *i.e.*, the region heats up. Likewise, if $\nabla^2 T$ is negative then T is locally greater than the average value, and heat flows out of the region: *i.e.*, $\partial T/\partial t < 0$. Thus, the above heat conduction equation makes physical sense.

2.18 CURL

Consider a vector field $\mathbf{A}(\mathbf{r})$, and a loop which lies in one plane. The integral of $\mathbf{A}$ around this loop is written $\oint \mathbf{A} \cdot d\mathbf{l}$, where $d\mathbf{l}$ is a line element of the loop. If $\mathbf{A}$ is a conservative field then $\mathbf{A} = \nabla\phi$ and $\oint \mathbf{A} \cdot d\mathbf{l} = 0$ for all loops. In general, for a non-conservative field, $\oint \mathbf{A} \cdot d\mathbf{l} \neq 0$.

For a small loop we expect $\oint \mathbf{A} \cdot d\mathbf{l}$ to be proportional to the area of the loop. Moreover, for a fixed-area loop we expect $\oint \mathbf{A} \cdot d\mathbf{l}$ to depend on the *orientation* of the loop. One particular orientation will give the maximum value: $\oint \mathbf{A} \cdot d\mathbf{l} = I_{max}$. If the loop subtends an angle θ with

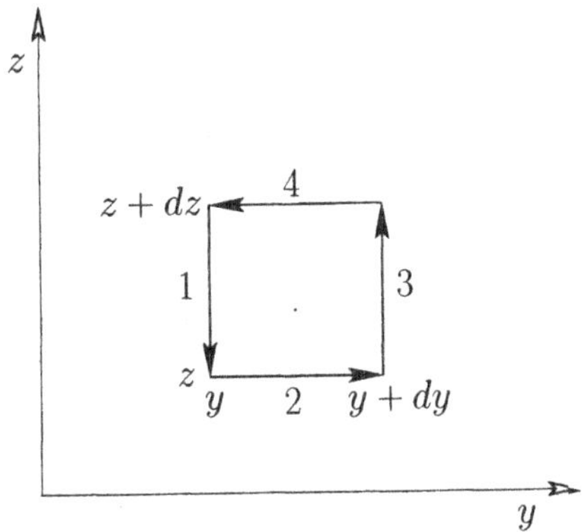

Figure 2.23: *A vector line integral around a small rectangular loop in the y-z plane.*

this optimum orientation then we expect $I = I_{max} \cos\theta$. Let us introduce the vector field **curl A** whose magnitude is

$$|\mathbf{curl\,A}| = \lim_{dS\to 0} \frac{\oint \mathbf{A}\cdot d\mathbf{l}}{dS} \tag{2.143}$$

for the orientation giving I_{max}. Here, dS is the area of the loop. The direction of **curl A** is perpendicular to the plane of the loop, when it is in the orientation giving I_{max}, with the sense given by the right-hand grip rule.

Let us now express **curl A** in terms of the components of **A**. First, we shall evaluate $\oint \mathbf{A}\cdot d\mathbf{l}$ around a small rectangle in the y-z plane—see Figure 2.23. The contribution from sides 1 and 3 is

$$A_z(y+dy)\,dz - A_z(y)\,dz = \frac{\partial A_z}{\partial y}\,dy\,dz. \tag{2.144}$$

The contribution from sides 2 and 4 is

$$-A_y(z+dz)\,dy + A_y(z)\,dy = -\frac{\partial A_y}{\partial y}\,dy\,dz. \tag{2.145}$$

So, the total of all contributions gives

$$\oint \mathbf{A}\cdot d\mathbf{l} = \left(\frac{\partial A_z}{\partial y} - \frac{\partial A_y}{\partial z}\right) dS, \tag{2.146}$$

where $dS = dy\,dz$ is the area of the loop.

Consider a non-rectangular (but still small) loop in the y-z plane. We can divide it into rectangular elements, and form $\oint \mathbf{A} \cdot d\mathbf{l}$ over all the resultant loops. The interior contributions cancel, so we are just left with the contribution from the outer loop. Also, the area of the outer loop is the sum of all the areas of the inner loops. We conclude that

$$\oint \mathbf{A} \cdot d\mathbf{l} = \left(\frac{\partial A_z}{\partial y} - \frac{\partial A_y}{\partial z} \right) dS_x \tag{2.147}$$

is valid for a small loop $d\mathbf{S} = (dS_x, 0, 0)$ of any shape in the y-z plane. Likewise, we can show that if the loop is in the x-z plane then $d\mathbf{S} = (0, dS_y, 0)$ and

$$\oint \mathbf{A} \cdot d\mathbf{l} = \left(\frac{\partial A_x}{\partial z} - \frac{\partial A_z}{\partial x} \right) dS_y. \tag{2.148}$$

Finally, if the loop is in the x-y plane then $d\mathbf{S} = (0, 0, dS_z)$ and

$$\oint \mathbf{A} \cdot d\mathbf{l} = \left(\frac{\partial A_y}{\partial x} - \frac{\partial A_x}{\partial y} \right) dS_z. \tag{2.149}$$

Imagine an arbitrary loop of vector area $d\mathbf{S} = (dS_x, dS_y, dS_z)$. We can construct this out of three vector areas, 1, 2, and 3, directed in the x-, y-, and z-directions, respectively, as indicated in Figure 2.24. If we form the line integral around all three loops then the interior contributions

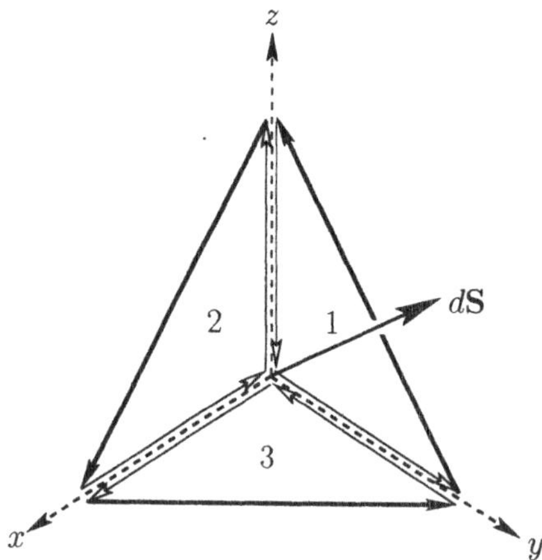

Figure 2.24: *Decomposition of a vector area into its Cartesian components.*

cancel, and we are left with the line integral around the original loop. Thus,

$$\oint \mathbf{A} \cdot d\mathbf{l} = \oint \mathbf{A} \cdot d\mathbf{l}_1 + \oint \mathbf{A} \cdot d\mathbf{l}_2 + \oint \mathbf{A} \cdot d\mathbf{l}_3, \tag{2.150}$$

giving

$$\oint \mathbf{A} \cdot d\mathbf{l} = \mathbf{curl}\,\mathbf{A} \cdot d\mathbf{S} = |\mathbf{curl}\,\mathbf{A}|\,|d\mathbf{S}|\,\cos\theta, \tag{2.151}$$

where

$$\mathbf{curl}\,\mathbf{A} = \left(\frac{\partial A_z}{\partial y} - \frac{\partial A_y}{\partial z}, \frac{\partial A_x}{\partial z} - \frac{\partial A_z}{\partial x}, \frac{\partial A_y}{\partial x} - \frac{\partial A_x}{\partial y}\right), \tag{2.152}$$

and θ is the angle subtended between the directions of $\mathbf{curl}\,\mathbf{A}$ and $d\mathbf{S}$. Note that

$$\mathbf{curl}\,\mathbf{A} = \nabla \times \mathbf{A} = \begin{vmatrix} \mathbf{e}_x & \mathbf{e}_y & \mathbf{e}_z \\ \partial/\partial x & \partial/\partial y & \partial/\partial z \\ A_x & A_y & A_z \end{vmatrix}. \tag{2.153}$$

This demonstrates that $\nabla \times \mathbf{A}$ is a good vector field, since it is the cross product of the ∇ operator (a good vector operator) and the vector field $\mathbf{A}$.

Consider a solid body rotating about the z-axis. The angular velocity is given by $\boldsymbol{\omega} = (0, 0, \omega)$, so the rotation velocity at position $\mathbf{r}$ is

$$\mathbf{v} = \boldsymbol{\omega} \times \mathbf{r} \tag{2.154}$$

[see Equation (2.44)]. Let us evaluate $\nabla \times \mathbf{v}$ on the axis of rotation. The x-component is proportional to the integral $\oint \mathbf{v} \cdot d\mathbf{l}$ around a loop in the y-z plane. This is plainly zero. Likewise, the y-component is also zero. The z-component is $\oint \mathbf{v} \cdot d\mathbf{l}/dS$ around some loop in the x-y plane. Consider a circular loop. We have $\oint \mathbf{v} \cdot d\mathbf{l} = 2\pi r\, \omega\, r$ with $dS = \pi r^2$. Here, r is the perpendicular distance from the rotation axis. It follows that $(\nabla \times \mathbf{v})_z = 2\,\omega$, which is independent of r. So, on the axis, $\nabla \times \mathbf{v} = (0\,,0\,,2\,\omega)$. Off the axis, at position $\mathbf{r}_0$, we can write

$$\mathbf{v} = \boldsymbol{\omega} \times (\mathbf{r} - \mathbf{r}_0) + \boldsymbol{\omega} \times \mathbf{r}_0. \tag{2.155}$$

The first part has the same curl as the velocity field on the axis, and the second part has zero curl, since it is constant. Thus, $\nabla \times \mathbf{v} = (0, 0, 2\,\omega)$ everywhere in the body. This allows us to form a physical picture of

$\nabla \times \mathbf{A}$. If we imagine $\mathbf{A}(\mathbf{r})$ as the velocity field of some fluid, then $\nabla \times \mathbf{A}$ at any given point is equal to twice the local angular rotation velocity: *i.e.*, $2\,\boldsymbol{\omega}$. Hence, a vector field with $\nabla \times \mathbf{A} = \mathbf{0}$ everywhere is said to be *irrotational*.

Another important result of vector field theory is the *curl theorem* or *Stokes' theorem*:

$$\oint_C \mathbf{A} \cdot d\mathbf{l} = \int_S \nabla \times \mathbf{A} \cdot d\mathbf{S}, \tag{2.156}$$

for some (non-planar) surface S bounded by a rim C. This theorem can easily be proved by splitting the loop up into many small rectangular loops, and forming the integral around all of the resultant loops. All of the contributions from the interior loops cancel, leaving just the contribution from the outer rim. Making use of Equation (2.151) for each of the small loops, we can see that the contribution from all of the loops is also equal to the integral of $\nabla \times \mathbf{A} \cdot d\mathbf{S}$ across the whole surface. This proves the theorem.

One immediate consequence of Stokes' theorem is that $\nabla \times \mathbf{A}$ is "incompressible." Consider any two surfaces, S_1 and S_2, which share the same rim—see Figure 2.21. It is clear from Stokes' theorem that $\int \nabla \times \mathbf{A} \cdot d\mathbf{S}$ is the same for both surfaces. Thus, it follows that $\oint \nabla \times \mathbf{A} \cdot d\mathbf{S} = 0$ for any closed surface. However, we have from the divergence theorem that $\oint \nabla \times \mathbf{A} \cdot d\mathbf{S} = \int \nabla \cdot (\nabla \times \mathbf{A})\, dV = 0$ for any volume. Hence,

$$\nabla \cdot (\nabla \times \mathbf{A}) \equiv 0. \tag{2.157}$$

So, $\nabla \times \mathbf{A}$ is a solenoidal field.

We have seen that for a conservative field $\oint \mathbf{A} \cdot d\mathbf{l} = 0$ for any loop. This is entirely equivalent to $\mathbf{A} = \nabla\phi$. However, the magnitude of $\nabla \times \mathbf{A}$ is $\lim_{dS\to 0} \oint \mathbf{A} \cdot d\mathbf{l}/dS$ for some particular loop. It is clear then that $\nabla \times \mathbf{A} = \mathbf{0}$ for a conservative field. In other words,

$$\nabla \times (\nabla\phi) \equiv \mathbf{0}. \tag{2.158}$$

Thus, a conservative field is also an irrotational one.

Finally, it can be shown that

$$\nabla \times (\nabla \times \mathbf{A}) = \nabla(\nabla \cdot \mathbf{A}) - \nabla^2\mathbf{A}, \tag{2.159}$$

where

$$\nabla^2\mathbf{A} = (\nabla^2 A_x, \nabla^2 A_y, \nabla^2 A_z). \tag{2.160}$$

It should be emphasized, however, that the above result is only valid in Cartesian coordinates.

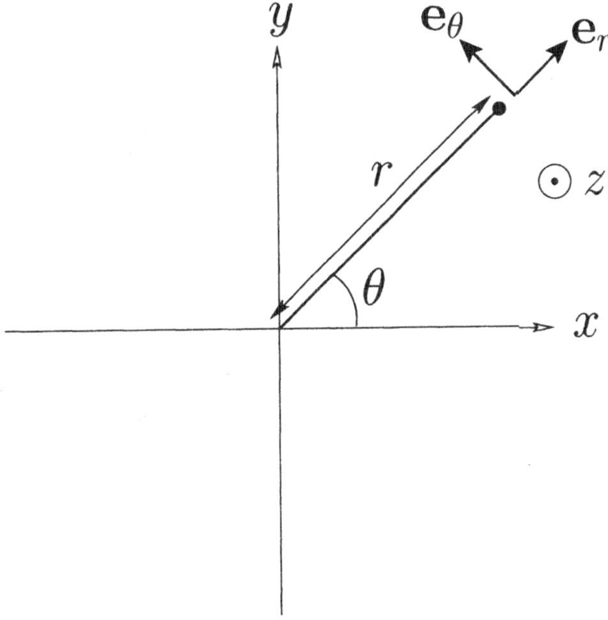

Figure 2.25: *Cylindrical polar coordinates.*

2.19 POLAR COORDINATES

In the *cylindrical* polar coordinate system the Cartesian coordinates x and y are replaced by $r = \sqrt{x^2 + y^2}$ and $\theta = \tan^{-1}(y/x)$. Here, r is the perpendicular distance from the z-axis, and θ the angle subtended between the perpendicular radius vector and the x-axis—see Figure 2.25. A general vector $\mathbf{A}$ is thus written

$$\mathbf{A} = A_r\,\mathbf{e}_r + A_\theta\,\mathbf{e}_\theta + A_z\,\mathbf{e}_z, \tag{2.161}$$

where $\mathbf{e}_r = \nabla r/|\nabla r|$ and $\mathbf{e}_\theta = \nabla\theta/|\nabla\theta|$—see Figure 2.25. Note that the unit vectors $\mathbf{e}_r$, $\mathbf{e}_\theta$, and $\mathbf{e}_z$ are mutually orthogonal. Hence, $A_r = \mathbf{A}\cdot\mathbf{e}_r$, *etc.* The volume element in this coordinate system is $d^3\mathbf{r} = r\,dr\,d\theta\,dz$. Moreover, gradient, divergence, and curl take the form

$$\nabla V = \frac{\partial V}{\partial r}\,\mathbf{e}_r + \frac{1}{r}\frac{\partial V}{\partial\theta}\,\mathbf{e}_\theta + \frac{\partial V}{\partial z}\,\mathbf{e}_z, \tag{2.162}$$

$$\nabla\cdot\mathbf{A} = \frac{1}{r}\frac{\partial}{\partial r}(r\,A_r) + \frac{1}{r}\frac{\partial A_\theta}{\partial\theta} + \frac{\partial A_z}{\partial z}, \tag{2.163}$$

$$\nabla\times\mathbf{A} = \left(\frac{1}{r}\frac{\partial A_z}{\partial\theta} - \frac{\partial A_\theta}{\partial z}\right)\mathbf{e}_r + \left(\frac{\partial A_r}{\partial z} - \frac{\partial A_z}{\partial r}\right)\mathbf{e}_\theta + \left(\frac{1}{r}\frac{\partial}{\partial r}(r\,A_\theta) - \frac{1}{r}\frac{\partial A_r}{\partial\theta}\right)\mathbf{e}_z, \tag{2.164}$$

respectively. Here, $V(\mathbf{r})$ is a general vector field, and $\mathbf{A}(\mathbf{r})$ a general scalar field. Finally, the Laplacian is written

$$\nabla^2 V = \frac{1}{r}\frac{\partial}{\partial r}\left(r\frac{\partial V}{\partial r}\right) + \frac{1}{r^2}\frac{\partial^2 V}{\partial\theta^2} + \frac{\partial^2 V}{\partial z^2}. \tag{2.165}$$

In the *spherical* polar coordinate system the Cartesian coordinates x, y, and z are replaced by $r = \sqrt{x^2+y^2+z^2}$, $\theta = \cos^{-1}(z/r)$, and $\phi = \tan^{-1}(y/x)$. Here, r is the distance from the origin, θ the angle subtended between the radius vector and the z-axis, and ϕ the angle subtended between the projection of the radius vector onto the x-y plane and the x-axis—see Figure 2.26. Note that r and θ in the spherical polar system are *not* the same as their counterparts in the cylindrical system. A general vector $\mathbf{A}$ is written

$$\mathbf{A} = A_r\,\mathbf{e}_r + A_\theta\,\mathbf{e}_\theta + A_\phi\,\mathbf{e}_\phi, \tag{2.166}$$

where $\mathbf{e}_r = \nabla r/|\nabla r|$, $\mathbf{e}_\theta = \nabla\theta/|\nabla\theta|$, and $\mathbf{e}_\phi = \nabla\phi/|\nabla\phi|$. The unit vectors $\mathbf{e}_r$, $\mathbf{e}_\theta$, and $\mathbf{e}_\phi$ are mutually orthogonal. Hence, $A_r = \mathbf{A}\cdot\mathbf{e}_r$, *etc.* The volume element in this coordinate system is $d^3\mathbf{r} = r^2\,\sin\theta\,dr\,d\theta\,d\phi$. Moreover, gradient, divergence, and curl take the form

$$\nabla V = \frac{\partial V}{\partial r}\,\mathbf{e}_r + \frac{1}{r}\frac{\partial V}{\partial\theta}\,\mathbf{e}_\theta + \frac{1}{r\,\sin\theta}\frac{\partial V}{\partial\phi}\,\mathbf{e}_\phi, \tag{2.167}$$

$$\nabla\cdot\mathbf{A} = \frac{1}{r^2}\frac{\partial}{\partial r}(r^2\,A_r) + \frac{1}{r\,\sin\theta}\frac{\partial}{\partial\theta}(\sin\theta\,A_\theta) + \frac{1}{r\,\sin\theta}\frac{\partial A_\phi}{\partial\phi}, \tag{2.168}$$

$$\nabla\times\mathbf{A} = \left(\frac{1}{r\,\sin\theta}\frac{\partial}{\partial\theta}(\sin\theta\,A_\phi) - \frac{1}{r\,\sin\theta}\frac{\partial A_\theta}{\partial\phi}\right)\mathbf{e}_r + \left(\frac{1}{r\,\sin\theta}\frac{\partial A_r}{\partial\phi} - \frac{1}{r}\frac{\partial}{\partial r}(r\,A_\phi)\right)\mathbf{e}_\theta + \left(\frac{1}{r}\frac{\partial}{\partial r}(r\,A_\theta) - \frac{1}{r}\frac{\partial A_r}{\partial\theta}\right)\mathbf{e}_\phi, \tag{2.169}$$

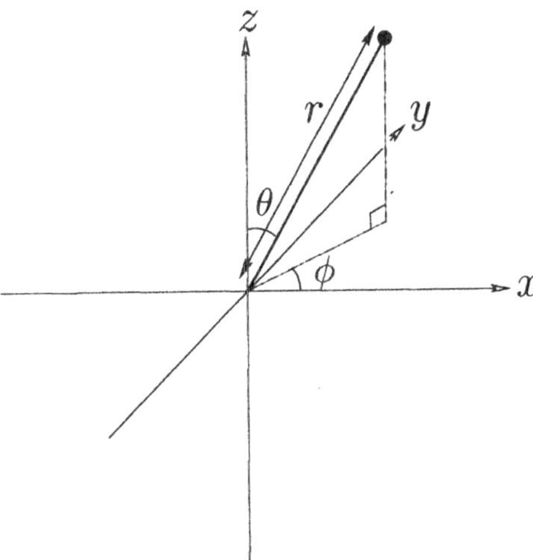

Figure 2.26: *Spherical polar coordinates.*

respectively. Here, $V(\mathbf{r})$ is a general vector field, and $\mathbf{A}(\mathbf{r})$ a general scalar field. Finally, the Laplacian is written

$$\nabla^2 V = \frac{1}{r^2}\frac{\partial}{\partial r}\left(r^2\frac{\partial V}{\partial r}\right) + \frac{1}{r^2\sin\theta}\frac{\partial}{\partial\theta}\left(\sin\theta\,\frac{\partial V}{\partial\theta}\right) + \frac{1}{r^2\sin^2\theta}\frac{\partial^2 V}{\partial\phi^2}. \tag{2.170}$$

2.20 EXERCISES

2.1. Prove the trigonometric law of sines

$$\frac{\sin a}{A} = \frac{\sin b}{B} = \frac{\sin c}{C}$$

using vector methods. Here, a, b, and c are the three angles of a plane triangle, and A, B, and C the lengths of the corresponding opposite sides.

2.2. Demonstrate using vectors that the diagonals of a parallelogram bisect one another. In addition, show that if the diagonals of a quadrilateral bisect one another then it is a parallelogram.

2.3. From the inequality

$$\mathbf{a}\cdot\mathbf{b} = |\mathbf{a}|\,|\mathbf{b}|\,\cos\theta \le |\mathbf{a}|\,|\mathbf{b}|$$

deduce the triangle inequality

$$|\mathbf{a}+\mathbf{b}| \le |\mathbf{a}| + |\mathbf{b}|.$$

2.4. Identify the following surfaces:

(a) $|\mathbf{r}| = a$,
(b) $\mathbf{r} \cdot \mathbf{n} = b$,
(c) $\mathbf{r} \cdot \mathbf{n} = c\,|\mathbf{r}|$,
(d) $|\mathbf{r} - (\mathbf{r} \cdot \mathbf{n})\,\mathbf{n}| = d$.

Here, $\mathbf{r}$ is the position vector, a, b, c, and d are positive constants, and $\mathbf{n}$ is a fixed unit vector.

2.5. Let $\mathbf{a}$, $\mathbf{b}$, and $\mathbf{c}$ be coplanar vectors related via

$$\alpha\,\mathbf{a} + \beta\,\mathbf{b} + \gamma\,\mathbf{c} = \mathbf{0},$$

where α, β, and γ are not all zero. Show that the condition for the points with position vectors $u\,\mathbf{a}$, $v\,\mathbf{b}$, and $w\,\mathbf{c}$ to be colinear is

$$\frac{\alpha}{u} + \frac{\beta}{v} + \frac{\gamma}{w} = 0.$$

2.6. If $\mathbf{p}$, $\mathbf{q}$, and $\mathbf{r}$ are any vectors, demonstrate that $\mathbf{a} = \mathbf{q} + \lambda\,\mathbf{r}$, $\mathbf{b} = \mathbf{r} + \mu\,\mathbf{p}$, and $\mathbf{c} = \mathbf{p} + \nu\,\mathbf{q}$ are coplanar provided that $\lambda\,\mu\,\nu = -1$, where λ, μ, and ν are scalars. Show that this condition is satisfied when $\mathbf{a}$ is perpendicular to $\mathbf{p}$, $\mathbf{b}$ to $\mathbf{q}$, and $\mathbf{c}$ to $\mathbf{r}$.

2.7. The vectors $\mathbf{a}$, $\mathbf{b}$, and $\mathbf{c}$ are not coplanar, and form a non-orthogonal vector base. The vectors $\mathbf{A}$, $\mathbf{B}$, and $\mathbf{C}$, defined by

$$\mathbf{A} = \frac{\mathbf{b} \times \mathbf{c}}{\mathbf{a} \cdot \mathbf{b} \times \mathbf{c}},$$

plus cyclic permutations, are said to be *reciprocal vectors*. Show that

$$\mathbf{a} = (\mathbf{B} \times \mathbf{C})/(\mathbf{A} \cdot \mathbf{B} \times \mathbf{C}),$$

plus cyclic permutations.

2.8. In the notation of the previous question, demonstrate that the plane passing through points $\mathbf{a}/\alpha$, $\mathbf{b}/\beta$, and $\mathbf{c}/\gamma$ is normal to the direction of the vector

$$\mathbf{h} = \alpha\,\mathbf{A} + \beta\,\mathbf{B} + \gamma\,\mathbf{C}.$$

In addition, show that the perpendicular distance of the plane from the origin is $|\mathbf{h}|^{-1}$.

2.9. A ray of light moving in the direction $\mathbf{k}$ impinges on a surface at a point where its normal is $\mathbf{n}$. If $\mathbf{k}'$ is the direction of the refracted ray, then Snell's law (in vector form) gives

$$\mu\,\mathbf{k} \times \mathbf{n} = \mu'\,\mathbf{k}' \times \mathbf{n},$$

where μ and μ' are the refractive indices of the media on either side of the surface. Note that $\mathbf{k}$, $\mathbf{k}'$, and $\mathbf{n}$ are unit vectors: *i.e.*, they all have unit length.

Show that $\mathbf{k}'$ is in the plane of $\mathbf{k}$ and $\mathbf{n}$, and that

$$\mathbf{k}' = \frac{\mu}{\mu'}\left[\mathbf{k} - (\mathbf{k}\cdot\mathbf{n})\,\mathbf{n} + \theta\,\mathbf{n}\right],$$

where

$$\theta^2 = (\mathbf{k}\cdot\mathbf{n})^2 + \left(\frac{\mu'}{\mu}\right)^2 - 1.$$

Which sign of the square-root must be taken? Under what circumstances is θ imaginary, and what does this imply physically?

2.10. Consider the following vector field:

$$\mathbf{A}(\mathbf{r}) = (8\,x^3 + 3\,x^2\,y^2,\ 2\,x^3\,y + 6\,y,\ 6).$$

Is this field conservative? Is it solenoidal? Is it irrotational? Justify your answers. Calculate $\oint_C \mathbf{A}\cdot d\mathbf{r}$, where the curve C is a unit circle in the x-y plane, centered on the origin, and the direction of integration is clockwise looking down the z-axis.

2.11. Consider the following vector field:

$$\mathbf{A}(\mathbf{r}) = (3\,x\,y^2\,z^2 - y^2,\ -y^3\,z^2 + x^2\,y,\ 3\,x^2 - x^2\,z).$$

Is this field conservative? Is it solenoidal? Is it irrotational? Justify your answers. Calculate the flux of $\mathbf{A}$ out of a unit sphere centered on the origin.

2.12. Find the gradients of the following scalar functions of the position vector $\mathbf{r} = (x, y, z)$:

(a) $\mathbf{k}\cdot\mathbf{r}$,
(b) $|\mathbf{r}|^n$,
(c) $|\mathbf{r} - \mathbf{k}|^{-n}$,
(d) $\cos(\mathbf{k}\cdot\mathbf{r})$.

Here, $\mathbf{k}$ is a fixed vector.

2.13. Find the divergences and curls of the following vector fields:

(a) $\mathbf{k}\times\mathbf{r}$,
(b) $|\mathbf{r}|^n\,\mathbf{r}$,
(c) $|\mathbf{r} - \mathbf{k}|^n\,(\mathbf{r} - \mathbf{k})$,
(d) $\mathbf{a}\,\cos(\mathbf{k}\cdot\mathbf{r})$.

Here, $\mathbf{k}$ and $\mathbf{a}$ are fixed vectors.

2.14. Calculate $\nabla^2\phi$ when $\phi = f(|\mathbf{r}|)$. Find f if $\nabla^2\phi = 0$.

2.15. Find the Cartesian components of the basis vectors $\mathbf{e}_r$, $\mathbf{e}_\theta$, and $\mathbf{e}_z$ of the cylindrical polar coordinate system. Verify that the vectors are mutually orthogonal. Do the same for the basis vectors $\mathbf{e}_r$, $\mathbf{e}_\theta$, and $\mathbf{e}_\phi$ of the spherical polar coordinate system.

Chapter 3

TIME-INDEPENDENT MAXWELL EQUATIONS

3.1 INTRODUCTION

In this chapter, we shall recast the familiar force laws of electrostatics and magnetostatics as vector field equations.

3.2 COULOMB'S LAW

Between 1785 and 1787, the French physicist Charles Augustine de Coulomb performed a series of experiments involving electric charges, and eventually established what is nowadays known as *Coulomb's law*. According to this law, the force acting between two static electric charges is central, inverse-square, and proportional to the product of the charges. Two like charges repel one another, whereas two unlike charges attract. Suppose that two charges, q_1 and q_2, are located at position vectors $\mathbf{r}_1$ and $\mathbf{r}_2$, respectively. The electrical force acting on the second charge is written

$$\mathbf{f}_2 = \frac{q_1\, q_2}{4\pi\epsilon_0} \frac{\mathbf{r}_2 - \mathbf{r}_1}{|\mathbf{r}_2 - \mathbf{r}_1|^3} \qquad (3.1)$$

in vector notation—see Figure 3.1. An equal and opposite force acts on the first charge, in accordance with Newton's third law of motion. The SI unit of electric charge is the coulomb (C). The magnitude of the charge on an electron is 1.6022×10^{-19} C. Finally, the universal constant ϵ_0 is called the *permittivity of free space*, and takes the value

$$\epsilon_0 = 8.8542 \times 10^{-12}\ \mathrm{C}^2\,\mathrm{N}^{-1}\mathrm{m}^{-2}. \qquad (3.2)$$

Suppose that two masses, m_1 and m_2, are located at position vectors $\mathbf{r}_1$ and $\mathbf{r}_2$, respectively. According to Newton's law of gravity, the gravitational force acting on the second mass is written

$$\mathbf{f}_2 = -G\, m_1\, m_2 \frac{\mathbf{r}_2 - \mathbf{r}_1}{|\mathbf{r}_2 - \mathbf{r}_1|^3} \qquad (3.3)$$

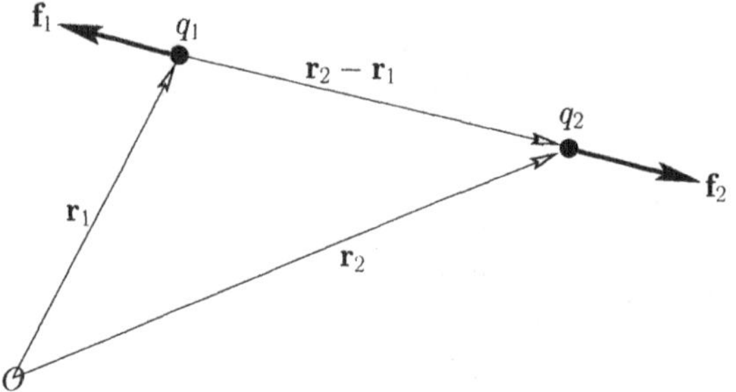

Figure 3.1: *Coulomb's law.*

in vector notation. The gravitational constant G takes the value

$$G = 6.6726 \times 10^{-11}\ \mathrm{N\,m^2\,kg^{-2}}. \tag{3.4}$$

Note that Coulomb's law has the same mathematical form as Newton's law of gravity. In particular, they are both *inverse-square* force laws: *i.e.*,

$$|\mathbf{f}_2| \propto \frac{1}{|\mathbf{r}_2 - \mathbf{r}_1|^2}. \tag{3.5}$$

However, these laws differ in two crucial respects. Firstly, the force due to gravity is always *attractive* (there is no such thing as a negative mass). Secondly, the magnitudes of the two forces are vastly different. Consider the ratio of the electrical and gravitational forces acting on two particles. This ratio is a constant, independent of the relative positions of the particles, and is given by

$$\frac{|\mathbf{f}_{\text{electrical}}|}{|\mathbf{f}_{\text{gravitational}}|} = \frac{|q_1|}{m_1}\frac{|q_2|}{m_2}\frac{1}{4\pi\epsilon_0\,G}. \tag{3.6}$$

For electrons, the charge-to-mass ratio is $|q|/m = 1.759 \times 10^{11}\ \mathrm{C\,kg^{-1}}$, so

$$\frac{|\mathbf{f}_{\text{electrical}}|}{|\mathbf{f}_{\text{gravitational}}|} = 4.17 \times 10^{42}. \tag{3.7}$$

This is a colossal number! Suppose we were studying a physics problem involving the motion of particles under the action of two forces with the same range, but differing in magnitude by a factor 10^{42}. It would seem a plausible approximation (to say the least) to start the investgation by neglecting the weaker force altogether. Applying this reasoning to the motion of particles in the Universe, we would expect the Universe to be governed entirely by electrical forces. However, this is not the case.

The force which holds us to the surface of the Earth, and prevents us from floating off into space, is gravity. The force which causes the Earth to orbit the Sun is also gravity. In fact, on astronomical length-scales gravity is the dominant force, and electrical forces are largely irrelevant. The key to understanding this paradox is that there are both positive and negative electric charges, whereas there are only positive gravitational "charges." This means that gravitational forces are always cumulative, whereas electrical forces can cancel one another out. Suppose, for the sake of argument, that the Universe starts out with randomly distributed electric charges. Initially, we expect electrical forces to completely dominate gravity. These forces try to make every positive charge get as far away as possible from the other positive charges in the Universe, and as close as possible to the other negative charges. After a while, we expect the positive and negative charges to form close pairs. Just how close is determined by Quantum Mechanics, but, in general, it is fairly close: *i.e.*, about 10^{-10} m. The electrical forces due to the charges in each pair effectively cancel one another out on length-scales much larger than the mutual spacing of the pair. However, it is only possible for gravity to be the dominant long-range force in the Universe if the number of positive charges is almost equal to the number of negative charges. In this situation, every positive charge can find a negative charge to team up with, and there are virtually no charges left over. In order for the cancellation of long-range electrical forces to be effective, the relative difference in the number of positive and negative charges in the Universe must be incredibly small. In fact, positive and negative charges have to cancel one another to such accuracy that most physicists believe that the net electrical charge of the Universe is *exactly* zero. But, it is not sufficient for the Universe to start out with zero charge. Suppose there were some elementary particle process which did not conserve electric charge. Even if this were to go on at a very low rate, it would not take long before the fine balance between positive and negative charges in the Universe was wrecked. So, it is important that electric charge is a *conserved* quantity (*i.e.*, the net charge of the Universe can neither increase or decrease). As far as we know, this is the case. To date, no elementary particle reactions have been discovered which create or destroy net electric charge.

In summary, there are two long-range forces in the Universe, electricity and gravity. The former is enormously stronger than the latter, but is usually "hidden" away inside neutral atoms. The fine balance of forces due to negative and positive electric charges starts to break down on atomic scales. In fact, interatomic and intermolecular forces are all electrical in nature. So, electrical forces are basically what prevent us from falling though the floor. But, this is electromagnetism on the microscopic or atomic scale—what is usually termed *Quantum Electromagnetism*. This book is about *Classical Electromagnetism*. That is, electromagnetism on length-scales much larger than the atomic scale. Classical Electromagnetism generally describes phenomena in which some sort of "violence" is done to matter, so that the close pairing of negative and positive charges is disrupted. This allows electrical forces to manifest themselves on macroscopic length-scales. Of course, very little disruption is necessary before gigantic forces are generated. Hence, it is no coincidence that the vast majority of useful machines which humankind has devised during the last century or so are electrical in nature.

Coulomb's law and Newton's law are both examples of what are usually referred to as *action-at-a-distance* laws. According to Equations (3.1) and (3.3), if the first charge or mass is moved then the force acting on the second charge or mass responds *immediately*. In particular, equal and opposite forces act on the two charges or masses at all times. However, this cannot be correct according to Einstein's Special Theory of Relativity, which implies that the maximum speed with which information can propagate through the Universe is the speed of light in vacuum. So, if the first charge or mass is moved then there must always be a time delay (*i.e.*, at least the time needed for a light signal to propagate between the two charges or masses) before the second charge or mass responds. Consider a rather extreme example. Suppose the first charge or mass is suddenly annihilated. The second charge or mass only finds out about this some time later. During this time interval, the second charge or mass experiences an electrical or gravitational force which is as if the first charge or mass were still there. So, during this period, there is an action but no reaction, which violates Newton's third law of motion. It is clear that action at a distance is not compatible with Relativity, and, consequently, that Newton's third law of motion is not strictly true. Of course, Newton's third law is intimately tied up with the conservation of linear momentum in the Universe. This is a concept which most physicists are loath to abandon. It turns out that we can "rescue" momentum conservation by abandoning action-at-a-distance theories, and instead adopting so-called *field theories* in which there is a medium, called a field, which transmits the force from one particle to another. Of course, in electromagnetism there are two fields—the electric field, and the magnetic field. Electromagnetic forces are transmitted via these fields at the speed of light, which implies that the laws of Relativity are never violated. Moreover, the fields can soak up energy and momentum. This means that even when the actions and reactions acting on charged particles are not quite equal and opposite, momentum is still conserved. We can bypass some of the problematic aspects of action at a distance by only considering *steady-state* situations. For the moment, this is how we shall proceed.

Consider N charges, q_1 though q_N, which are located at position vectors $\mathbf{r}_1$ through $\mathbf{r}_N$, respectively. Electrical forces obey what is known as the *principle of superposition*: *i.e.*, the electrical force acting on a test charge q at position vector **r** is simply the vector sum of all of the Coulomb law forces exerted on it by each of the N charges taken in isolation. In other words, the electrical force exerted by the ith charge (say) on the test charge is the same as if all of the other charges were not there. Thus, the force acting on the test charge is given by

$$\mathbf{f}(\mathbf{r}) = q \sum_{i=1}^{N} \frac{q_i}{4\pi\epsilon_0} \frac{\mathbf{r} - \mathbf{r}_i}{|\mathbf{r} - \mathbf{r}_i|^3}. \tag{3.8}$$

It is helpful to define a vector field $\mathbf{E}(\mathbf{r})$, called the *electric field*, which is the force exerted on a unit test charge located at position vector **r**. So, the force on a test charge is written

$$\mathbf{f} = q\,\mathbf{E}, \tag{3.9}$$

and the electric field is given by

$$\mathbf{E}(\mathbf{r}) = \sum_{i=1}^{N} \frac{q_i}{4\pi\epsilon_0} \frac{\mathbf{r} - \mathbf{r}_i}{|\mathbf{r} - \mathbf{r}_i|^3}. \tag{3.10}$$

At this point, we have no reason to believe that the electric field has any real physical existence. It is just a useful device for calculating the force which acts on test charges placed at various locations.

The electric field from a single charge q located at the origin is purely radial, points outward if the charge is positive, inward if it is negative, and has magnitude

$$E_r(r) = \frac{q}{4\pi\epsilon_0\, r^2}, \tag{3.11}$$

where $r = |\mathbf{r}|$. We can represent an electric field by *field-lines*. The direction of the lines indicates the direction of the local electric field, and the density of the lines perpendicular to this direction is proportional to the magnitude of the local electric field. It follows from Equation (3.11) that the number of field-lines crossing the surface of a sphere centered on a point charge (which is equal to E_r times the area, $4\pi r^2$, of the surface) is *independent* of the radius of the sphere. Thus, the field of a point positive charge is represented by a group of equally spaced, unbroken, straight

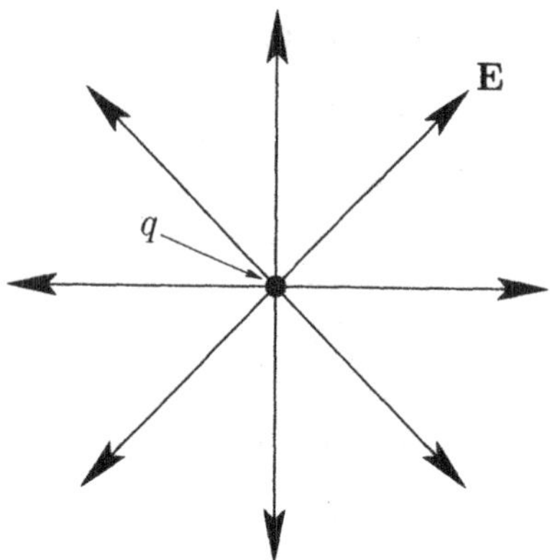

Figure 3.2: *Electric field-lines generated by a positive charge.*

lines radiating from the charge—see Figure 3.2. Likewise, field of a point negative charge is represented by a group of equally spaced, unbroken, straight-lines converging on the charge.

The electric field from a collection of charges is simply the vector sum of the fields from each of the charges taken in isolation. In other words, electric fields are completely *superposable*. Suppose that, instead of having discrete charges, we have a continuous distribution of charge represented by a *charge density* $\rho(\mathbf{r})$. Thus, the charge at position vector $\mathbf{r}'$ is $\rho(\mathbf{r}')\,d^3\mathbf{r}'$, where $d^3\mathbf{r}'$ is the volume element at $\mathbf{r}'$. It follows from a simple extension of Equation (3.10) that the electric field generated by this charge distribution is

$$\mathbf{E}(\mathbf{r}) = \frac{1}{4\pi\epsilon_0}\int \rho(\mathbf{r}')\,\frac{\mathbf{r}-\mathbf{r}'}{|\mathbf{r}-\mathbf{r}'|^3}\,d^3\mathbf{r}', \tag{3.12}$$

where the volume integral is over all space, or, at least, over all space for which $\rho(\mathbf{r}')$ is non-zero. We shall sometimes refer to the above result as *Coulomb's law*, since it is essentially equivalent to Equation (3.1).

3.3 THE ELECTRIC SCALAR POTENTIAL

Suppose that $\mathbf{r} = (x, y, z)$ and $\mathbf{r}' = (x', y', z')$ in Cartesian coordinates. The x component of $(\mathbf{r}-\mathbf{r}')/|\mathbf{r}-\mathbf{r}'|^3$ is written

$$\frac{x-x'}{[(x-x')^2+(y-y')^2+(z-z')^2]^{3/2}}. \tag{3.13}$$

However, it is easily demonstrated that

$$\frac{x-x'}{[(x-x')^2+(y-y')^2+(z-z')^2]^{3/2}} = -\frac{\partial}{\partial x}\left(\frac{1}{[(x-x')^2+(y-y')^2+(z-z')^2]^{1/2}}\right). \tag{3.14}$$

Since there is nothing special about the x-axis, we can write

$$\frac{\mathbf{r}-\mathbf{r}'}{|\mathbf{r}-\mathbf{r}'|^3} = -\nabla\left(\frac{1}{|\mathbf{r}-\mathbf{r}'|}\right), \tag{3.15}$$

where $\nabla \equiv (\partial/\partial x,\ \partial/\partial y,\ \partial/\partial z)$ is a differential operator which involves the components of $\mathbf{r}$ but not those of $\mathbf{r}'$. It follows from Equation (3.12) that

$$\mathbf{E} = -\nabla\phi, \tag{3.16}$$

where

$$\phi(\mathbf{r}) = \frac{1}{4\pi\epsilon_0}\int \frac{\rho(\mathbf{r}')}{|\mathbf{r}-\mathbf{r}'|}\, d^3\mathbf{r}'. \tag{3.17}$$

Thus, the electric field generated by a collection of fixed charges can be written as the gradient of a scalar field—known as the *electric scalar potential*—and this field can be expressed as a simple volume integral involving the charge distribution.

The scalar potential generated by a charge q located at the origin is

$$\phi(r) = \frac{q}{4\pi\epsilon_0\, r}. \tag{3.18}$$

According to Equation (3.10), the scalar potential generated by a set of N discrete charges q_i, located at $\mathbf{r}_i$, is

$$\phi(\mathbf{r}) = \sum_{i=1}^{N} \phi_i(\mathbf{r}), \tag{3.19}$$

where

$$\phi_i(\mathbf{r}) = \frac{q_i}{4\pi\epsilon_0\, |\mathbf{r}-\mathbf{r}_i|}. \tag{3.20}$$

Thus, the scalar potential is just the sum of the potentials generated by each of the charges taken in isolation.

Suppose that a particle of charge q is taken along some path from point P to point Q. The net work done on the particle by electrical forces is

$$\mathcal{W} = \int_P^Q \mathbf{f} \cdot d\mathbf{l}, \tag{3.21}$$

where $\mathbf{f}$ is the electrical force, and $d\mathbf{l}$ is a line element along the path. Making use of Equations (3.9) and (3.16), we obtain

$$\mathcal{W} = q \int_P^Q \mathbf{E} \cdot d\mathbf{l} = -q \int_P^Q \nabla\phi \cdot d\mathbf{l} = -q\,[\phi(Q) - \phi(P)]. \tag{3.22}$$

Thus, the work done on the particle is simply minus its charge times the difference in electric potential between the end point and the beginning point. This quantity is clearly *independent* of the path taken between P and Q. So, an electric field generated by stationary charges is an example of a *conservative* field. In fact, this result follows immediately from vector field theory once we are told, in Equation (3.16), that the electric field is the gradient of a scalar potential. The work done on the particle when it is taken around a closed loop is zero, so

$$\oint_C \mathbf{E} \cdot d\mathbf{l} = 0 \tag{3.23}$$

for any closed loop C. This implies from Stokes' theorem that

$$\nabla \times \mathbf{E} = \mathbf{0} \tag{3.24}$$

for any electric field generated by stationary charges. Equation (3.24) also follows directly from Equation (3.16), since $\nabla \times \nabla\phi \equiv \mathbf{0}$ for any scalar potential ϕ.

The SI unit of electric potential is the volt, which is equivalent to a joule per coulomb. Thus, according to Equation (3.22), the electrical work done on a particle when it is taken between two points is the product of minus its charge and the voltage difference between the points.

We are familiar with the idea that a particle moving in a gravitational field possesses potential energy as well as kinetic energy. If the particle moves from point P to a lower point Q then the gravitational field does work on the particle causing its kinetic energy to increase. The increase in kinetic energy of the particle is balanced by an equal decrease in its potential energy, so that the overall energy of the particle is a conserved quantity. Therefore, the work done on the particle as it moves from P to Q is *minus* the difference in its gravitational potential energy between points Q and P. Of course, it only makes sense to talk about gravitational potential energy because the gravitational field is *conservative*. Thus, the work done in taking a particle between two points is *path independent*, and, therefore, well-defined. This means that the difference in potential energy of the particle between the beginning and end points is also well-defined. We have already seen that an electric field generated by stationary charges is a conservative field. In follows that we can define an electrical potential energy of a particle moving in such a field. By analogy with gravitational fields, the work done in taking a particle from point P to point Q is equal to minus the difference in potential energy of the particle between points Q and P. It follows from Equation (3.22) that the potential energy of the particle at a general point Q, relative to some reference point P (where the potential energy is set to zero), is given by

$$W(Q) = q\,\phi(Q). \tag{3.25}$$

Free particles try to move down gradients of potential energy, in order to attain a minimum potential energy state. Thus, free particles in the Earth's gravitational field tend to fall downward. Likewise, positive charges moving in an electric field tend to migrate toward regions with the most negative voltage, and *vice versa* for negative charges.

The scalar electric potential is undefined to an additive constant. So, the transformation

$$\phi(\mathbf{r}) \to \phi(\mathbf{r}) + c \tag{3.26}$$

leaves the electric field unchanged according to Equation (3.16). The potential can be fixed unambiguously by specifying its value at a single point. The usual convention is to say that the potential is zero at infinity. This convention is implicit in Equation (3.17), where it can be seen that $\phi \to 0$ as $|\mathbf{r}| \to \infty$, provided that the total charge $\int \rho(\mathbf{r}')\, d^3\mathbf{r}'$ is finite.

3.4 GAUSS' LAW

Consider a single charge q located at the origin. The electric field generated by such a charge is given by Equation (3.11). Suppose that we surround the charge by a concentric spherical surface S of radius r—see

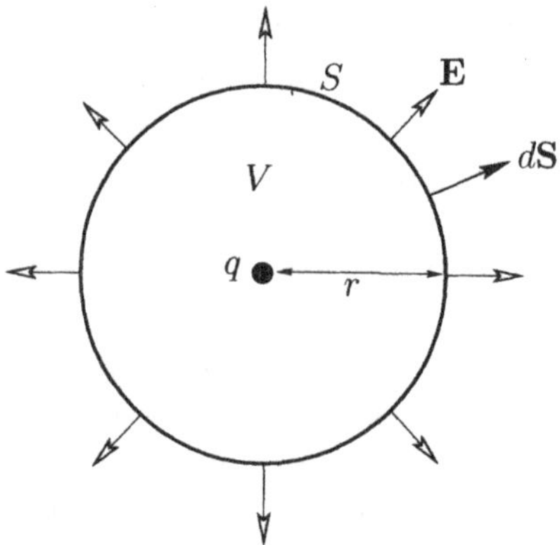

Figure 3.3: *Gauss' law.*

Figure 3.3. The flux of the electric field through this surface is given by

$$\oint_S \mathbf{E}\cdot d\mathbf{S} = \oint_S E_r\, dS_r = E_r(r)\, 4\pi r^2 = \frac{q}{4\pi\epsilon_0\, r^2}\, 4\pi r^2 = \frac{q}{\epsilon_0}, \tag{3.27}$$

since the normal to the surface is always parallel to the local electric field. However, we also know from Gauss' theorem that

$$\oint_S \mathbf{E}\cdot d\mathbf{S} = \int_V \nabla\cdot\mathbf{E}\; d^3\mathbf{r}, \tag{3.28}$$

where V is the volume enclosed by surface S. Let us evaluate $\nabla\cdot\mathbf{E}$ directly. In Cartesian coordinates, the field is written

$$\mathbf{E} = \frac{q}{4\pi\epsilon_0}\left(\frac{x}{r^3}, \frac{y}{r^3}, \frac{z}{r^3}\right), \tag{3.29}$$

where $r^2 = x^2 + y^2 + z^2$. So,

$$\frac{\partial E_x}{\partial x} = \frac{q}{4\pi\epsilon_0}\left(\frac{1}{r^3} - \frac{3\,x}{r^4}\frac{x}{r}\right) = \frac{q}{4\pi\epsilon_0}\frac{r^2 - 3\,x^2}{r^5}. \tag{3.30}$$

Here, use has been made of

$$\frac{\partial r}{\partial x} = \frac{x}{r}. \tag{3.31}$$

Formulae analogous to Equation (3.30) can be obtained for $\partial E_y/\partial y$ and $\partial E_z/\partial z$. The divergence of the field is thus given by

$$\nabla\cdot\mathbf{E} = \frac{\partial E_x}{\partial x} + \frac{\partial E_y}{\partial y} + \frac{\partial E_z}{\partial z} = \frac{q}{4\pi\epsilon_0}\frac{3\,r^2 - 3\,x^2 - 3\,y^2 - 3\,z^2}{r^5} = 0. \tag{3.32}$$

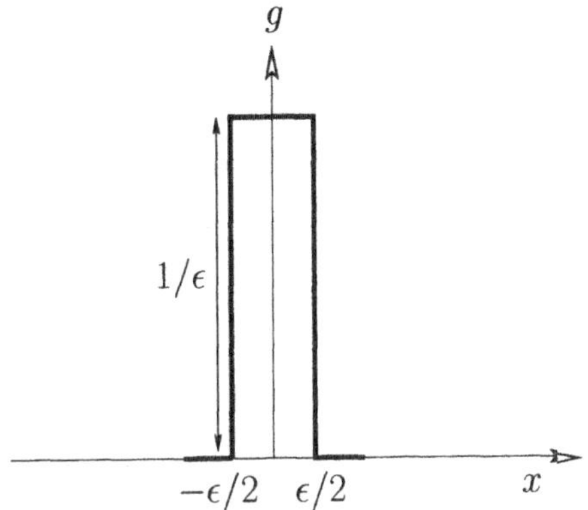

Figure 3.4: *A box-car function.*

This is a puzzling result! We have from Equations (3.27) and (3.28) that

$$\int_V \nabla \cdot \mathbf{E}\, d^3\mathbf{r} = \frac{q}{\epsilon_0}, \tag{3.33}$$

and yet we have just proved that $\nabla \cdot \mathbf{E} = 0$. This paradox can be resolved after a close examination of Equation (3.32). At the origin ($r = 0$) we find that $\nabla \cdot \mathbf{E} = 0/0$, which means that $\nabla \cdot \mathbf{E}$ can take any value at this point. Thus, Equations (3.32) and (3.33) can be reconciled if $\nabla \cdot \mathbf{E}$ is some sort of "spike" function: *i.e.*, it is zero everywhere except arbitrarily close to the origin, where it becomes very large. This must occur in such a manner that the volume integral over the spike is finite.

Let us examine how we might construct a one-dimensional spike function. Consider the "box-car" function

$$g(x, \epsilon) = \begin{cases} 1/\epsilon & \text{for } |x| < \epsilon/2 \\ 0 & \text{otherwise} \end{cases} \tag{3.34}$$

—see Figure 3.4. It is clear that

$$\int_{-\infty}^{\infty} g(x, \epsilon)\, dx = 1. \tag{3.35}$$

Now consider the function

$$\delta(x) = \lim_{\epsilon \to 0} g(x, \epsilon). \tag{3.36}$$

This is zero everywhere except arbitrarily close to $x = 0$. According to Equation (3.35), it also possess a finite integral;

$$\int_{-\infty}^{\infty} \delta(x)\, dx = 1. \tag{3.37}$$

Thus, $\delta(x)$ has all of the required properties of a spike function. The one-dimensional spike function $\delta(x)$ is called the *Dirac delta-function*, after the Cambridge physicist Paul Dirac who invented it in 1927 whilst investigating Quantum Mechanics. The delta-function is an example of what mathematicians call a *generalized function*: it is not well-defined at $x = 0$, but its integral is nevertheless well-defined. Consider the integral

$$\int_{-\infty}^{\infty} f(x)\, \delta(x)\, dx, \tag{3.38}$$

where $f(x)$ is a function which is well-behaved in the vicinity of $x = 0$. Since the delta-function is zero everywhere apart from very close to $x = 0$, it is clear that

$$\int_{-\infty}^{\infty} f(x)\, \delta(x)\, dx = f(0) \int_{-\infty}^{\infty} \delta(x)\, dx = f(0), \tag{3.39}$$

where use has been made of Equation (3.37). The above equation, which is valid for any well-behaved function, $f(x)$, is effectively the definition of a delta-function. A simple change of variables allows us to define $\delta(x - x_0)$, which is a spike function centered on $x = x_0$. Equation (3.39) gives

$$\int_{-\infty}^{\infty} f(x)\, \delta(x - x_0)\, dx = f(x_0). \tag{3.40}$$

We actually want a three-dimensional spike function: *i.e.*, a function which is zero everywhere apart from arbitrarily close to the origin, and whose volume integral is unity. If we denote this function by $\delta(\mathbf{r})$ then it is easily seen that the three-dimensional delta-function is the product of three one-dimensional delta-functions:

$$\delta(\mathbf{r}) = \delta(x)\, \delta(y)\, \delta(z). \tag{3.41}$$

This function is clearly zero everywhere except the origin. But is its volume integral unity? Let us integrate over a cube of dimension $2\,a$ which is centered on the origin, and aligned along the Cartesian axes. This volume integral is obviously separable, so that

$$\int \delta(\mathbf{r})\, d^3\mathbf{r} = \int_{-a}^{a} \delta(x)\, dx \int_{-a}^{a} \delta(y)\, dy \int_{-a}^{a} \delta(z)\, dz. \tag{3.42}$$

The integral can be turned into an integral over all space by taking the limit $a \to \infty$. However, we know that for one-dimensional delta-functions $\int_{-\infty}^{\infty} \delta(x)\, dx = 1$, so it follows from the above equation that

$$\int \delta(\mathbf{r})\, d^3\mathbf{r} = 1, \tag{3.43}$$

which is the desired result. A simple generalization of previous arguments yields

$$\int f(\mathbf{r})\, \delta(\mathbf{r})\, d^3\mathbf{r} = f(\mathbf{0}), \tag{3.44}$$

where $f(\mathbf{r})$ is any well-behaved scalar field. Finally, we can change variables and write

$$\delta(\mathbf{r} - \mathbf{r}') = \delta(x - x')\, \delta(y - y')\, \delta(z - z'), \tag{3.45}$$

which is a three-dimensional spike function centered on $\mathbf{r} = \mathbf{r}'$. It is easily demonstrated that

$$\int f(\mathbf{r})\, \delta(\mathbf{r} - \mathbf{r}')\, d^3\mathbf{r} = f(\mathbf{r}'). \tag{3.46}$$

Up to now, we have only considered volume integrals taken over all space. However, it should be obvious that the above result also holds for integrals over any finite volume V which contains the point $\mathbf{r} = \mathbf{r}'$. Likewise, the integral is zero if V does not contain $\mathbf{r} = \mathbf{r}'$.

Let us now return to the problem in hand. The electric field generated by a charge q located at the origin has $\nabla \cdot \mathbf{E} = 0$ everywhere apart from the origin, and also satisfies

$$\int_V \nabla \cdot \mathbf{E}\, d^3\mathbf{r} = \frac{q}{\epsilon_0} \tag{3.47}$$

for a spherical volume V centered on the origin. These two facts imply that

$$\nabla \cdot \mathbf{E} = \frac{q}{\epsilon_0}\, \delta(\mathbf{r}), \tag{3.48}$$

where use has been made of Equation (3.43).

At this stage, vector field theory has yet to show its worth. After all, we have just spent an inordinately long time proving something using vector field theory which we previously proved in one line [see Equation (3.27)] using conventional analysis. It is time to demonstrate the power of vector field theory. Consider, again, a charge q at the origin surrounded by a spherical surface S which is centered on the origin. Suppose that we now displace the surface S, so that it is no longer centered on the origin. What is the flux of the electric field out of S? This is no longer a simple problem for conventional analysis, because the normal to the surface is not parallel to the local electric field. However, using vector field theory this problem is no more difficult than the previous one. We have

$$\oint_S \mathbf{E}\cdot d\mathbf{S} = \int_V \nabla\cdot\mathbf{E}\, d^3\mathbf{r} \tag{3.49}$$

from Gauss' theorem, plus Equation (3.48). From these equations, it is clear that the flux of $\mathbf{E}$ out of S is q/ϵ_0 for a spherical surface displaced from the origin. However, the flux becomes zero when the displacement is sufficiently large that the origin is no longer enclosed by the sphere. It is possible to prove this via conventional analysis, but it is certainly not easy. Suppose that the surface S is not spherical but is instead highly distorted. What now is the flux of $\mathbf{E}$ out of S? This is a virtually impossible problem in conventional analysis, but it is still easy using vector field theory. Gauss' theorem and Equation (3.48) tell us that the flux is q/ϵ_0 provided that the surface contains the origin, and that the flux is zero otherwise. This result is completely independent of the shape of S.

Consider N charges q_i located at $\mathbf{r}_i$. A simple generalization of Equation (3.48) gives

$$\nabla\cdot\mathbf{E} = \sum_{i=1}^{N} \frac{q_i}{\epsilon_0}\,\delta(\mathbf{r}-\mathbf{r}_i). \tag{3.50}$$

Thus, Gauss' theorem (3.49) implies that

$$\oint_S \mathbf{E}\cdot d\mathbf{S} = \int_V \nabla\cdot\mathbf{E}\, d^3\mathbf{r} = \frac{Q}{\epsilon_0}, \tag{3.51}$$

where Q is the total charge enclosed by the surface S. This result is called *Gauss' law*, and does not depend on the shape of the surface.

Suppose, finally, that instead of having a set of discrete charges, we have a continuous charge distribution described by a charge density $\rho(\mathbf{r})$. The charge contained in a small rectangular volume of dimensions dx, dy, and dz located at position $\mathbf{r}$ is $Q = \rho(\mathbf{r})\, dx\, dy\, dz$. However, if we integrate $\nabla\cdot\mathbf{E}$ over this volume element we obtain

$$\nabla\cdot\mathbf{E}\, dx\, dy\, dz = \frac{Q}{\epsilon_0} = \frac{\rho\, dx\, dy\, dz}{\epsilon_0}, \tag{3.52}$$

where use has been made of Equation (3.51). Here, the volume element is assumed to be sufficiently small that $\nabla \cdot \mathbf{E}$ does not vary significantly across it. Thus, we get

$$\nabla \cdot \mathbf{E} = \frac{\rho}{\epsilon_0}. \tag{3.53}$$

This is the first of four field equations, called Maxwell's equations, which together form a complete description of electromagnetism. Of course, our derivation of Equation (3.53) is only valid for electric fields generated by stationary charge distributions. In principle, additional terms might be required to describe fields generated by moving charge distributions. However, it turns out that this is not the case, and that Equation (3.53) is universally valid.

Equation (3.53) is a differential equation describing the electric field generated by a set of charges. We already know the solution to this equation when the charges are stationary: it is given by Equation (3.12),

$$\mathbf{E}(\mathbf{r}) = \frac{1}{4\pi\epsilon_0} \int \rho(\mathbf{r}') \, \frac{\mathbf{r} - \mathbf{r}'}{|\mathbf{r} - \mathbf{r}'|^3} \, d^3\mathbf{r}'. \tag{3.54}$$

Equations (3.53) and (3.54) can be reconciled provided

$$\nabla \cdot \left(\frac{\mathbf{r} - \mathbf{r}'}{|\mathbf{r} - \mathbf{r}'|^3} \right) = -\nabla^2 \left(\frac{1}{|\mathbf{r} - \mathbf{r}'|} \right) = 4\pi \, \delta(\mathbf{r} - \mathbf{r}'), \tag{3.55}$$

where use has been made of Equation (3.15). It follows that

$$\begin{aligned} \nabla \cdot \mathbf{E}(\mathbf{r}) &= \frac{1}{4\pi\epsilon_0} \int \rho(\mathbf{r}') \, \nabla \cdot \left(\frac{\mathbf{r} - \mathbf{r}'}{|\mathbf{r} - \mathbf{r}'|^3} \right) d^3\mathbf{r}' \\ &= \int \frac{\rho(\mathbf{r}')}{\epsilon_0} \, \delta(\mathbf{r} - \mathbf{r}') \, d^3\mathbf{r}' = \frac{\rho(\mathbf{r})}{\epsilon_0}, \end{aligned} \tag{3.56}$$

which is the desired result. The most general form of Gauss' law, Equation (3.51), is obtained by integrating Equation (3.53) over a volume V surrounded by a surface S, and making use of Gauss' theorem:

$$\oint_S \mathbf{E} \cdot d\mathbf{S} = \frac{1}{\epsilon_0} \int_V \rho(\mathbf{r}) \, d^3\mathbf{r}. \tag{3.57}$$

One particularly interesting application of Gauss' law is *Earnshaw's theorem*, which states that it is impossible for a collection of charged particles to remain in static equilibrium solely under the influence of (classical) electrostatic forces. For instance, consider the motion of the ith particle in the electric field, $\mathbf{E}(\mathbf{r})$, generated by all of the other static particles. The equilibrium position of the ith particle corresponds to some point $\mathbf{r}_i$, where $\mathbf{E}(\mathbf{r}_i) = \mathbf{0}$. By implication, $\mathbf{r}_i$ does not correspond to the equilibrium position of any other particle. However, in order for $\mathbf{r}_i$ to be a *stable* equilibrium point, the particle must experience a *restoring force* when it moves a small distance away from $\mathbf{r}_i$ in *any* direction. Assuming that the ith particle is positively charged, this means that the electric field must point radially toward $\mathbf{r}_i$ at all neighboring points. Hence, if we apply Gauss' law to a small sphere centered on $\mathbf{r}_i$ then there must be a negative flux of $\mathbf{E}$ through the surface of the sphere, implying the presence of a negative charge at $\mathbf{r}_i$. However, there is no such charge at $\mathbf{r}_i$. Hence, we conclude that $\mathbf{E}$ cannot point radially toward $\mathbf{r}_i$ at all neighboring points. In other words, there must be some neighboring points at which $\mathbf{E}$ is directed *away* from $\mathbf{r}_i$. Hence, a positively charged particle placed at $\mathbf{r}_i$ can always escape by moving to such points. One corollary of Earnshaw's theorem is that classical electrostatics cannot account for the stability of atoms and molecules.

As an example of the use of Gauss' law, let us calculate the electric field generated by a spherically symmetric charge annulus of inner radius a, and outer radius b, centered on the origin, and carrying a uniformly distributed charge Q. Now, from symmetry, we expect a spherically symmetric charge distribution to generate a spherically symmetric potential, $\phi(r)$. It therefore follows that the electric field is both spherically symmetric and radial: *i.e.*, $\mathbf{E} = E_r(r)\,\mathbf{e}_r$. Let us apply Gauss' law to an imaginary spherical surface, of radius r, centered on the origin—see Figure 3.5. Such a surface is generally known as a *Gaussian surface*. Now, according to Gauss' law, the flux of the electric field out of the surface is equal to the enclosed charge, divided by ϵ_0. The flux is easy to calculate since the electric field is everywhere perpendicular to the surface. We obtain

$$4\pi\, r^2\, E_r(r) = \frac{Q(r)}{\epsilon_0},$$

where $Q(r)$ is the charge enclosed by a Gaussian surface of radius r. Now, it is evident that

$$Q(r) = \begin{cases} 0 & r < a \\ [(r^3 - a^3)/(b^3 - a^3)]\,Q & a \le r \le b \\ Q & b < r \end{cases}. \tag{3.58}$$

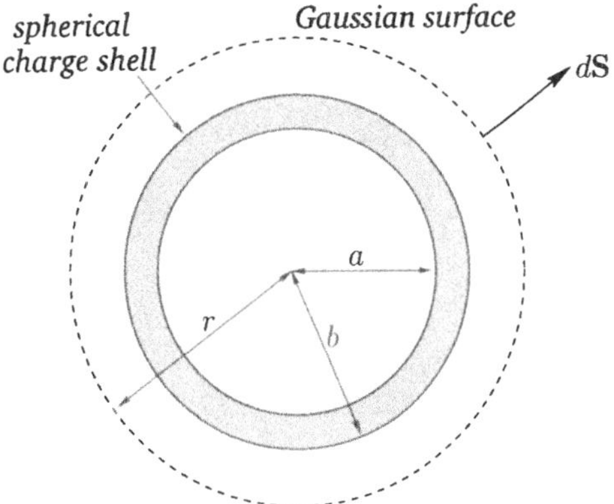

Figure 3.5: *An example use of Gauss' law.*

Hence,

$$E_r(r) = \begin{cases} 0 & r < a \\ [Q/(4\pi\epsilon_0\, r^2)]\,[(r^3 - a^3)/(b^3 - a^3)] & a \leq r \leq b \\ Q/(4\pi\epsilon_0\, r^2) & b < r \end{cases}. \tag{3.59}$$

The above electric field distribution illustrates two important points. Firstly, the electric field generated outside a spherically symmetric charge distribution is the same that which would be generated if all of the charge in the distribution were concentrated at its center. Secondly, zero electric field is generated inside an empty cavity surrounded by a spherically symmetric charge distribution.

We can easily determine the electric potential associated with the above electric field using

$$\frac{\partial\phi(r)}{\partial r} = -E_r(r). \tag{3.60}$$

The boundary conditions are that $\phi(\infty) = 0$, and that ϕ is *continuous* at $r = a$ and $r = b$. (Of course, a discontinuous potential would lead to an infinite electric field, which is unphysical.) It follows that

$$\phi(r) = \begin{cases} [Q/(4\pi\epsilon_0)]\,(3/2)\,[(b^2 - a^2)/(b^3 - a^3)] & r < a \\ [Q/(4\pi\epsilon_0\, r)]\,[(3b^3\, r - r^3 - 2\, a^3)/2\,(b^3 - a^3)] & a \leq r \leq b \\ Q/(4\pi\epsilon_0\, r) & b < r \end{cases}. \tag{3.61}$$

Hence, the work done in slowly moving a charge from infinity to the center of the distribution (which is minus the work done by the electric field) is

$$W = q\,[\phi(0) - \phi(\infty)] = \frac{q\,Q}{4\pi\epsilon_0}\,\frac{3}{2}\left(\frac{b^2 - a^2}{b^3 - a^3}\right). \tag{3.62}$$

3.5 POISSON'S EQUATION

We have seen that the electric field generated by a set of stationary charges can be written as the gradient of a scalar potential, so that

$$\mathbf{E} = -\nabla\phi. \tag{3.63}$$

This equation can be combined with the field equation (3.53) to give a partial differential equation for the scalar potential:

$$\nabla^2\phi = -\frac{\rho}{\epsilon_0}. \tag{3.64}$$

This is an example of a very famous type of partial differential equation known as *Poisson's equation*.

In its most general form, Poisson's equation is written

$$\nabla^2 u = v, \tag{3.65}$$

where $u(\mathbf{r})$ is some scalar potential which is to be determined, and $v(\mathbf{r})$ is a known "source function." The most common boundary condition applied to this equation is that the potential u is zero at infinity. The solutions to Poisson's equation are completely superposable. Thus, if u_1 is the potential generated by the source function v_1, and u_2 is the potential generated by the source function v_2, so that

$$\nabla^2 u_1 = v_1, \qquad \nabla^2 u_2 = v_2, \tag{3.66}$$

then the potential generated by $v_1 + v_2$ is $u_1 + u_2$, since

$$\nabla^2(u_1 + u_2) = \nabla^2 u_1 + \nabla^2 u_2 = v_1 + v_2. \tag{3.67}$$

Poisson's equation has this property because it is *linear* in both the potential and the source term.

The fact that the solutions to Poisson's equation are superposable suggests a general method for solving this equation. Suppose that we could construct all of the solutions generated by point sources. Of course, these solutions must satisfy the appropriate boundary conditions. Any general source function can be built up out of a set of suitably weighted point sources, so the general solution of Poisson's equation must be expressible as a similarly weighted sum over the point source solutions. Thus, once we know all of the point source solutions we can construct any other solution. In mathematical terminology, we require the solution to

$$\nabla^2 G(\mathbf{r}, \mathbf{r}') = \delta(\mathbf{r} - \mathbf{r}') \tag{3.68}$$

which goes to zero as $|\mathbf{r}| \to \infty$. The function $G(\mathbf{r}, \mathbf{r}')$ is the solution generated by a unit point source located at position $\mathbf{r}'$. This function is known to mathematicians as a *Green's function*. The solution generated by a general source function $v(\mathbf{r})$ is simply the appropriately weighted sum of all of the Green's function solutions:

$$u(\mathbf{r}) = \int G(\mathbf{r}, \mathbf{r}')\, v(\mathbf{r}')\, d^3\mathbf{r}'. \tag{3.69}$$

We can easily demonstrate that this is the correct solution:

$$\nabla^2 u(\mathbf{r}) = \int \left[\nabla^2 G(\mathbf{r}, \mathbf{r}')\right] v(\mathbf{r}')\, d^3\mathbf{r}' = \int \delta(\mathbf{r} - \mathbf{r}')\, v(\mathbf{r}')\, d^3\mathbf{r}' = v(\mathbf{r}). \tag{3.70}$$

Let us return to Equation (3.64):

$$\nabla^2 \phi = -\frac{\rho}{\epsilon_0}. \tag{3.71}$$

The Green's function for this equation satisfies Equation (3.68) with $|G| \to \infty$ as $|\mathbf{r}| \to 0$. It follows from Equation (3.55) that

$$G(\mathbf{r}, \mathbf{r}') = -\frac{1}{4\pi} \frac{1}{|\mathbf{r} - \mathbf{r}'|}. \tag{3.72}$$

Note, from Equation (3.20), that the Green's function has the same form as the potential generated by a point charge. This is hardly surprising, given the definition of a Green's function. It follows from Equation (3.69) and (3.72) that the general solution to Poisson's equation, (3.71), is written

$$\phi(\mathbf{r}) = \frac{1}{4\pi\epsilon_0} \int \frac{\rho(\mathbf{r}')}{|\mathbf{r} - \mathbf{r}'|}\, d^3\mathbf{r}'. \tag{3.73}$$

In fact, we have already obtained this solution by another method [see Equation (3.17)].

3.6 AMPÈRE'S EXPERIMENTS

As legend has it, in 1820 the Danish physicist Hans Christian Ørsted was giving a lecture demonstration of various electrical and magnetic effects. Suddenly, much to his surprise, he noticed that the needle of a compass he was holding was deflected when he moved it close to a current-carrying wire. Up until then, magnetism has been thought of as solely a property of some rather unusual rocks called loadstones. Word of this discovery spread quickly along the scientific grapevine, and the French physicist Andre Marie Ampère immediately decided to investigate further. Ampère's apparatus consisted (essentially) of a long straight wire carrying an electric current I. Ampère quickly discovered that the needle of a small compass maps out a series of concentric circular loops in the plane perpendicular to a current-carrying wire—see Figure 3.6. The direction of circulation around these magnetic loops is conventionally taken to be the direction in which the *North* pole of the compass needle points. Using this convention, the circulation of the loops is given by a right-hand rule: if the thumb of the right-hand points along the direction of the current then the fingers of the right-hand circulate in the same sense as the magnetic loops.

Ampère's next series of experiments involved bringing a short test wire, carrying a current I', close to the original wire, and investigating the force exerted on the test wire—see Figure 3.7. This experiment is

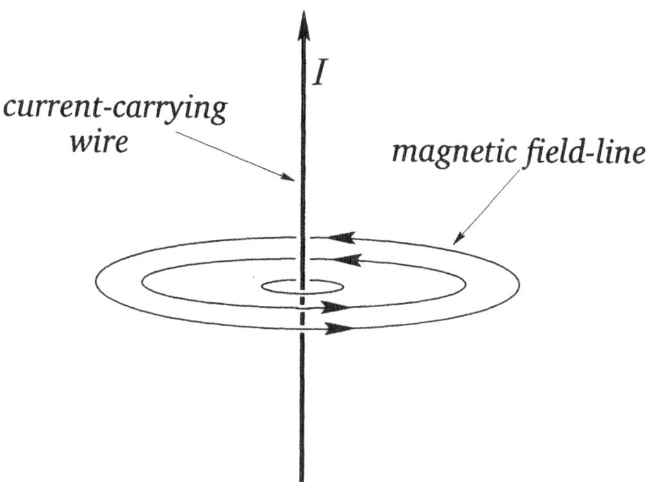

Figure 3.6: *Magnetic loops around a current-carrying wire.*

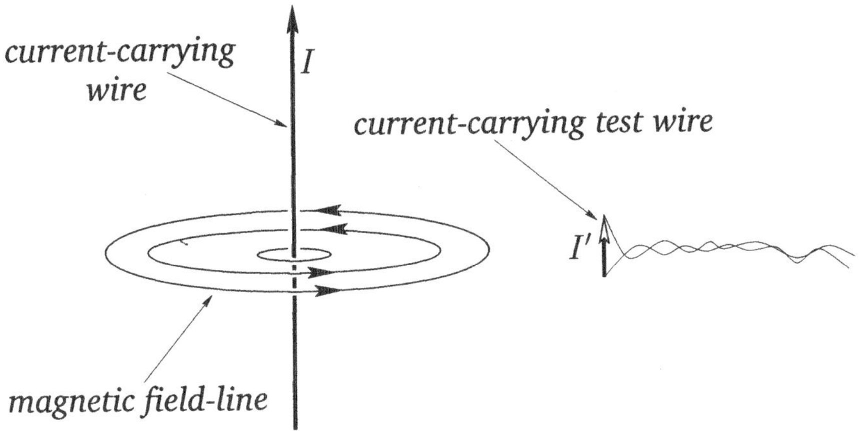

Figure 3.7: *Ampère's experiment.*

not quite as clear cut as Coulomb's experiment because, unlike electric charges, electric currents cannot exist as point entities—they have to flow in complete circuits. We must imagine that the circuit which connects with the central wire is sufficiently far away that it has no appreciable influence on the outcome of the experiment. The circuit which connects with the test wire is more problematic. Fortunately, if the feed wires are twisted around each other, as indicated in Figure 3.7, then they effectively cancel one another out, and also do not influence the outcome of the experiment.

Ampère discovered that the force exerted on the test wire is directly proportional to its length. He also made the following observations. If the current in the test wire (*i.e.*, the test current) flows parallel to the current in the central wire then the two wires attract one another. If the current in the test wire is reversed then the two wires repel one another. If the test current points radially toward the central wire (and the current in the central wire flows upward) then the test wire is subject to a downward force. If the test current is reversed then the force is upward. If the test current is rotated in a single plane, so that it starts parallel to the central current and ends up pointing radially toward it, then the force on the test wire is of constant magnitude, and is always at right-angles to the test current. If the test current is parallel to a magnetic loop then there is no force exerted on the test wire. If the test current is rotated in a single plane, so that it starts parallel to the central current and ends up pointing along a magnetic loop, then the magnitude of the force on the test wire attenuates like $\cos\theta$ (where θ is the angle the current is turned through—$\theta = 0$ corresponds to the case where the test current is parallel to the central current), and its direction is again always at right-angles to the test current. Finally, Ampère was able to establish that the attractive force between two parallel current-carrying wires is proportional to the product of the two currents, and falls off like the inverse of the perpendicular distance between the wires.

This rather complicated force law can be summed up succinctly in vector notation provided that we define a vector field $\mathbf{B}(\mathbf{r})$, called the *magnetic field*, whose direction is always parallel to the loops mapped out by a small compass. The dependence of the force per unit length, $\mathbf{F}$, acting on a test wire with the different possible orientations of the test current is described by

$$\mathbf{F} = \mathbf{I}' \times \mathbf{B}, \tag{3.74}$$

where $\mathbf{I}'$ is a vector whose direction and magnitude are the same as those of the test current. Incidentally, the SI unit of electric current is the ampere (A), which is the same as a coulomb per second. The SI unit of magnetic field-strength is the tesla (T), which is the same as a newton per ampere per meter. The variation of the force per unit length acting on a test wire with the strength of the central current and the perpendicular distance r to the central wire is summed up by saying that the magnetic field-strength is proportional to I and inversely proportional to r. Thus, defining cylindrical polar coordinates aligned along the axis of the central current, we have

$$B_\theta = \frac{\mu_0 I}{2\pi r}, \tag{3.75}$$

with $B_r = B_z = 0$. The constant of proportionality μ_0 is called the *permeability of free space*, and takes the value

$$\mu_0 = 4\pi \times 10^{-7}\,\mathrm{N\,A^{-2}}. \tag{3.76}$$

The concept of a magnetic field allows the calculation of the force on a test wire to be conveniently split into two parts. In the first part, we calculate the magnetic field generated by the current flowing in the central wire. This field circulates in the plane normal to the wire: its magnitude is proportional to the central current, and inversely proportional to the perpendicular distance from the wire. In the second part, we use Equation (3.74) to calculate the force per unit length acting on a short current-carrying wire located in the magnetic field generated by the central current. This force is perpendicular to both the magnetic field and the direction of the test current. Note that, at this stage, we have no reason to suppose that the magnetic field has any real physical existence. It is introduced merely to facilitate the calculation of the force exerted on the test wire by the central wire.

3.7 THE LORENTZ FORCE

The flow of an electric current down a conducting wire is ultimately due to the motion of electrically charged particles (in most cases, electrons) through the conducting medium. It seems reasonable, therefore, that the force exerted on the wire when it is placed in a magnetic field is really the resultant of the forces exerted on these moving charges. Let us suppose that this is the case.

Let A be the (uniform) cross-sectional area of the (cylindrical) wire, and let n be the number density of mobile charges in the conductor. Suppose that the mobile charges each have charge q and velocity $\mathbf{v}$. We must assume that the conductor also contains stationary charges, of charge $-q$ and number density n (say), so that the net charge density in the wire is zero. In most conductors, the mobile charges are electrons, and the stationary charges are ions. The magnitude of the electric current flowing through the wire is simply the number of coulombs per second which flow past a given point. In one second, a mobile charge moves a distance v, so all of the charges contained in a cylinder of cross-sectional area A and length v flow past a given point. Thus, the magnitude of the current is $q\,n\,A\,v$. The direction of the current is the same as the direction of motion of the charges, so the vector current is $\mathbf{I}' = q\,n\,A\,\mathbf{v}$. According to Equation (3.74), the force per unit length acting on the wire is

$$\mathbf{F} = q\,n\,A\,\mathbf{v} \times \mathbf{B}. \tag{3.77}$$

However, a unit length of the wire contains $n\,A$ moving charges. So, assuming that each charge is subject to an equal force from the magnetic field (we have no reason to suppose otherwise), the force acting on an individual charge is

$$\mathbf{f} = q\,\mathbf{v} \times \mathbf{B}. \tag{3.78}$$

We can combine this with Equation (3.9) to give the force acting on a charge q moving with velocity $\mathbf{v}$ in an electric field $\mathbf{E}$ and a magnetic field $\mathbf{B}$:

$$\mathbf{f} = q\,\mathbf{E} + q\,\mathbf{v} \times \mathbf{B}. \tag{3.79}$$

This is called the *Lorentz force law*, after the Dutch physicist Hendrik Antoon Lorentz who first formulated it. The electric force on a charged particle is parallel to the local electric field. The magnetic force, however, is perpendicular to both the local magnetic field and the particle's direction of motion. No magnetic force is exerted on a stationary charged particle.

The equation of motion of a free particle of charge q and mass m moving in electric and magnetic fields is

$$m\frac{d\mathbf{v}}{dt} = q\,\mathbf{E} + q\,\mathbf{v}\times\mathbf{B}, \tag{3.80}$$

according to the Lorentz force law. This equation of motion was first verified in a famous experiment carried out by the Cambridge physicist J.J. Thompson in 1897. Thompson was investigating *cathode rays*, a then mysterious form of radiation emitted by a heated metal element held at a large negative voltage (*i.e.*, a cathode) with respect to another metal element (*i.e.*, an anode) in an evacuated tube. German physicists held that cathode rays were a form of electromagnetic radiation, whilst British and French physicists suspected that they were, in reality, a stream of charged particles. Thompson was able to demonstrate that the latter view was correct. In Thompson's experiment, the cathode rays passed though a region of "crossed" electric and magnetic fields (still in vacuum). The fields were perpendicular to the original trajectory of the rays, and were also mutually perpendicular.

Let us analyze Thompson's experiment. Suppose that the rays are originally traveling in the x-direction, and are subject to a uniform electric field E in the z-direction and a uniform magnetic field B in the $-y$-direction—see Figure 3.8. Let us assume, as Thompson did, that cathode rays are a stream of particles of mass m and charge q. The equation of motion of the particles in the z-direction is

$$m\,\frac{d^2z}{dt^2} = q\,(E - v\,B), \tag{3.81}$$

where v is the velocity of the particles in the x-direction. Thompson started off his experiment by only turning on the electric field in his apparatus, and measuring the deflection d of the ray in the z-direction after it had traveled a distance l through the electric field. It is clear from the equation of motion that

$$d = \frac{q}{m}\frac{E\,t^2}{2} = \frac{q}{m}\frac{E\,l^2}{2\,v^2}, \tag{3.82}$$

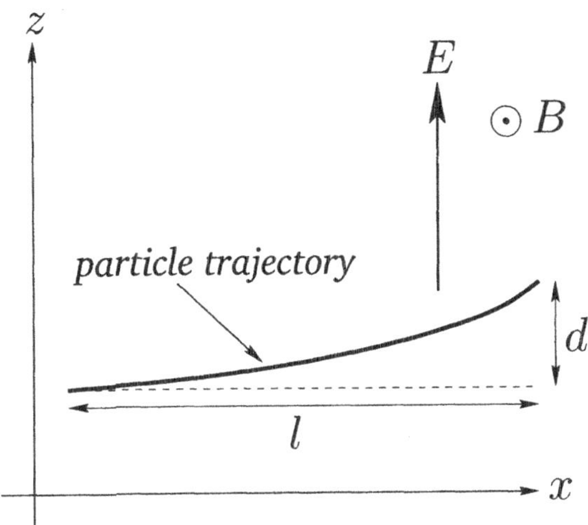

Figure 3.8: *Thompson's experiment.*

where the "time of flight" t is replaced by l/v. This formula is only valid if $d \ll l$, which is assumed to be the case. Next, Thompson turned on the magnetic field in his apparatus, and adjusted it so that the cathode ray was no longer deflected. The lack of deflection implies that the net force on the particles in the z-direction was zero. In other words, the electric and magnetic forces balanced exactly. It follows from Equation (3.81) that with a properly adjusted magnetic field-strength

$$v = \frac{E}{B}. \tag{3.83}$$

Equations (3.82) and (3.83) can be combined and rearranged to give the charge-to-mass ratio of the particles in terms of measured quantities:

$$\frac{q}{m} = \frac{2\,d\,E}{l^2\,B^2}. \tag{3.84}$$

Using this method, Thompson inferred that cathode rays were made up of negatively charged particles (the sign of the charge is obvious from the direction of the deflection in the electric field) with a charge-to-mass ratio of -1.7×10^{11} C/kg. A decade later, in 1908, the American Robert Millikan performed his famous "oil drop" experiment, and discovered that mobile electric charges are quantized in units of -1.6×10^{-19} C. Assuming that mobile electric charges and the particles which make up cathode rays are one and the same thing, Thompson's and Millikan's experiments imply that the mass of these particles is 9.4×10^{-31} kg. Of course, this is the mass of an electron (the modern value is 9.1×10^{-31} kg), and -1.6×10^{-19} C is the charge of an electron. Thus, cathode rays are, in fact, streams of electrons which are emitted from a heated cathode, and then accelerated because of the large voltage difference between the cathode and anode.

Consider, now, a particle of mass m and charge q moving in a uniform magnetic field, $\mathbf{B} = B\,\mathbf{e}_z$. According to Equation (3.80), the particle's equation of motion can be written:

$$m\,\frac{d\mathbf{v}}{dt} = q\,\mathbf{v} \times \mathbf{B}. \tag{3.85}$$

This reduces to

$$\frac{dv_x}{dt} = \Omega\,v_y, \tag{3.86}$$

$$\frac{dv_y}{dt} = -\Omega\,v_x, \tag{3.87}$$

$$\frac{dv_z}{dt} = 0. \tag{3.88}$$

Here, $\Omega = q\,B/m$ is called the *cyclotron frequency*. The above equations can easily be solved to give

$$v_x = v_\perp\,\cos(\Omega\,t), \tag{3.89}$$

$$v_y = -\,v_\perp\,\sin(\Omega\,t), \tag{3.90}$$

$$v_z = v_\parallel, \tag{3.91}$$

and

$$x = \frac{v_\perp}{\Omega}\,\sin(\Omega\,t), \tag{3.92}$$

$$y = \frac{v_\perp}{\Omega}\,\cos(\Omega\,t), \tag{3.93}$$

$$z = v_\parallel\,t. \tag{3.94}$$

According to these equations, the particle trajectory is a *spiral* whose axis is parallel to the magnetic field—see Figure 3.9. The radius of the spiral is $\rho = v_\perp/\Omega$, where $v_\perp$ is the particle's constant speed in the plane perpendicular to the magnetic field. Here, ρ is termed the *Larmor radius*. The particle drifts parallel to the magnetic field at a constant velocity, $v_\parallel$. Finally, the particle gyrates in the plane perpendicular to the magnetic

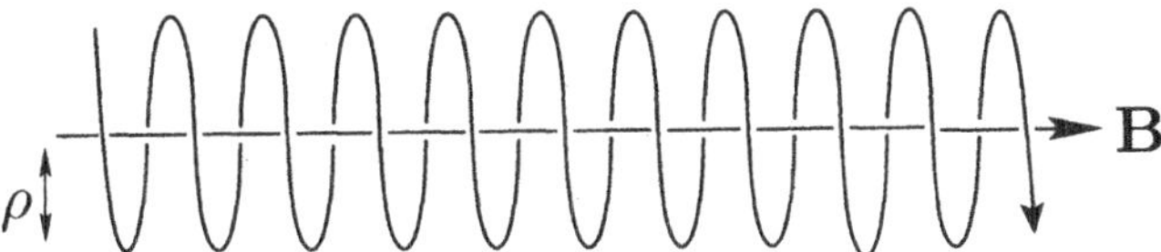

Figure 3.9: *Trajectory of a charged particle in a uniform magnetic field.*

field at the cyclotron frequency. Oppositely charged particles gyrate in opposite directions.

Finally, if a particle is subject to a force $\mathbf{f}$, and moves a distance $\delta\mathbf{r}$ in a time interval δt, then the work done on the particle by the force is

$$\delta W = \mathbf{f} \cdot \delta\mathbf{r}. \tag{3.95}$$

Thus, the power input to the particle from the force field is

$$P = \lim_{\delta t \to 0} \frac{\delta W}{\delta t} = \mathbf{f} \cdot \mathbf{v}, \tag{3.96}$$

where $\mathbf{v}$ is the particle's velocity. It follows from the Lorentz force law, Equation (3.79), that the power input to a particle moving in electric and magnetic fields is

$$P = q\,\mathbf{v} \cdot \mathbf{E}. \tag{3.97}$$

Note that a charged particle can gain (or lose) energy from an electric field, but not from a magnetic field. This is because the magnetic force is always perpendicular to the particle's direction of motion, and, therefore, does no work on the particle [see Equation (3.95)]. Thus, in particle accelerators, magnetic fields are often used to guide particle motion (*e.g.*, in a circle), but the actual acceleration is always performed by electric fields.

3.8 AMPÈRE'S LAW

Magnetic fields, like electric fields, are completely superposable. So, if a field $\mathbf{B}_1$ is generated by a current I_1 flowing through some circuit, and a field $\mathbf{B}_2$ is generated by a current I_2 flowing through another circuit, then when the currents I_1 and I_2 flow through both circuits simultaneously the generated magnetic field is $\mathbf{B}_1 + \mathbf{B}_2$.

Consider two parallel wires separated by a perpendicular distance r and carrying electric currents I_1 and I_2, respectively—see Figure 3.10.

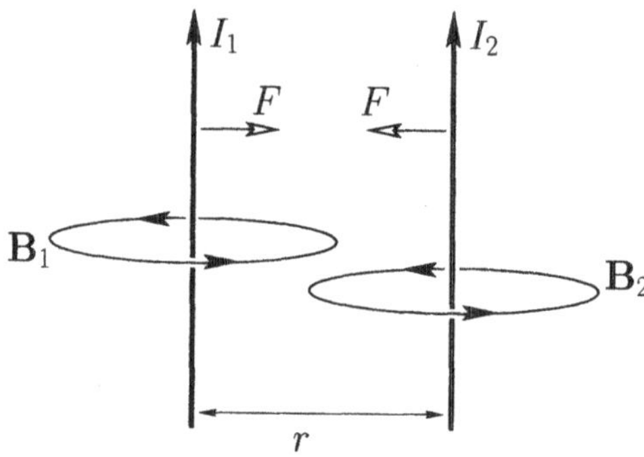

Figure 3.10: *Two parallel current-carrying wires.*

The magnetic field strength at the second wire due to the current flowing in the first wire is $B = \mu_0 I_1/2\pi r$. This field is orientated at right-angles to the second wire, so the force per unit length exerted on the second wire is

$$F = \frac{\mu_0 I_1 I_2}{2\pi r}. \tag{3.98}$$

This follows from Equation (3.74), which is valid for continuous wires as well as short test wires. The force acting on the second wire is directed radially inward toward the first wire (assuming that $I_1 I_2 > 0$). The magnetic field strength at the first wire due to the current flowing in the second wire is $B = \mu_0 I_2/2\pi r$. This field is orientated at right-angles to the first wire, so the force per unit length acting on the first wire is equal and opposite to that acting on the second wire, according to Equation (3.74). Equation (3.98) is sometimes called *Ampère's law*, and is clearly another example of an action-at-a-distance law: *i.e.*, if the current in the first wire is suddenly changed then the force on the second wire immediately adjusts. In reality, there should be a short time delay, at least as long as the propagation time for a light signal between the two wires. Clearly, Ampère's law is not strictly correct. However, as long as we restrict our investigations to *steady* currents it is perfectly adequate.

3.9 MAGNETIC MONOPOLES?

Suppose that we have an infinite straight wire carrying an electric current I. Let the wire be aligned along the z-axis. The magnetic field generated by such a wire is written

$$\mathbf{B} = \frac{\mu_0 I}{2\pi}\left(\frac{-y}{r^2}, \frac{x}{r^2}, 0\right) \tag{3.99}$$

in Cartesian coordinates, where $r = \sqrt{x^2 + y^2}$. The divergence of this field is

$$\nabla \cdot \mathbf{B} = \frac{\mu_0 I}{2\pi}\left(\frac{2\,y\,x}{r^4} - \frac{2\,x\,y}{r^4}\right) = 0, \tag{3.100}$$

where use has been made of $\partial r/\partial x = x/r$, *etc.* We saw in Section 3.4 that the divergence of the electric field appeared, at first sight, to be zero, but, was, in reality, a delta-function, because the volume integral of $\nabla \cdot \mathbf{E}$ was non-zero. Does the same sort of thing happen for the divergence of the magnetic field? Well, if we could find a closed surface S for which $\oint_S \mathbf{B}\cdot d\mathbf{S} \neq 0$ then, according to Gauss' theorem, $\int_V \nabla \cdot \mathbf{B}\, dV \neq 0$, where V is the volume enclosed by S. This would certainly imply that $\nabla \cdot \mathbf{B}$ is some sort of delta-function. So, can we find such a surface? Consider a cylindrical surface aligned with the wire. The magnetic field is everywhere tangential to the outward surface element, so this surface certainly has zero magnetic flux coming out of it. In fact, it is impossible to invent any closed surface for which $\oint_S \mathbf{B} \cdot d\mathbf{S} \neq 0$ with $\mathbf{B}$ given by Equation (3.99) (if you do not believe this, just try and find one!). This suggests that the divergence of a magnetic field generated by steady electric currents really is zero. Admittedly, we have only proved this for infinite straight currents, but, as will be demonstrated presently, it is true in general.

If $\nabla \cdot \mathbf{B} = 0$ then $\mathbf{B}$ is a *solenoidal* vector field. In other words, field-lines of $\mathbf{B}$ never begin or end. This is certainly the case in Equation (3.99) where the field-lines are a set of concentric circles centered on the z-axis. What about magnetic fields generated by permanent magnets (the modern equivalent of loadstones)? Do they also never begin or end? Well, we know that a conventional bar magnet has both a North and South magnetic pole (like the Earth). If we track the magnetic field-lines with a small compass they all emanate from the North pole, spread out, and eventually reconverge on the South pole—see Figure 3.11. It appears likely (but we cannot prove it with a compass) that the field-lines inside the magnet connect from the South to the North pole so as to form closed loops which never begin or end.

Can we produce an isolated North or South magnetic pole: for instance, by snapping a bar magnet in two? A compass needle would always point away from an isolated North pole, so this would act like a

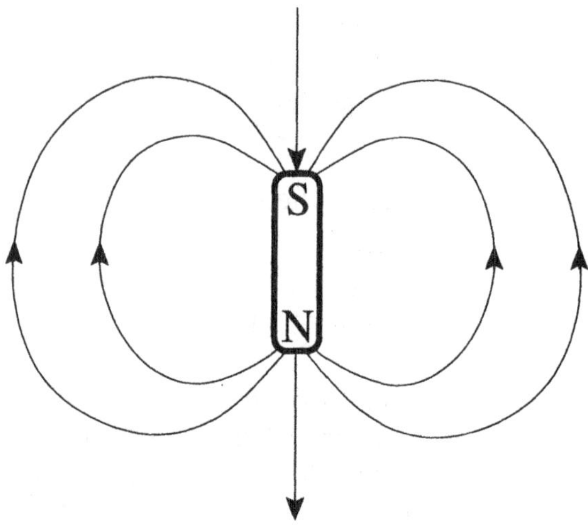

Figure 3.11: *Magnetic field-lines generated by a bar magnet.*

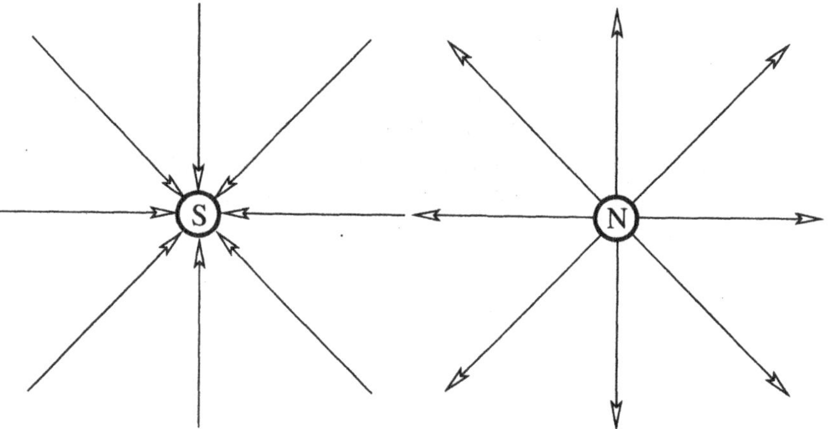

Figure 3.12: *Magnetic field-lines generated by magnetic monopoles.*

positive "magnetic charge." Likewise, a compass needle would always point towards an isolated South pole, so this would act like a negative "magnetic charge." It is clear, from Figure 3.12, that if we take a closed surface S containing an isolated magnetic pole, which is usually termed a *magnetic monopole*, then $\oint_S \mathbf{B} \cdot d\mathbf{S} \neq 0$: the flux will be positive for an isolated North pole, and negative for an isolated South pole. It follows from Gauss' theorem that if $\oint_S \mathbf{B} \cdot d\mathbf{S} \neq 0$ then $\nabla \cdot \mathbf{B} \neq 0$. Thus, the statement that magnetic fields are solenoidal, or that $\nabla \cdot \mathbf{B} = 0$, is equivalent to the statement that *there are no magnetic monopoles*. It is not clear, *a priori*, that this is a true statement. In fact, it is quite possible to formulate electromagnetism so as to allow for magnetic monopoles. However, as far as we are aware, there are no magnetic monopoles in the Universe. At least, if there are any then they are all hiding from us! We know that if we try to make a magnetic monopole by snapping a bar magnet in two then we just end up with two smaller bar magnets. If we snap one of these smaller magnets in two then we end up with two even smaller bar magnets. We can continue this process down to the atomic level without ever producing a magnetic monopole. In fact, permanent magnetism is generated by electric currents circulating on the atomic scale, and so this type of magnetism is not fundamentally different to the magnetism generated by macroscopic currents.

In conclusion, *all* steady magnetic fields in the Universe are generated by circulating electric currents of some description. Such fields are solenoidal: that is, they never begin or end, and satisfy the field equation

$$\nabla \cdot \mathbf{B} = 0. \tag{3.101}$$

This, incidentally, is the second of Maxwell's equations. Essentially, it says that there is no such thing as a magnetic monopole. We have only proved that $\nabla \cdot \mathbf{B} = 0$ for steady magnetic fields, but, in fact, this is also the case for time-dependent fields (see later).

3.10 AMPÈRE'S CIRCUITAL LAW

Consider, again, an infinite straight wire aligned along the z-axis and carrying a current I. The field generated by such a wire is written

$$B_\theta = \frac{\mu_0 I}{2\pi r} \tag{3.102}$$

in cylindrical polar coordinates. Consider a circular loop C in the x-y plane which is centered on the wire. Suppose that the radius of this loop is r. Let us evaluate the line integral $\oint_C \mathbf{B} \cdot d\mathbf{l}$. This integral is easy to perform because the magnetic field is always parallel to the line element. We have

$$\oint_C \mathbf{B} \cdot d\mathbf{l} = \oint B_\theta \, r \, d\theta = \mu_0 I. \tag{3.103}$$

However, we know from Stokes' theorem that

$$\oint_C \mathbf{B}\cdot d\mathbf{l} = \int_S \nabla\times\mathbf{B}\cdot d\mathbf{S}, \tag{3.104}$$

where S is any surface attached to the loop C.

Let us evaluate $\nabla\times\mathbf{B}$ directly. According to Equation (3.99),

$$(\nabla\times\mathbf{B})_x = \frac{\partial B_z}{\partial y} - \frac{\partial B_y}{\partial z} = 0, \tag{3.105}$$

$$(\nabla\times\mathbf{B})_y = \frac{\partial B_x}{\partial z} - \frac{\partial B_z}{\partial x} = 0, \tag{3.106}$$

$$\begin{aligned}(\nabla\times\mathbf{B})_z &= \frac{\partial B_y}{\partial x} - \frac{\partial B_x}{\partial y}\\ &= \frac{\mu_0\, I}{2\pi}\left(\frac{1}{r^2} - \frac{2\,x^2}{r^4} + \frac{1}{r^2} - \frac{2\,y^2}{r^4}\right) = 0,\end{aligned} \tag{3.107}$$

where use has been made of $\partial r/\partial x = x/r$, *etc.* We now have a problem. Equations (3.103) and (3.104) imply that

$$\int_S \nabla\times\mathbf{B}\cdot d\mathbf{S} = \mu_0\, I. \tag{3.108}$$

But, we have just demonstrated that $\nabla\times\mathbf{B} = \mathbf{0}$. This problem is very reminiscent of the difficulty we had earlier with $\nabla\cdot\mathbf{E}$. Recall that $\int_V \nabla\cdot\mathbf{E}\, dV = q/\epsilon_0$ for a volume V containing a discrete charge q, but that $\nabla\cdot\mathbf{E} = 0$ at a general point. We got around this problem by saying that $\nabla\cdot\mathbf{E}$ is a three-dimensional delta-function whose spike is coincident with the location of the charge. Likewise, we can get around our present difficulty by saying that $\nabla\times\mathbf{B}$ is a two-dimensional delta-function. A three-dimensional delta-function is a singular (but integrable) *point* in space, whereas a two-dimensional delta-function is a singular *line* in space. It is clear from an examination of Equations (3.105)–(3.107) that the only component of $\nabla\times\mathbf{B}$ which can be singular is the z-component, and that this can only be singular on the z-axis (*i.e.*, $r = 0$). Thus, the singularity coincides with the location of the current, and we can write

$$\nabla\times\mathbf{B} = \mu_0\, I\,\delta(x)\,\delta(y)\,\mathbf{e}_z. \tag{3.109}$$

The above equation certainly gives $(\nabla\times\mathbf{B})_x = (\nabla\times\mathbf{B})_y = 0$, and $(\nabla\times\mathbf{B})_z = 0$ everywhere apart from the z-axis, in accordance with Equations (3.105)–(3.107). Suppose that we integrate over a plane surface S connected to the loop C. The surface element is $\mathbf{dS} = dx\, dy\, \mathbf{e}_z$, so

$$\int_S \nabla \times \mathbf{B} \cdot \mathbf{dS} = \mu_0\, I \int\!\!\int \delta(x)\, \delta(y)\, dx\, dy \tag{3.110}$$

where the integration is performed over the region $\sqrt{x^2 + y^2} \leq r$. However, since the only part of S which actually contributes to the surface integral is the bit which lies infinitesimally close to the z-axis, we can integrate over all x and y without changing the result. Thus, we obtain

$$\int_S \nabla \times \mathbf{B} \cdot \mathbf{dS} = \mu_0\, I \int_{-\infty}^{\infty} \delta(x)\, dx \int_{-\infty}^{\infty} \delta(y)\, dy = \mu_0\, I, \tag{3.111}$$

which is in agreement with Equation (3.108).

But, why have we gone to so much trouble to prove something using vector field theory which can be demonstrated in one line via conventional analysis [see Equation (3.103)]? The answer, of course, is that the vector field result is easily generalized, whereas the conventional result is just a special case. For instance, it is clear that Equation (3.111) is true for *any* surface attached to the loop C, not just a plane surface. Moreover, suppose that we distort our simple circular loop C so that it is no longer circular or even lies in one plane. What now is the line integral of **B** around the loop? This is no longer a simple problem for conventional analysis, because the magnetic field is not parallel to a line element of the loop. However, according to Stokes' theorem,

$$\oint_C \mathbf{B} \cdot \mathbf{dl} = \int_S \nabla \times \mathbf{B} \cdot \mathbf{dS}, \tag{3.112}$$

with $\nabla \times \mathbf{B}$ given by Equation (3.109). Note that the only part of S which contributes to the surface integral is an infinitesimal region centered on the z-axis. So, as long as S actually intersects the z-axis, it does not matter what shape the rest of the surface is, and we always get the same answer for the surface integral: namely,

$$\oint_C \mathbf{B} \cdot \mathbf{dl} = \int_S \nabla \times \mathbf{B} \cdot \mathbf{dS} = \mu_0\, I. \tag{3.113}$$

Thus, provided the curve C circulates the z-axis, and, therefore, any surface S attached to C intersects the z-axis (an odd number of times), the line integral $\oint_C \mathbf{B} \cdot \mathbf{dl}$ is equal to $\mu_0\, I$. Of course, if C does not circulate the z-axis then an attached surface S does not intersect the z-axis (an odd number of times) and $\oint_C \mathbf{B} \cdot \mathbf{dl}$ is zero. There is one more proviso. The line integral $\oint_C \mathbf{B} \cdot d\mathbf{l}$ is $\mu_0 I$ for a loop which circulates the z-axis in a clockwise direction (looking up the z-axis). However, if the loop circulates in a counterclockwise direction then the integral is $-\mu_0 I$. This follows because in the latter case the z-component of the surface element $d\mathbf{S}$ is oppositely directed to the current flow at the point where the surface intersects the wire.

Let us now consider N wires directed parallel to the z-axis, with coordinates (x_i, y_i) in the x-y plane, each carrying a current I_i in the positive z-direction. It is fairly obvious that Equation (3.109) generalizes to

$$\nabla \times \mathbf{B} = \mu_0 \sum_{i=1}^{N} I_i\, \delta(x - x_i)\, \delta(y - y_i)\, \mathbf{e}_z. \tag{3.114}$$

If we integrate the magnetic field around some closed curve C, which can have any shape and does not necessarily lie in one plane, then Stokes' theorem and the above equation imply that

$$\oint_C \mathbf{B} \cdot d\mathbf{l} = \int_S \nabla \times \mathbf{B} \cdot d\mathbf{S} = \mu_0 \mathcal{I}, \tag{3.115}$$

where $\mathcal{I}$ is the total current enclosed by the curve. Again, if the curve circulates the *i*th wire in a clockwise direction (looking down the direction of current flow) then the wire contributes I_i to the aggregate current $\mathcal{I}$. On the other hand, if the curve circulates in a counterclockwise direction then the wire contributes $-I_i$. Finally, if the curve does not circulate the wire at all then the wire contributes nothing to $\mathcal{I}$.

Equation (3.114) is a field equation describing how a set of z-directed current-carrying wires generate a magnetic field. These wires have zero-thickness, which implies that we are trying to squeeze a finite amount of current into an infinitesimal region. This accounts for the delta-functions on the right-hand side of the equation. Likewise, we obtained delta-functions in Section 3.4 because we were dealing with point charges. Let us now generalize to the more realistic case of diffuse currents. Suppose that the z-current flowing through a small rectangle in the x-y plane, centered on coordinates (x, y) and of dimensions dx and dy, is $j_z(x, y)\, dx\, dy$. Here, j_z is termed the *current density* in the z-direction. Let us integrate $(\nabla \times \mathbf{B})_z$ over this rectangle. The rectangle is assumed to be sufficiently small that $(\nabla \times \mathbf{B})_z$ does not vary appreciably across it. According to Equation (3.115), this integral is equal to μ_0 times the total z-current flowing through the rectangle. Thus,

$$(\nabla \times \mathbf{B})_z\, dx\, dy = \mu_0\, j_z\, dx\, dy, \tag{3.116}$$

which implies that

$$(\nabla \times \mathbf{B})_z = \mu_0\, j_z. \tag{3.117}$$

Of course, there is nothing special about the z-axis. Hence, we can obtain analogous equations for diffuse currents flowing in the y- and z-directions. We can combine all of these equations to form a single vector field equation which describes how electric currents generate magnetic fields: *i.e.*,

$$\nabla \times \mathbf{B} = \mu_0\, \mathbf{j}, \tag{3.118}$$

where $\mathbf{j} = (j_x, j_y, j_z)$ is the vector current density. This is the third Maxwell equation. The electric current flowing through a small area $d\mathbf{S}$ located at position $\mathbf{r}$ is $\mathbf{j}(\mathbf{r}) \cdot d\mathbf{S}$. Suppose that space is filled with particles of charge q, number density $n(\mathbf{r})$, and velocity $\mathbf{v}(\mathbf{r})$. The charge density is given by $\rho(\mathbf{r}) = q\, n$. The current density is given by $\mathbf{j}(\mathbf{r}) = q\, n\, \mathbf{v}$, and is obviously a proper vector field (velocities are proper vectors since they are ultimately derived from displacements).

If we form the line integral of $\mathbf{B}$ around some general closed curve C, making use of Stokes' theorem and the field equation (3.118), then we obtain

$$\oint_C \mathbf{B} \cdot d\mathbf{l} = \mu_0 \int_S \mathbf{j} \cdot d\mathbf{S}. \tag{3.119}$$

In other words, the line integral of the magnetic field around any closed loop C is equal to μ_0 times the flux of the current density through C. This result is called *Ampère's circuital law*. If the currents flow in zero-thickness wires then Ampère's circuital law reduces to Equation (3.115).

The flux of the current density through C is evaluated by integrating $\mathbf{j} \cdot d\mathbf{S}$ over any surface S attached to C. Suppose that we take two different surfaces S_1 and S_2. It is clear that if Ampère's circuital law is to make any sense then the surface integral $\int_{S_1} \mathbf{j} \cdot d\mathbf{S}$ had better equal the integral $\int_{S_2} \mathbf{j} \cdot d\mathbf{S}$. That is, when we work out the flux of the current density though C using two different attached surfaces then we had better get the same answer, otherwise Equation (3.119) is wrong (since the left-hand side is clearly independent of the surface spanning C). We saw in Chapter 2 that if the integral of a vector field $\mathbf{A}$ over some surface attached to a loop depends only on the loop, and is independent of the surface which spans it, then this implies that $\nabla \cdot \mathbf{A} = 0$. Hence, we require that $\nabla \cdot \mathbf{j} = 0$ in order to make the flux of the current density through C a well-defined quantity. We can also see this directly from the field equation (3.118). We know that the divergence of a curl is automatically zero, so taking the divergence of Equation (3.118), we obtain

$$\nabla \cdot \mathbf{j} = 0. \tag{3.120}$$

We have shown that if Ampère's circuital law is to make any sense then we need $\nabla \cdot \mathbf{j} = 0$. Physically, this implies that the net current flowing through any closed surface S is zero. Up to now, we have only considered stationary charges and steady currents. It is clear that if all charges are stationary and all currents are steady then there can be no net current flowing through a closed surface S, since this would imply a build up of charge in the volume V enclosed by S. In other words, as long as we restrict our investigation to stationary charges, and steady currents, then we expect $\nabla \cdot \mathbf{j} = 0$, and Ampère's circuital law makes sense. However, suppose that we now relax this restriction. Suppose that some of the charges in a volume V decide to move outside V. Clearly, there will be a non-zero net flux of electric current through the bounding surface S whilst this is happening. This implies from Gauss' theorem that $\nabla \cdot \mathbf{j} \neq 0$. Under these circumstances Ampère's circuital law collapses in a heap. We shall see later that we can rescue Ampère's circuital law by adding an extra term involving a time derivative to the right-hand side of the field equation (3.118). For steady-state situations (*i.e.*, $\partial/\partial t = 0$), this extra term can be neglected. Thus, the field equation $\nabla \times \mathbf{B} = \mu_0 \mathbf{j}$ is, in fact, only two-thirds of Maxwell's third equation: there is a term missing from the right-hand side.

We have now derived two field equations involving magnetic fields (strictly speaking, we have only derived one and two-thirds equations):

$$\nabla \cdot \mathbf{B} = 0, \tag{3.121}$$

$$\nabla \times \mathbf{B} = \mu_0 \mathbf{j}. \tag{3.122}$$

We obtained these equations by looking at the fields generated by infinitely long, straight, steady currents. This, of course, is a rather special class of currents. We should now go back and repeat the process for general currents. In fact, if we did this we would find that the above field equations still hold (provided that the currents are steady). Unfortunately, this demonstration is rather messy and extremely tedious. There is a better approach. Let us *assume* that the above field equations are valid for any set of steady currents. We can then, with relatively little effort, use these equations to generate the correct formula for the magnetic field induced by a general set of steady currents, thus proving that our assumption is correct. More of this later.

As an example of the use of Ampère's circuital law, let us calculate the magnetic field generated by a cylindrical current annulus of inner

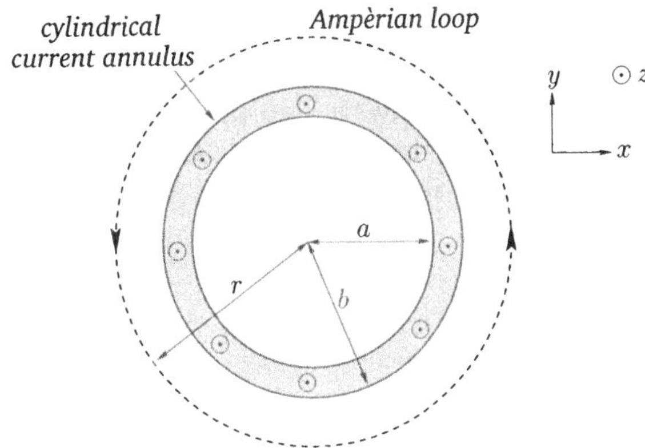

Figure 3.13: *An example use of Ampère's circuital law.*

radius a, and outer radius b, co-axial with the z-axis, and carrying a uniformly distributed z-directed current I. Now, from symmetry, and by analogy with the magnetic field generated by a straight wire, we expect the current distribution to generate a magnetic field of the form $\mathbf{B} = B_\theta(r)\, \mathbf{e}_\theta$. Let us apply Ampère's circuital law to an imaginary circular loop in the x-y plane, of radius r, centered on the z-axis—see Figure 3.13. Such a loop is generally known as an *Ampèrian loop*. Now, according to Ampère's circuital law, the line integral of the magnetic field around the loop is equal to the current enclosed by the loop, multiplied by μ_0. The line integral is easy to calculate since the magnetic field is everywhere tangential to the loop. We obtain

$$2\pi r\, B_\theta(r) = \mu_0\, I(r),$$

where $I(r)$ is the current enclosed by an Ampèrian loop of radius r. Now, it is evident that

$$I(r) = \begin{cases} 0 & r < a \\ [(r^2 - a^2)/(b^2 - a^2)]\, I & a \le r \le b \\ I & b < r \end{cases}. \tag{3.123}$$

Hence,

$$B_\theta(r) = \begin{cases} 0 & r < a \\ [\mu_0\, I/(2\pi r)]\, [(r^2 - a^2)/(b^2 - a^2)] & a \le r \le b \\ \mu_0\, I/(2\pi r) & b < r \end{cases}. \tag{3.124}$$

3.11 HELMHOLTZ'S THEOREM

Up to now, we have only studied the electric and magnetic fields generated by stationary charges and steady currents. We have found that these fields are describable in terms of four field equations: *i.e.*,

$$\nabla \cdot \mathbf{E} = \frac{\rho}{\epsilon_0}, \tag{3.125}$$

$$\nabla \times \mathbf{E} = \mathbf{0} \tag{3.126}$$

for electric fields, and

$$\nabla \cdot \mathbf{B} = 0, \tag{3.127}$$

$$\nabla \times \mathbf{B} = \mu_0 \mathbf{j} \tag{3.128}$$

for magnetic fields. There are no other field equations. This strongly suggests that if we know the divergence and the curl of a vector field then we know everything there is to know about the field. In fact, this is the case. There is a mathematical theorem which sums this up. It is called *Helmholtz's theorem*, after the German polymath Hermann Ludwig Ferdinand von Helmholtz.

Let us start with scalar fields. Field equations are a type of differential equation: *i.e.*, they deal with the infinitesimal differences in quantities between neighbouring points. So what kind of differential equation completely specifies a scalar field? This is easy. Suppose that we have a scalar field $\phi(\mathbf{r})$ and a field equation which tells us the gradient of this field at all points: something like

$$\nabla\phi = \mathbf{A}, \tag{3.129}$$

where $\mathbf{A}(\mathbf{r})$ is a vector field. Note that we need $\nabla \times \mathbf{A} = \mathbf{0}$ for self-consistency, since the curl of a gradient is automatically zero. The above equation completely specifies $\phi(\mathbf{r})$ once we are given the value of the field at a single point, P (say). Thus,

$$\phi(\mathrm{Q}) = \phi(\mathrm{P}) + \int_{\mathrm{P}}^{\mathrm{Q}} \nabla\phi \cdot d\mathbf{l} = \phi(\mathrm{P}) + \int_{\mathrm{P}}^{\mathrm{Q}} \mathbf{A} \cdot d\mathbf{l}, \tag{3.130}$$

where Q is a general point. The fact that $\nabla \times \mathbf{A} = \mathbf{0}$ means that $\mathbf{A}$ is a conservative field, which guarantees that the above equation gives a *unique* value for ϕ at a general point in space.

Suppose that we have a vector field $\mathbf{F}(\mathbf{r})$. How many differential equations do we need to completely specify this field? Hopefully, we only need two: one giving the divergence of the field, and one giving its curl.

Let us test this hypothesis. Suppose that we have two field equations:

$$\nabla \cdot \mathbf{F} = D, \tag{3.131}$$

$$\nabla \times \mathbf{F} = \mathbf{C}, \tag{3.132}$$

where $D(\mathbf{r})$ is a scalar field and $\mathbf{C}(\mathbf{r})$ is a vector field. For self-consistency, we need

$$\nabla \cdot \mathbf{C} = 0, \tag{3.133}$$

since the divergence of a curl is automatically zero. So, do these two field equations, plus some suitable boundary conditions, completely specify **F**? Suppose that we write

$$\mathbf{F} = -\nabla U + \nabla \times \mathbf{W}. \tag{3.134}$$

In other words, we are saying that a general vector field **F** is the sum of a conservative field, ∇U, and a solenoidal field, $\nabla \times \mathbf{W}$. This sounds plausible, but it remains to be proved. Let us start by taking the divergence of the above equation, and making use of Equation (3.131). We get

$$\nabla^2 U = -D. \tag{3.135}$$

Note that the vector field **W** does not figure in this equation, because the divergence of a curl is automatically zero. Let us now take the curl of Equation (3.134):

$$\nabla \times \mathbf{F} = \nabla \times \nabla \times \mathbf{W} = \nabla(\nabla \cdot \mathbf{W}) - \nabla^2 \mathbf{W} = -\nabla^2 \mathbf{W}. \tag{3.136}$$

Here, we assume that the divergence of **W** is zero. This is another thing which remains to be proved. Note that the scalar field U does not figure in this equation, because the curl of a divergence is automatically zero. Using Equation (3.132), we get

$$\nabla^2 W_x = -C_x, \tag{3.137}$$

$$\nabla^2 W_y = -C_y, \tag{3.138}$$

$$\nabla^2 W_z = -C_z, \tag{3.139}$$

So, we have transformed our problem into four differential equations, Equation (3.135) and Equations (3.137)–(3.139), which we need to solve. Let us look at these equations. We immediately notice that they all have exactly the same form. In fact, they are all versions of Poisson's equation. We can now make use of a principle made famous by Richard P. Feynman: "the same equations have the same solutions." Recall that earlier on we came across the following equation:

$$\nabla^2 \phi = -\frac{\rho}{\epsilon_0}, \tag{3.140}$$

where ϕ is the electrostatic potential and ρ is the charge density. We proved that the solution to this equation, with the boundary condition that ϕ goes to zero at infinity, is

$$\phi(\mathbf{r}) = \frac{1}{4\pi\epsilon_0} \int \frac{\rho(\mathbf{r}')}{|\mathbf{r} - \mathbf{r}'|} \, d^3\mathbf{r}'. \tag{3.141}$$

Well, if the same equations have the same solutions, and Equation (3.141) is the solution to Equation (3.140), then we can immediately write down the solutions to Equation (3.135) and Equations (3.137)–(3.139). We get

$$U(\mathbf{r}) = \frac{1}{4\pi} \int \frac{D(\mathbf{r}')}{|\mathbf{r} - \mathbf{r}'|} \, d^3\mathbf{r}', \tag{3.142}$$

and

$$W_x(\mathbf{r}) = \frac{1}{4\pi} \int \frac{C_x(\mathbf{r}')}{|\mathbf{r} - \mathbf{r}'|} \, d^3\mathbf{r}', \tag{3.143}$$

$$W_y(\mathbf{r}) = \frac{1}{4\pi} \int \frac{C_y(\mathbf{r}')}{|\mathbf{r} - \mathbf{r}'|} \, d^3\mathbf{r}', \tag{3.144}$$

$$W_z(\mathbf{r}) = \frac{1}{4\pi} \int \frac{C_z(\mathbf{r}')}{|\mathbf{r} - \mathbf{r}'|} \, d^3\mathbf{r}'. \tag{3.145}$$

The last three equations can be combined to form a single vector equation:

$$\mathbf{W}(\mathbf{r}) = \frac{1}{4\pi} \int \frac{\mathbf{C}(\mathbf{r}')}{|\mathbf{r} - \mathbf{r}'|} \, d^3\mathbf{r}'. \tag{3.146}$$

We assumed earlier that $\nabla \cdot \mathbf{W} = 0$. Let us check to see if this is true. Note that

$$\frac{\partial}{\partial x}\left(\frac{1}{|\mathbf{r} - \mathbf{r}'|}\right) = -\frac{x - x'}{|\mathbf{r} - \mathbf{r}'|^3} = \frac{x' - x}{|\mathbf{r} - \mathbf{r}'|^3} = -\frac{\partial}{\partial x'}\left(\frac{1}{|\mathbf{r} - \mathbf{r}'|}\right), \tag{3.147}$$

which implies that

$$\nabla\left(\frac{1}{|\mathbf{r} - \mathbf{r}'|}\right) = -\nabla'\left(\frac{1}{|\mathbf{r} - \mathbf{r}'|}\right), \tag{3.148}$$

where ∇' is the operator $(\partial/\partial x', \partial/\partial y', \partial/\partial z')$. Taking the divergence of Equation (3.146), and making use of the above relation, we obtain

$$\begin{aligned}\nabla \cdot \mathbf{W} &= \frac{1}{4\pi}\int \mathbf{C}(\mathbf{r}') \cdot \nabla\left(\frac{1}{|\mathbf{r}-\mathbf{r}'|}\right) d^3\mathbf{r}' \\ &= -\frac{1}{4\pi}\int \mathbf{C}(\mathbf{r}') \cdot \nabla'\left(\frac{1}{|\mathbf{r}-\mathbf{r}'|}\right) d^3\mathbf{r}'.\end{aligned} \tag{3.149}$$

Now

$$\int_{-\infty}^{\infty} g\,\frac{\partial f}{\partial x}\,dx = [g\,f]_{-\infty}^{\infty} - \int_{-\infty}^{\infty} f\,\frac{\partial g}{\partial x}\,dx. \tag{3.150}$$

However, if $g\,f \to 0$ as $x \to \pm\infty$ then we can neglect the first term on the right-hand side of the above equation and write

$$\int_{-\infty}^{\infty} g\,\frac{\partial f}{\partial x}\,dx = -\int_{-\infty}^{\infty} f\,\frac{\partial g}{\partial x}\,dx. \tag{3.151}$$

A simple generalization of this result yields

$$\int \mathbf{g} \cdot \nabla f\, d^3\mathbf{r} = -\int f\,\nabla \cdot \mathbf{g}\, d^3\mathbf{r}, \tag{3.152}$$

provided that $g_x\, f \to 0$ as $|\mathbf{r}| \to \infty$, *etc.* Thus, we can deduce that

$$\nabla \cdot \mathbf{W} = \frac{1}{4\pi}\int \frac{\nabla' \cdot \mathbf{C}(\mathbf{r}')}{|\mathbf{r}-\mathbf{r}'|}\, d^3\mathbf{r}' \tag{3.153}$$

from Equation (3.149), provided $|\mathbf{C}(\mathbf{r})|$, is bounded as $|\mathbf{r}| \to \infty$. However, we have already shown that $\nabla \cdot \mathbf{C} = 0$ from self-consistency arguments, so the above equation implies that $\nabla \cdot \mathbf{W} = 0$, which is the desired result.

We have constructed a vector field $\mathbf{F}(\mathbf{r})$ which satisfies Equations (3.131) and (3.132) and behaves sensibly at infinity: *i.e.*, $|\mathbf{F}| \to 0$ as $|\mathbf{r}| \to \infty$. But, is our solution the only possible solution of Equations (3.131) and (3.132) with sensible boundary conditions at infinity? Another way of posing this question is to ask whether there are any solutions of

$$\nabla^2 U = 0, \quad \nabla^2 W_i = 0, \tag{3.154}$$

where i denotes x, y, or z, which are bounded at infinity. If there are then we are in trouble, because we can take our solution and add to it an arbitrary amount of a vector field with zero divergence and zero curl, and thereby obtain another solution which also satisfies physical boundary conditions. This would imply that our solution is not unique.

In other words, it is not possible to unambiguously reconstruct a vector field given its divergence, its curl, and physical boundary conditions. Fortunately, the equation

$$\nabla^2\phi = 0, \tag{3.155}$$

which is called *Laplace's equation*, has a very interesting property: its solutions are *unique*. That is, if we can find a solution to Laplace's equation which satisfies the boundary conditions then we are guaranteed that this is the *only* solution. We shall prove this later on in Section 5.8. Well, let us invent some solutions to Equations (3.154) which are bounded at infinity. How about

$$u = w_i = 0\,? \tag{3.156}$$

These solutions certainly satisfy Laplace's equation, and are well-behaved at infinity. Because the solutions to Laplace's equations are unique, we know that Equations (3.156) are the only solutions to Equations (3.154). This means that there is no vector field which satisfies physical boundary equations at infinity and has zero divergence and zero curl. In other words, our solution to Equations (3.131) and (3.132) is the *only* solution. Thus, we have unambiguously reconstructed the vector field **F** given its divergence, its curl, and sensible boundary conditions at infinity. This is Helmholtz's theorem.

We have just demonstrated a number of very useful, and also very important, points. First, according to Equation (3.134), a general vector field can be written as the sum of a conservative field and a solenoidal field. Thus, we ought to be able to write electric and magnetic fields in this form. Second, a general vector field which is zero at infinity is completely specified once its divergence and its curl are given. Thus, we can guess that the laws of electromagnetism can be written as four field equations,

$$\nabla\cdot\mathbf{E} = \textit{something}, \tag{3.157}$$

$$\nabla\times\mathbf{E} = \textbf{something}, \tag{3.158}$$

$$\nabla\cdot\mathbf{B} = \textit{something}, \tag{3.159}$$

$$\nabla\times\mathbf{B} = \textbf{something}, \tag{3.160}$$

without knowing the first thing about electromagnetism (other than the fact that it deals with two vector fields). Of course, Equations (3.125)–(3.128) are of exactly this form. We also know that there are only four field equations, since the above equations are sufficient to completely reconstruct both **E** and **B**. Furthermore, we know that we can solve the field equations without even knowing what the right-hand sides look like. After all, we solved Equations (3.131)–(3.132) for completely general right-hand sides. [Actually, the right-hand sides have to go to zero at infinity, otherwise integrals like Equation (3.142) blow up.] We also know that any solutions we find are unique. In other words, there is only one possible steady electric and magnetic field which can be generated by a given set of stationary charges and steady currents. The third thing which we proved was that if the right-hand sides of the above field equations are all zero then the only physical solution is $\mathbf{E}(\mathbf{r}) = \mathbf{B}(\mathbf{r}) = \mathbf{0}$. This implies that steady electric and magnetic fields cannot generate themselves. Instead, they have to be generated by stationary charges and steady currents. So, if we come across a steady electric field then we know that if we trace the field-lines back we shall eventually find a charge. Likewise, a steady magnetic field implies that there is a steady current flowing somewhere. All of these results follow from vector field theory (*i.e.*, from the mathematical properties of vector fields in three-dimensional space), prior to any investigation of electromagnetism.

3.12 THE MAGNETIC VECTOR POTENTIAL

Electric fields generated by stationary charges obey

$$\nabla \times \mathbf{E} = \mathbf{0}. \tag{3.161}$$

This immediately allows us to write

$$\mathbf{E} = -\nabla\phi, \tag{3.162}$$

since the curl of a gradient is automatically zero. In fact, whenever we come across an irrotational vector field in Physics we can always write it as the gradient of some scalar field. This is clearly a useful thing to do, since it enables us to replace a vector field by a much simpler scalar field. The quantity ϕ in the above equation is known as the *electric scalar potential*.

Magnetic fields generated by steady currents (and unsteady currents, for that matter) satisfy

$$\nabla \cdot \mathbf{B} = 0. \tag{3.163}$$

This immediately allows us to write

$$\mathbf{B} = \nabla \times \mathbf{A}, \tag{3.164}$$

since the divergence of a curl is automatically zero. In fact, whenever we come across a solenoidal vector field in Physics we can always write it as the curl of some other vector field. This is not an obviously useful thing to do, however, since it only allows us to replace one vector field by another. Nevertheless, Equation (3.164) is one of the most useful equations we shall come across in this book. The quantity **A** is known as the *magnetic vector potential*.

We know from Helmholtz's theorem that a vector field is fully specified by its divergence and its curl. The curl of the vector potential gives us the magnetic field via Equation (3.164). However, the divergence of **A** has no physical significance. In fact, we are completely free to choose $\nabla \cdot \mathbf{A}$ to be whatever we like. Note that, according to Equation (3.164), the magnetic field is invariant under the transformation

$$\mathbf{A} \to \mathbf{A} - \nabla\psi. \tag{3.165}$$

In other words, the vector potential is undetermined to the gradient of a scalar field. This is just another way of saying that we are free to choose $\nabla \cdot \mathbf{A}$. Recall that the electric scalar potential is undetermined to an arbitrary additive constant, since the transformation

$$\phi \to \phi + c \tag{3.166}$$

leaves the electric field invariant in Equation (3.162). The transformations (3.165) and (3.166) are examples of what mathematicians call *gauge transformations*. The choice of a particular function ψ or a particular constant c is referred to as a choice of the gauge. We are free to fix the gauge to be whatever we like. The most sensible choice is the one which makes our equations as simple as possible. The usual gauge for the scalar potential ϕ is such that $\phi \to 0$ at infinity. The usual gauge for **A** (in steady-state situations) is such that

$$\nabla \cdot \mathbf{A} = 0. \tag{3.167}$$

This particular choice is known as the *Coulomb gauge*.

It is obvious that we can always add a constant to ϕ so as to make it zero at infinity. But it is not at all obvious that we can always perform a gauge transformation such as to make $\nabla \cdot \mathbf{A}$ zero. Suppose that we have found some vector field $\mathbf{A}(\mathbf{r})$ whose curl gives the magnetic field but whose divergence in non-zero. Let

$$\nabla \cdot \mathbf{A} = \nu(\mathbf{r}). \tag{3.168}$$

So, can we find a scalar field ψ such that after we perform the gauge transformation (3.165) we are left with $\nabla \cdot \mathbf{A} = 0$? Taking the divergence of Equation (3.165) it is clear that we need to find a function ψ which satisfies

$$\nabla^2\psi = \nu. \tag{3.169}$$

But this is just Poisson's equation. We know that we can always find a unique solution of this equation (see Section 3.11). This proves that, in practice, we can always set the divergence of **A** equal to zero.

Let us again consider an infinite straight wire directed along the z-axis and carrying a current I. The magnetic field generated by such a wire is written

$$\mathbf{B} = \frac{\mu_0 I}{2\pi}\left(\frac{-y}{r^2}, \frac{x}{r^2}, 0\right). \tag{3.170}$$

We wish to find a vector potential **A** whose curl is equal to the above magnetic field, and whose divergence is zero. It is not difficult to see that

$$\mathbf{A} = -\frac{\mu_0 I}{4\pi}\left(0, 0, \ln[x^2 + y^2]\right) \tag{3.171}$$

fits the bill. Note that the vector potential is parallel to the direction of the current. This would seem to suggest that there is a more direct relationship between the vector potential and the current than there is between the magnetic field and the current. The potential is not very well-behaved on the z-axis, but this is just because we are dealing with an infinitely thin current.

Let us take the curl of Equation (3.164). We find that

$$\nabla \times \mathbf{B} = \nabla \times \nabla \times \mathbf{A} = \nabla(\nabla \cdot \mathbf{A}) - \nabla^2\mathbf{A} = -\nabla^2\mathbf{A}, \tag{3.172}$$

where use has been made of the Coulomb gauge condition (3.167). We can combine the above relation with the field equation (3.118) to give

$$\nabla^2\mathbf{A} = -\mu_0\,\mathbf{j}. \tag{3.173}$$

Writing this in component form, we obtain

$$\nabla^2 A_x = -\mu_0\, j_x, \tag{3.174}$$

$$\nabla^2 A_y = -\mu_0\, j_y, \tag{3.175}$$

$$\nabla^2 A_z = -\mu_0\, j_z. \tag{3.176}$$

But, this is just Poisson's equation three times over. We can immediately write the unique solutions to the above equations:

$$A_x(\mathbf{r}) = \frac{\mu_0}{4\pi}\int \frac{j_x(\mathbf{r}')}{|\mathbf{r}-\mathbf{r}'|}\,d^3\mathbf{r}', \tag{3.177}$$

$$A_y(\mathbf{r}) = \frac{\mu_0}{4\pi}\int \frac{j_y(\mathbf{r}')}{|\mathbf{r}-\mathbf{r}'|}\,d^3\mathbf{r}', \tag{3.178}$$

$$A_z(\mathbf{r}) = \frac{\mu_0}{4\pi}\int \frac{j_z(\mathbf{r}')}{|\mathbf{r}-\mathbf{r}'|}\,d^3\mathbf{r}'. \tag{3.179}$$

These solutions can be recombined to form a single vector solution

$$\mathbf{A}(\mathbf{r}) = \frac{\mu_0}{4\pi}\int \frac{\mathbf{j}(\mathbf{r}')}{|\mathbf{r}-\mathbf{r}'|}\,d^3\mathbf{r}'. \tag{3.180}$$

Of course, we have seen a equation like this before:

$$\phi(\mathbf{r}) = \frac{1}{4\pi\epsilon_0}\int \frac{\rho(\mathbf{r}')}{|\mathbf{r}-\mathbf{r}'|}\,d^3\mathbf{r}'. \tag{3.181}$$

Equations (3.180) and (3.181) are the unique solutions (given the arbitrary choice of gauge) to the field equations (3.125)–(3.128): they specify the magnetic vector and electric scalar potentials generated by a set of stationary charges, of charge density $\rho(\mathbf{r})$, and a set of steady currents, of current density $\mathbf{j}(\mathbf{r})$. Incidentally, we can prove that Equation (3.180) satisfies the gauge condition $\nabla\cdot\mathbf{A} = 0$ by repeating the analysis of Equations (3.146)–(3.153) (with $\mathbf{W}\to\mathbf{A}$ and $\mathbf{C}\to\mu_0\,\mathbf{j}$), and using the fact that $\nabla\cdot\mathbf{j} = 0$ for steady currents.

As an example, let us find the vector potential associated with the magnetic field distribution (3.124). By symmetry, and by analogy with the vector potential generated by a straight wire, we expect that $\mathbf{A} = A_z(r)\,\mathbf{e}_z$. Note that this form for $\mathbf{A}$ satisfies the Coulomb gauge. Hence, using $\nabla\times\mathbf{A} = \mathbf{B}$, we get

$$\frac{\partial A_z}{\partial r} = -B_\theta(r). \tag{3.182}$$

The boundary conditions are that A_z be continuous at $r = a$ and $r = b$, since a discontinuous A_z would generate an infinite magnetic field, which is unphysical. Thus, we obtain

$$A_z(r) = \begin{cases} (\mu_0\,I/2\pi)\,[1/2 - a^2\,\ln(b/a)/(b^2-a^2)] & r < a \\ (\mu_0\,I/2\pi)\,([b^2/2 - r^2/2 - a^2\,\ln(b/r)]/[b^2-a^2]) & a \le r \le b\,. \\ -(\mu_0\,I/2\pi)\,\ln(r/b) & b < r \end{cases} \tag{3.183}$$

3.13 THE BIOT-SAVART LAW

According to Equation (3.162), we can obtain an expression for the electric field generated by stationary charges by taking minus the gradient of Equation (3.181). This yields

$$\mathbf{E}(\mathbf{r}) = \frac{1}{4\pi\epsilon_0} \int \rho(\mathbf{r}') \frac{\mathbf{r} - \mathbf{r}'}{|\mathbf{r} - \mathbf{r}'|^3} \, d^3\mathbf{r}', \tag{3.184}$$

which we recognize as Coulomb's law written for a continuous charge distribution. According to Equation (3.164), we can obtain an equivalent expression for the magnetic field generated by steady currents by taking the curl of Equation (3.180). This gives

$$\mathbf{B}(\mathbf{r}) = \frac{\mu_0}{4\pi} \int \frac{\mathbf{j}(\mathbf{r}') \times (\mathbf{r} - \mathbf{r}')}{|\mathbf{r} - \mathbf{r}'|^3} \, d^3\mathbf{r}', \tag{3.185}$$

where use has been made of the vector identity $\nabla \times (\phi \mathbf{A}) = \phi \nabla \times \mathbf{A} + \nabla\phi \times \mathbf{A}$. Equation (3.185) is known as the *Biot-Savart law* after the French physicists Jean Baptiste Biot and Felix Savart: it completely specifies the magnetic field generated by a steady (but otherwise quite general) distributed current.

Let us reduce our distributed current to an idealized zero-thickness wire. We can do this by writing

$$\mathbf{j}(\mathbf{r}) \, d^3\mathbf{r} = \mathbf{I}(\mathbf{r}) \, dl, \tag{3.186}$$

where $\mathbf{I}$ is the vector current (*i.e.*, its direction and magnitude specify the direction and magnitude of the current) and dl is an element of length along the wire. Equations (3.185) and (3.186) can be combined to give

$$\mathbf{B}(\mathbf{r}) = \frac{\mu_0}{4\pi} \int \frac{\mathbf{I}(\mathbf{r}') \times (\mathbf{r} - \mathbf{r}')}{|\mathbf{r} - \mathbf{r}'|^3} \, dl, \tag{3.187}$$

which is the form in which the Biot-Savart law is most usually written. This law is to magnetostatics (*i.e.*, the study of magnetic fields generated by steady currents) what Coulomb's law is to electrostatics (*i.e.*, the study of electric fields generated by stationary charges). Furthermore, it can be experimentally verified given a set of currents, a compass, a test wire, and a great deal of skill and patience. This justifies our earlier assumption that the field equations (3.121) and (3.122) are valid for general current distributions (recall that we derived them by studying the fields generated by infinite straight wires). Note that both Coulomb's law and the Biot-Savart law are *gauge independent*: *i.e.*, they do not depend on the particular choice of gauge.

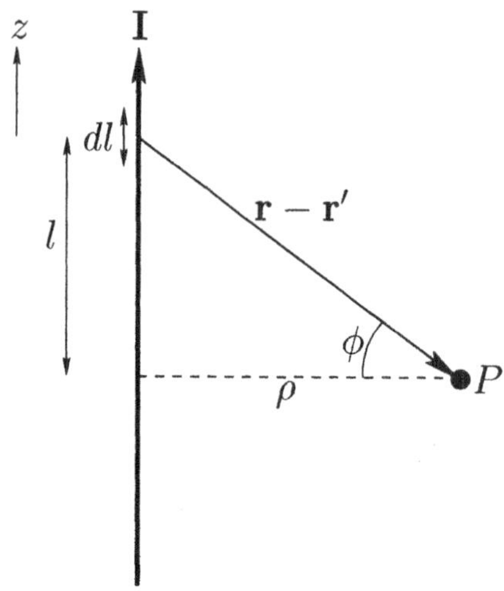

Figure 3.14: *A Biot-Savart law calculation.*

Consider an infinite straight wire, directed along the z-axis, and carrying a current I—see Figure 3.14. Let us reconstruct the magnetic field generated by the wire at point P using the Biot-Savart law. Suppose that the perpendicular distance to the wire is ρ. It is easily seen that

$$\mathbf{I} \times (\mathbf{r} - \mathbf{r}') = I\,\rho\,\mathbf{e}_\theta, \tag{3.188}$$

$$l = \rho \tan\phi, \tag{3.189}$$

$$dl = \frac{\rho}{\cos^2\phi}\,d\phi, \tag{3.190}$$

$$|\mathbf{r} - \mathbf{r}'| = \frac{\rho}{\cos\phi}, \tag{3.191}$$

where θ is a cylindrical polar coordinate. Thus, according to Equation (3.187), we have

$$\begin{aligned} B_\theta &= \frac{\mu_0}{4\pi}\int_{-\pi/2}^{\pi/2}\frac{I\,\rho}{\rho^3\,(\cos\phi)^{-3}}\frac{\rho}{\cos^2\phi}\,d\phi \\ &= \frac{\mu_0\,I}{4\pi\,\rho}\int_{-\pi/2}^{\pi/2}\cos\phi\,d\phi = \frac{\mu_0\,I}{4\pi\,\rho}\,[\sin\phi]_{-\pi/2}^{\pi/2}, \end{aligned} \tag{3.192}$$

which gives the familiar result

$$B_\theta = \frac{\mu_0\,I}{2\pi\,\rho}. \tag{3.193}$$

So, we have come full circle in our investigation of magnetic fields. Note that the simple result (3.193) can only be obtained from the Biot-Savart law after some non-trivial algebra. Examination of more complicated current distributions using this law invariably leads to lengthy, involved, and extremely unpleasant calculations.

3.14 ELECTROSTATICS AND MAGNETOSTATICS

We have now completed our theoretical investigation of electrostatics and magnetostatics. Our next task is to incorporate time variation into our analysis. However, before we start this, let us briefly review our progress so far. We have found that the electric fields generated by stationary charges, and the magnetic fields generated by steady currents, are describable in terms of four field equations:

$$\nabla \cdot \mathbf{E} = \frac{\rho}{\epsilon_0}, \tag{3.194}$$

$$\nabla \times \mathbf{E} = \mathbf{0}, \tag{3.195}$$

$$\nabla \cdot \mathbf{B} = 0, \tag{3.196}$$

$$\nabla \times \mathbf{B} = \mu_0 \, \mathbf{j}. \tag{3.197}$$

The boundary conditions are that the fields are zero at infinity, assuming that the generating charges and currents are localized to some region in space. According to Helmholtz's theorem, the above field equations, plus the boundary conditions, are sufficient to *uniquely* specify the electric and magnetic fields. The physical significance of this is that divergence and curl are the only *rotationally invariant* first-order differential properties of a general vector field: *i.e.*, the only quantities which do not change their physical characteristics when the coordinate axes are rotated. Since Physics does not depend on the orientation of the coordinate axes (which is, after all, quite arbitrary), it follows that divergence and curl are the *only* quantities which can appear in first-order differential field equations which claim to describe physical phenomena.

The field equations can be integrated to give:

$$\oint_S \mathbf{E} \cdot d\mathbf{S} = \frac{1}{\epsilon_0} \int_V \rho \, dV, \tag{3.198}$$

$$\oint_C \mathbf{E} \cdot d\mathbf{l} = 0, \tag{3.199}$$

$$\oint_S \mathbf{B}\cdot d\mathbf{S} = 0, \tag{3.200}$$

$$\oint_C \mathbf{B}\cdot d\mathbf{l} = \mu_0 \int_{S'} \mathbf{j}\cdot d\mathbf{S}. \tag{3.201}$$

Here, S is a closed surface enclosing a volume V. Also, C is a closed loop, and S′ is some surface attached to this loop. The field equations (3.194)–(3.197) can be deduced from Equations (3.198)–(3.201) using Gauss' theorem and Stokes' theorem. Equation (3.198) is called Gauss' law, and says that the flux of the electric field out of a closed surface is proportional to the enclosed electric charge. Equation (3.200) has no particular name, and says that there is no such things as a magnetic monopole. Equation (3.201) is called Ampère's circuital law, and says that the line integral of the magnetic field around any closed loop is proportional to the flux of the current density through the loop.

The field equation (3.195) is automatically satisfied if we write

$$\mathbf{E} = -\nabla\phi. \tag{3.202}$$

Likewise, the field equation (3.196) is automatically satisfied if we write

$$\mathbf{B} = \nabla\times\mathbf{A}. \tag{3.203}$$

Here, ϕ is the electric scalar potential, and **A** is the magnetic vector potential. The electric field is clearly unchanged if we add a constant to the scalar potential:

$$\mathbf{E}\rightarrow\mathbf{E} \quad \text{as} \quad \phi\rightarrow\phi + c. \tag{3.204}$$

The magnetic field is similarly unchanged if we subtract the gradient of a scalar field from the vector potential:

$$\mathbf{B}\rightarrow\mathbf{B} \quad \text{as} \quad \mathbf{A}\rightarrow\mathbf{A} - \nabla\psi. \tag{3.205}$$

The above transformations, which leave the **E** and **B** fields invariant, are called gauge transformations. We are free to choose c and ψ to be whatever we like: *i.e.*, we are free to choose the gauge. The most sensible gauge is the one which makes our equations as simple and symmetric as possible. This corresponds to the choice

$$\phi(\mathbf{r})\rightarrow 0 \quad \text{as} \quad |\mathbf{r}|\rightarrow\infty, \tag{3.206}$$

and

$$\nabla\cdot\mathbf{A} = 0. \tag{3.207}$$

The latter convention is known as the Coulomb gauge.

Taking the divergence of Equation (3.202) and the curl of Equation (3.203), and making use of the Coulomb gauge, we find that the four field equations (3.194)–(3.197) can be reduced to Poisson's equation written four times over:

$$\nabla^2\phi = -\frac{\rho}{\epsilon_0}, \tag{3.208}$$

$$\nabla^2\mathbf{A} = -\mu_0\,\mathbf{j}. \tag{3.209}$$

Poisson's equation is just about the simplest *rotationally invariant* second-order partial differential equation it is possible to write. Note that ∇^2 is clearly rotationally invariant, since it is the divergence of a gradient, and both divergence and gradient are rotationally invariant. We can always construct the solution to Poisson's equation, given the boundary conditions. Furthermore, we have a *uniqueness theorem* which tells us that our solution is the only possible solution. Physically, this means that there is only one electric and magnetic field which is consistent with a given set of stationary charges and steady currents. This sounds like an obvious, almost trivial, statement. But there are many areas of Physics (for instance, Fluid Mechanics and Plasma Physics) where we also believe, for physical reasons, that for a given set of boundary conditions the solution should be unique. The difficulty is that in most cases when we reduce a given problem to a partial differential equation we end up with something far nastier than Poisson's equation. In general, we cannot solve this equation. In fact, we usually cannot even prove that it possess a solution for general boundary conditions, let alone that the solution is unique. So, we are very fortunate indeed that in Electrostatics and Magnetostatics a general problem always boils down to solving a tractable partial differential equation. When physicists make statements to the effect that "Electromagnetism is the best understood theory in Physics," which they often do, what they are really saying is that the partial differential equations which crop up in this theory are soluble and have unique solutions.

Poisson's equation

$$\nabla^2 u = v \tag{3.210}$$

is *linear*, which means that its solutions are superposable. We can exploit this fact to construct a general solution to this equation. Suppose that we can find the solution to

$$\nabla^2 G(\mathbf{r}, \mathbf{r}') = \delta(\mathbf{r} - \mathbf{r}') \tag{3.211}$$

which satisfies the boundary conditions. This is the solution driven by a unit amplitude point source located at position vector $\mathbf{r}'$. Since any general source can be built up out of a weighted sum of point sources, it follows that a general solution to Poisson's equation can be built up out of a similarly weighted superposition of point source solutions. Mathematically, we can write

$$u(\mathbf{r}) = \int G(\mathbf{r}, \mathbf{r}')\, v(\mathbf{r}')\, d^3\mathbf{r}'. \tag{3.212}$$

The function $G(\mathbf{r}, \mathbf{r}')$ is called the Green's function. The Green's function for Poisson's equation is

$$G(\mathbf{r}, \mathbf{r}') = -\frac{1}{4\pi}\frac{1}{|\mathbf{r} - \mathbf{r}'|}. \tag{3.213}$$

Note that this Green's function is proportional to the scalar potential of a point charge located at $\mathbf{r}'$: this is hardly surprising, given the definition of a Green's function.

According to Equations (3.208), (3.209), (3.210), (3.212), and (3.213), the scalar and vector potentials generated by a set of stationary charges and steady currents take the form

$$\phi(\mathbf{r}) = \frac{1}{4\pi\epsilon_0}\int \frac{\rho(\mathbf{r}')}{|\mathbf{r} - \mathbf{r}'|}\, d^3\mathbf{r}', \tag{3.214}$$

$$\mathbf{A}(\mathbf{r}) = \frac{\mu_0}{4\pi}\int \frac{\mathbf{j}(\mathbf{r}')}{|\mathbf{r} - \mathbf{r}'|}\, d^3\mathbf{r}'. \tag{3.215}$$

Making use of Equations (3.202), (3.203), (3.214), and (3.215), we obtain Coulomb's law,

$$\mathbf{E}(\mathbf{r}) = \frac{1}{4\pi\epsilon_0}\int \rho(\mathbf{r}')\,\frac{\mathbf{r} - \mathbf{r}'}{|\mathbf{r} - \mathbf{r}'|^3}\, d^3\mathbf{r}', \tag{3.216}$$

and the Biot-Savart law,

$$\mathbf{B}(\mathbf{r}) = \frac{\mu_0}{4\pi}\int \frac{\mathbf{j}(\mathbf{r}') \times (\mathbf{r} - \mathbf{r}')}{|\mathbf{r} - \mathbf{r}'|^3}\, d^3\mathbf{r}'. \tag{3.217}$$

Of course, both of these laws are examples of action-at-a-distance laws (in that the electric and magnetic fields at a given point respond instantaneously to changes in the charge and current densities at distant points), and, therefore, violate the Special Theory of Relativity. However, this is not a problem as long as we restrict ourselves to fields generated by *time-independent* charge and current distributions.

The next question is by how much is this scheme which we have just worked out going to be disrupted when we take time variation into account. The answer, somewhat surprisingly, is by very little indeed. So, in Equations (3.194)–(3.217) we can already discern the basic outline of Classical Electromagnetism. Let us continue our investigation.

3.15 EXERCISES

3.1. A charge Q is uniformly distributed in a sphere of radius a centered on the origin. Use symmetry and Gauss' law to find the electric field generated inside and outside the sphere. What is the corresponding electric potential inside and outside the sphere?

3.2. A charge per unit length Q is uniformly distributed in an infinitely long cylinder of radius a whose axis corresponds to the z-axis. Use symmetry and Gauss' law to find the electric field generated inside and outside the cylinder. What is the corresponding electric potential inside and outside the cylinder.

3.3. Find the electric charge distribution which generates the Yukawa potential

$$\phi(r) = \frac{q}{4\pi\epsilon_0}\,\frac{e^{-r/a}}{r},$$

where r is a spherical polar coordinate, and a a positive constant. Why must the total charge in the distribution be zero?

3.4. An electric dipole consists of two equal and opposite charges, q and $-q$, separated by a *small* distance d. The strength and orientation of the dipole is measured by its vector moment $\mathbf{p}$, which is of magnitude $q\,d$, and points in the direction of the displacement of the positive charge from the negative. Use the principle of superposition to demonstrate that the electric potential generated by an electric dipole of moment $\mathbf{p}$ situated at the origin is

$$\phi(\mathbf{r}) = \frac{\mathbf{p}\cdot\mathbf{r}}{4\pi\epsilon_0\,r^3}.$$

Show that the corresponding electric field distribution is

$$\mathbf{E}(\mathbf{r}) = \frac{3\,(\mathbf{p}\cdot\mathbf{r})\,\mathbf{r} - r^2\,\mathbf{p}}{4\pi\epsilon_0\,r^5}.$$

3.5. An electric dipole of fixed moment $\mathbf{p}$ is situated at position $\mathbf{r}$ in a non-uniform external electric field $\mathbf{E}(\mathbf{r})$. Demonstrate that the net force on the dipole can be written $\mathbf{f} = -\nabla W$, where

$$W = -\mathbf{p} \cdot \mathbf{E}.$$

Hence, show that the potential energy of an electric dipole of moment $\mathbf{p}_1$ in the electric field generated by a second dipole of moment $\mathbf{p}_2$ is

$$W = \frac{r^2\,(\mathbf{p}_1 \cdot \mathbf{p}_2) - 3\,(\mathbf{p}_1 \cdot \mathbf{r})\,(\mathbf{p}_2 \cdot \mathbf{r})}{4\pi\epsilon_0\, r^5},$$

where $\mathbf{r}$ is the displacement of one dipole from another.

3.6. Show that the torque on an electric dipole of moment $\mathbf{p}$ in a uniform external electric field $\mathbf{E}$ is

$$\boldsymbol{\tau} = \mathbf{p} \times \mathbf{E}.$$

Hence, deduce that the potential energy of the dipole is

$$W = -\mathbf{p} \cdot \mathbf{E}.$$

3.7. A charge distribution $\rho(\mathbf{r})$ is localized in the vicinity of the origin in a region of radius a. Consider the electric potential generated at position $\mathbf{r}$, where $|\mathbf{r}| \gg a$. Demonstrate via a suitable expansion in a/r that

$$\phi(\mathbf{r}) = \frac{Q}{4\pi\epsilon_0\, r} + \frac{\mathbf{p} \cdot \mathbf{r}}{4\pi\epsilon_0\, r^3} + \cdots,$$

where $Q = \int \rho(\mathbf{r})\, d^3\mathbf{r}$ is the total charge contained in the distribution, and $\mathbf{p} = \int \rho(\mathbf{r})\, \mathbf{r}\, d^3\mathbf{r}$ its electric dipole moment.

3.8. Consider a scalar potential field $\phi(\mathbf{r})$ generated by a set of stationary charges. Demonstrate that the mean potential over any spherical surface which does not contain a charge is equal to the potential at the center. Hence, deduce that there can be no maxima or minima of the scalar potential in a charge-free region. Hint: The solution to this problem is more intuitive than mathematical, and depends on the fact that the potential generated outside a uniform spherical charge shell is the same as that generated when all of the charge is collected at its center.

3.9. Demonstrate that the Green's function for Poisson's equation in two dimensions (*i.e.*, $\partial/\partial z \equiv 0$) is

$$G(\mathbf{r}, \mathbf{r}') = -\frac{\ln|\mathbf{r} - \mathbf{r}'|}{2\pi},$$

where $\mathbf{r} = (x, y)$, *etc.* Hence, deduce that the scalar potential field generated by the two-dimensional charge distribution $\rho(\mathbf{r})$ is

$$\phi(\mathbf{r}) = -\frac{1}{2\pi\epsilon_0} \int \rho(\mathbf{r}')\, \ln|\mathbf{r} - \mathbf{r}'|\, d^3\mathbf{r}'.$$

3.10. A particle of mass m and charge q starts at rest from the origin at $t = 0$ in a uniform electric field E directed along the y-axis, and a uniform magnetic field

B directed along the z-axis. Find the particle's subsequent motion in the x-y plane. Sketch the particle's trajectory.

3.11. In a parallel-plate magnetron the cathode and the anode are flat parallel plates, and a uniform magnetic field B is applied in a direction parallel to the plates. Electrons are emitted from the cathode with essentially zero velocity. If the separation between the anode and cathode is d, and if the anode is held at a constant positive potential V with respect to the cathode, show that no current will flow between the plates when

$$V \leq \frac{e\,B^2\,d^2}{2\,m_e}.$$

Here, e is the magnitude of the electron charge, and m_e the electron mass.

3.12. An infinite, straight, circular cross-section wire of radius a runs along the z-axis and carries a uniformly distributed z-directed current I. Use symmetry and Ampère's circuital law to find the magnetic field distribution inside and outside the wire. What is the corresponding magnetic vector potential inside and outside the wire? Use the Coulomb gauge.

3.13. An infinite cylindrical current annulus of inner radius a, outer radius b, and axis running along the z-axis carries a uniformly distributed current per unit length I in the θ direction: *i.e.*, $\mathbf{j} \propto \mathbf{e}_\theta$. Use symmetry and Ampère's circuital law to find the magnetic field distribution inside and outside the annulus. What is the corresponding magnetic vector potential inside and outside the annulus? Use the Coulomb gauge.

3.14. A thick slab extends from $z = -a$ to $z = a$, and is infinite in the x-y plane. The slab carries a uniform current density $\mathbf{j} = J\,\mathbf{e}_x$. Find the magnetic field and magnetic vector potential inside and outside the slab. Use the Coulomb gauge.

3.15. Show that the magnetic vector potential due to two long, straight, z-directed wires, the first carrying a current I, and the second a current $-I$, is

$$\mathbf{A} = \frac{\mu_0\,I}{2\pi}\,\ln\left(\frac{r_1}{r_2}\right)\mathbf{e}_z,$$

where r_1 and r_2 are the perpendicular distances to the two wires.

3.16. Use the Biot-Savart law to:

(a) Find the magnetic field at the center of a circular loop of radius r carrying a current I.

(b) Find the magnetic field at the center of a square loop carrying a current I. Let r be the perpendicular distance from the center to one of the sides of the loop.

(c) Find the magnetic field at the center of a regular n-sided polygon carrying a current I. Let r be the perpendicular distance from the center to one of the sides of the loop. Check that your answer reduces to the answer from part (a) in the limit $n \to \infty$.

3.17. Use the Biot-Savart law to show that the magnetic field generated along the axis of a circular current loop of radius a lying in the x-y plane and centered on the origin is

$$B_z = \frac{\mu_0\, I}{2} \frac{a^2}{(a^2 + z^2)^{3/2}}.$$

Here, I is the current circulating counterclockwise (looking down the z-axis) around the loop. Demonstrate that

$$\int_{-\infty}^{\infty} B_z(z)\, dz = \mu_0\, I.$$

Derive this result from Ampère's circuital law.

3.18. A Helmholtz coil consists of two identical, single turn, circular coils, of radius a, carrying the same current, I, in the same sense, which are coaxial with one another, and are separated by a distance d. Show that the variation of the magnetic field-strength in the vicinity of the axial midpoint is minimized when $d = a$. Demonstrate that, in this optimal case, the magnetic field-strength at the axial midpoint is

$$B = \frac{8\, \mu_0\, I}{5\sqrt{5}\, a}.$$

3.19. A force-free magnetic field is such that $\mathbf{j} \times \mathbf{B} = \mathbf{0}$. Demonstrate that such a field satisfies

$$\nabla^2 \mathbf{B} = -\alpha^2\, \mathbf{B},$$

where α is some constant. Find the force-free field with the lowest value of α (excluding $\alpha = 0$) in a cubic volume of dimension a bounded by superconducting walls (in which $\mathbf{B} = \mathbf{0}$).

3.20. Consider a small circular current loop of radius a lying in the x-y plane and centered on the origin. Such a loop constitutes a magnetic dipole of moment $\mathbf{m}$, where $m = \pi\, a^2\, I$, and I is the circulating current. The direction of $\mathbf{m}$ is conventionally taken to be normal to the plane of the loop, in the sense given by the right-hand grip rule. Demonstrate that the magnetic vector potential generated at position $\mathbf{r}$, where $|\mathbf{r}| \gg a$, is

$$\mathbf{A}(\mathbf{r}) = \frac{\mu_0}{4\pi} \frac{\mathbf{m} \times \mathbf{r}}{r^3}.$$

Show that the corresponding magnetic field is

$$\mathbf{B}(\mathbf{r}) = \frac{\mu_0}{4\pi}\left(\frac{3\,(\mathbf{r}\cdot\mathbf{m})\,\mathbf{r} - r^2\,\mathbf{m}}{r^5}\right).$$

3.21. Demonstrate that the torque on a magnetic dipole of moment $\mathbf{m}$ placed in a uniform external magnetic field $\mathbf{B}$ is

$$\boldsymbol{\tau} = \mathbf{m}\times\mathbf{B}.$$

Hence, deduce that the potential energy of the magnetic dipole is

$$W = -\mathbf{m}\cdot\mathbf{B}.$$

3.22. Consider two magnetic dipoles, $\mathbf{m}_1$ and $\mathbf{m}_2$. Suppose that $\mathbf{m}_1$ is fixed, whereas $\mathbf{m}_2$ can rotate freely in any direction. Demonstrate that the equilibrium configuration of the second dipole is such that

$$\tan\theta_1 = -2\,\tan\theta_2,$$

where θ_1 and θ_2 are the angles subtended by $\mathbf{m}_1$ and $\mathbf{m}_2$, respectively, with the radius vector joining them.

Chapter 4 TIME-DEPENDENT MAXWELL EQUATIONS

4.1 INTRODUCTION

In this chapter, we shall generalize the time-independent Maxwell equations, derived in the previous chapter, to obtain the full set of time-dependent Maxwell equations.

4.2 FARADAY'S LAW

The history of humankind's development of Physics can be thought of as the history of the synthesis of ideas. Physicists keep finding that apparently disparate phenomena can be understood as different aspects of some more fundamental phenomenon. This process has continued until, today, all physical phenomena can be described in terms of three fundamental forces: *Gravity*, the *Electroweak Force*, and the *Strong Force*. One of the main goals of modern physics is to find some way of combining these three forces so that all of physics can be described in terms of a single unified force.

The first great synthesis of ideas in Physics took place in 1666 when Issac Newton realised that the force which causes apples to fall downward is the same as that which maintains the Planets in elliptical orbits around the Sun. The second great synthesis, which we are about to study in more detail, took place in 1830 when Michael Faraday discovered that Electricity and Magnetism are two aspects of the same phenomenon, usually referred to as *Electromagnetism*. The third great synthesis, which we shall discuss presently, took place in 1873 when James Clerk Maxwell demonstrated that light and electromagnetism are intimately related. The last (but, hopefully, not the final) great synthesis took place in 1967 when Steve Weinberg and Abdus Salam showed that the electromagnetic force and the weak nuclear force (*i.e.*, the one which is responsible for β decays) can be combined to give the electroweak force.

Let us now consider Faraday's experiments, having put them in their proper historical context. Prior to 1830, the only known way in which to make an electric current flow through a conducting wire was to connect the ends of the wire to the positive and negative terminals of a battery. We measure a battery's ability to push current down a wire in terms of its *voltage*, by which we mean the voltage difference between its positive and negative terminals. What does voltage correspond to in Physics? Well, volts are the units used to measure electric scalar potential, so when we talk about a 6 V battery, what we are really saying is that the difference in electric scalar potential between its positive and negative terminals is six volts. This insight allows us to write

$$V = \phi(\oplus) - \phi(\ominus) = -\int_{\oplus}^{\ominus} \nabla\phi \cdot \mathbf{dl} = \int_{\oplus}^{\ominus} \mathbf{E} \cdot \mathbf{dl}, \qquad (4.1)$$

where V is the battery voltage, $\oplus$ denotes the positive terminal, $\ominus$ the negative terminal, and $\mathbf{dl}$ is an element of length along the wire. Of course, the above equation is a direct consequence of $\mathbf{E} = -\nabla\phi$. Clearly, a voltage difference between two ends of a wire attached to a battery implies the presence of a longitudinal electric field which pushes charges through the wire. This field is directed from the positive terminal of the battery to the negative terminal, and is, therefore, such as to force electrons (which are negatively charged) to flow through the wire from the negative to the positive terminal. As expected, this means that a net positive current flows from the positive to the negative terminal. The fact that $\mathbf{E}$ is a conservative field ensures that the voltage difference V is independent of the path of the wire between the terminals. In other words, two different wires attached to the same battery develop identical voltage differences.

Let us now consider a closed loop of wire (with no battery). The voltage around such a loop, which is sometimes called the *electromotive force* or *emf*, is

$$V = \oint \mathbf{E} \cdot \mathbf{dl} = 0. \qquad (4.2)$$

The fact that the right-hand side of the above equation is zero is a direct consequence of the field equation $\nabla \times \mathbf{E} = \mathbf{0}$. We conclude that, because $\mathbf{E}$ is a conservative field, the electromotive force around a closed loop of wire is automatically zero, and so there is no current flow around such a loop. This all seems to make sense. However, in 1830 Michael Faraday discovered that a changing magnetic field can cause a current to flow around a closed loop of wire (in the absence of a battery). Well, if current flows around the loop then there must be an electromotive force. So,

$$V = \oint \mathbf{E} \cdot d\mathbf{l} \neq 0, \tag{4.3}$$

which immediately implies that $\mathbf{E}$ is *not* a conservative field, and that $\nabla \times \mathbf{E} \neq \mathbf{0}$. Clearly, we are going to have to modify some of our ideas regarding electric fields.

Faraday continued his experiments, and found that another way of generating an electromotive force around a loop of wire is to keep the magnetic field constant and move the loop. Eventually, Faraday was able to formulate a law which accounted for all of his experiments—the emf generated around a loop of wire in a magnetic field is proportional to the rate of change of the flux of the magnetic field through the loop. So, if the loop is denoted C, and S is some surface attached to the loop, then Faraday's experiments can be summed up by writing

$$V = \oint_C \mathbf{E} \cdot d\mathbf{l} = A \frac{\partial}{\partial t} \int_S \mathbf{B} \cdot d\mathbf{S}, \tag{4.4}$$

where A is a constant of proportionality. Thus, the changing flux of the magnetic field through the loop generates an electric field directed around the loop. This process is know as *magnetic induction*.

SI units have been carefully chosen so as to make $|A| = 1$ in the above equation. The only thing we now have to decide is whether $A = +1$ or $A = -1$. In other words, we need to decide which way around the loop the induced emf wants to drive the current. We possess a general principle which allows us to decide questions like this. It is called *Le Chatelier's principle*. According to Le Chatelier's principle, every change generates a reaction which tries to minimize the change. Essentially, this means that the Universe is stable to small perturbations. When this principle is applied to the special case of magnetic induction, it is usually called *Lenz's law*. According to Lenz's law, the current induced around a closed loop is always such that the magnetic field it produces tries to counteract the change in magnetic flux which generates the electromotive force. From Figure 4.1, it is clear that if the magnetic field $\mathbf{B}$ is increasing and the current I circulates clockwise (as seen from above) then it generates a field $\mathbf{B}'$ which opposes the increase in magnetic flux through the loop, in accordance with Lenz's law. The direction of the current is opposite to the sense of the current loop C, as determined by the right-hand grip rule (assuming that the flux of $\mathbf{B}$ through the loop is positive), so this implies that $A = -1$ in Equation (4.4). Thus, Faraday's law takes

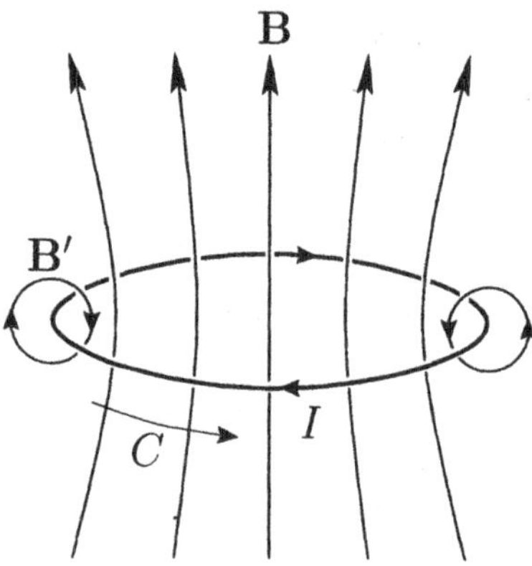

Figure 4.1: *Lenz's law.*

the form

$$\oint_C \mathbf{E} \cdot d\mathbf{l} = -\frac{\partial}{\partial t}\int_S \mathbf{B} \cdot d\mathbf{S}. \tag{4.5}$$

Experimentally, Faraday's law is found to correctly predict the emf (*i.e.*, $\oint \mathbf{E} \cdot d\mathbf{l}$) generated around any wire loop, irrespective of the position or shape of the loop. It is reasonable to assume that the same emf would be generated in the absence of the wire (of course, no current would flow in this case). We conclude that Equation (4.5) is valid for *any* closed loop C. Now, if Faraday's law is to make any sense then it must also be true for all surfaces S attached to the loop C. Clearly, if the flux of the magnetic field through the loop depends on the surface upon which it is evaluated then Faraday's law is going to predict different emfs for different surfaces. Since there is no preferred surface for a general non-coplanar loop, this would not make very much sense. The condition for the flux of the magnetic field, $\int_S \mathbf{B} \cdot d\mathbf{S}$, to depend only on the loop C to which the surface S is attached, and not on the nature of the surface itself, is

$$\oint_{S'} \mathbf{B} \cdot d\mathbf{S}' = 0, \tag{4.6}$$

for any closed surface S′.

Faraday's law, Equation (4.5), can be converted into a field equation using Stokes' theorem. We obtain

$$\nabla \times \mathbf{E} = -\frac{\partial \mathbf{B}}{\partial t}. \tag{4.7}$$

This is the final Maxwell equation. It describes how a changing magnetic field can generate, or induce, an electric field. Gauss' theorem applied to Equation (4.6) gives the familiar field equation

$$\nabla \cdot \mathbf{B} = 0. \tag{4.8}$$

This ensures that the magnetic flux through a loop is a well-defined quantity.

The divergence of Equation (4.7) yields

$$\frac{\partial \nabla \cdot \mathbf{B}}{\partial t} = 0. \tag{4.9}$$

Thus, the field equation (4.7) actually demands that the divergence of the magnetic field be constant in time for self-consistency (this means that the flux of the magnetic field through a loop need not be a well-defined quantity, as long as its time derivative is well-defined). However, a constant non-solenoidal magnetic field can only be generated by magnetic monopoles, and magnetic monopoles do not exist (as far as we are aware). Hence, $\nabla \cdot \mathbf{B} = 0$. Note that the absence of magnetic monopoles is an observational fact—it cannot be predicted by any theory. If magnetic monopoles were discovered tomorrow then this would not cause physicists any great difficulties, since they know how to generalize Maxwell's equations to include both magnetic monopoles and currents of magnetic monopoles. In this generalized formalism, Maxwell's equations are completely symmetric with respect to electric and magnetic fields, and $\nabla \cdot \mathbf{B} \neq 0$. However, an extra term (involving the current of magnetic monopoles) must be added to the right-hand side of Equation (4.7) in order to make it mathematically self-consistent.

As an example of the use of Faraday's law, let us calculate the electric field generated by a decaying magnetic field of the form $\mathbf{B} = B_z(r, t)\, \mathbf{e}_z$, where

$$B_z(r, t) = \begin{cases} B_0 \exp(-t/\tau) & r \leq a \\ 0 & r > a \end{cases}. \tag{4.10}$$

Here, B_0 and τ are positive constants. By symmetry, we expect an induced electric field of the form $\mathbf{E}(r, t)$. We also expect $\nabla \cdot \mathbf{E} = 0$, since there are no electric charges in the problem. This rules out a radial electric field. We can also rule out a z-directed electric field, since $\nabla \times E_z(r)\, \mathbf{e}_z = -(\partial E_z/\partial r)\, \mathbf{e}_\theta$, and we require $\nabla \times \mathbf{E} \propto \mathbf{B} \propto \mathbf{e}_z$. Hence, the induced electric field must be of the form $\mathbf{E}(r, t) = E_\theta(r, t)\, \mathbf{e}_\theta$. Now, according to Faraday's law, the line integral of the electric field around some closed loop is equal to minus the rate of change of the magnetic flux through the loop. If we choose a loop which is a circle of radius r in the x-y plane, then we have

$$2\pi r\, E_\theta(r,t) = -\frac{d\Phi}{dt}, \tag{4.11}$$

where Φ is the magnetic flux (in the $+z$ direction) through a circular loop of radius r. It is evident that

$$\Phi(r,t) = \begin{cases} \pi r^2\, B_0 \exp(-t/\tau) & r \leq a \\ \pi a^2\, B_0 \exp(-t/\tau) & r > a \end{cases}. \tag{4.12}$$

Hence,

$$E_\theta(r,t) = \begin{cases} (B_0/2\tau)\, r \exp(-t/\tau) & r \leq a \\ (B_0/2\tau)\,(a^2/r) \exp(-t/\tau) & r > a \end{cases}. \tag{4.13}$$

4.3 ELECTRIC SCALAR POTENTIAL?

We now have a problem. We can only write the electric field in terms of a scalar potential (*i.e.*, $\mathbf{E} = -\nabla\phi$) provided that $\nabla \times \mathbf{E} = \mathbf{0}$. However, we have just found that the curl of the electric field is non-zero in the presence of a changing magnetic field. In other words, $\mathbf{E}$ is not, in general, a conservative field. Does this mean that we have to abandon the concept of electric scalar potential? Fortunately, it does not. It is still possible to define a scalar potential which is physically meaningful.

Let us start from the equation

$$\nabla \cdot \mathbf{B} = 0, \tag{4.14}$$

which is valid for both time-varying and non-time-varying magnetic fields. Since the magnetic field is solenoidal, we can write it as the curl of a vector potential:

$$\mathbf{B} = \nabla \times \mathbf{A}. \tag{4.15}$$

So, there is no problem with the vector potential in the presence of time-varying fields. Let us substitute Equation (4.15) into the field equation (4.7). We obtain

$$\nabla \times \mathbf{E} = -\frac{\partial\, \nabla \times \mathbf{A}}{\partial t}, \tag{4.16}$$

which can be written

$$\nabla \times \left(\mathbf{E} + \frac{\partial \mathbf{A}}{\partial t} \right) = \mathbf{0}. \tag{4.17}$$

Now, we know that a curl-free vector field can always be expressed as the gradient of a scalar potential, so let us write

$$\mathbf{E} + \frac{\partial \mathbf{A}}{\partial t} = -\nabla \phi, \tag{4.18}$$

or

$$\mathbf{E} = -\nabla \phi - \frac{\partial \mathbf{A}}{\partial t}. \tag{4.19}$$

This equation tells us that the scalar potential ϕ only describes the conservative electric field generated by electric charges. The electric field induced by time-varying magnetic fields is non-conservative, and is described by the magnetic vector potential **A**.

4.4 GAUGE TRANSFORMATIONS

Electric and magnetic fields can be written in terms of scalar and vector potentials, as follows:

$$\mathbf{E} = -\nabla \phi - \frac{\partial \mathbf{A}}{\partial t}, \tag{4.20}$$

$$\mathbf{B} = \nabla \times \mathbf{A}. \tag{4.21}$$

However, this prescription is not unique. There are many different potentials which can generate the same fields. We have come across this problem before. It is called *gauge invariance*. The most general transformation which leaves the **E** and **B** fields unchanged in Equations (4.20) and (4.21) is

$$\phi \to \phi + \frac{\partial \psi}{\partial t}, \tag{4.22}$$

$$\mathbf{A} \to \mathbf{A} - \nabla \psi. \tag{4.23}$$

This is clearly a generalization of the gauge transformation which we found earlier for static fields:

$$\phi \to \phi + c, \tag{4.24}$$

$$\mathbf{A} \to \mathbf{A} - \nabla \psi, \tag{4.25}$$

where c is a constant. In fact, if $\psi(\mathbf{r},t) \to \psi(\mathbf{r}) + c\,t$ then Equations (4.22) and (4.23) reduce to Equations (4.24) and (4.25).

We are free to choose the gauge so as to make our equations as simple as possible. As before, the most sensible gauge for the scalar potential is to make it go to zero at infinity:

$$\phi(\mathbf{r},t) \to 0 \qquad \text{as } |\mathbf{r}| \to \infty. \tag{4.26}$$

For steady fields, we found that the optimum gauge for the vector potential was the so-called Coulomb gauge:

$$\nabla \cdot \mathbf{A} = 0. \tag{4.27}$$

We can still use this gauge for non-steady fields. The argument, which we gave earlier (see Section 3.12), that it is always possible to transform away the divergence of a vector potential remains valid. One of the nice features of the Coulomb gauge is that when we write the electric field,

$$\mathbf{E} = -\nabla\phi - \frac{\partial \mathbf{A}}{\partial t}, \tag{4.28}$$

we find that the part which is generated by charges (*i.e.*, the first term on the right-hand side) is conservative, and the part induced by magnetic fields (*i.e.*, the second term on the right-hand side) is purely solenoidal. Earlier on, we proved mathematically that a general vector field can be written as the sum of a conservative field and a solenoidal field (see Section 3.11). Now we are finding that when we split up the electric field in this manner the two fields have different physical origins—the conservative part of the field emanates from electric charges, whereas the solenoidal part is induced by magnetic fields.

Equation (4.28) can be combined with the field equation

$$\nabla \cdot \mathbf{E} = \frac{\rho}{\epsilon_0} \tag{4.29}$$

(which remains valid for non-steady fields) to give

$$-\nabla^2\phi - \frac{\partial\, \nabla \cdot \mathbf{A}}{\partial t} = \frac{\rho}{\epsilon_0}. \tag{4.30}$$

With the Coulomb gauge condition, $\nabla \cdot \mathbf{A} = 0$, the above expression reduces to

$$\nabla^2\phi = -\frac{\rho}{\epsilon_0}, \tag{4.31}$$

which is just Poisson's equation. Thus, we can immediately write down an expression for the scalar potential generated by non-steady fields. It is exactly analogous to our previous expression for the scalar potential generated by steady fields: *i.e.*,

$$\phi(\mathbf{r},t) = \frac{1}{4\pi\epsilon_0}\int \frac{\rho(\mathbf{r}',t)}{|\mathbf{r}-\mathbf{r}'|}\,d^3\mathbf{r}'. \tag{4.32}$$

However, this apparently simple result is *extremely* deceptive. Equation (4.32) is a typical action at a distance law. If the charge density changes suddenly at $\mathbf{r}'$ then the potential at $\mathbf{r}$ responds *immediately*. However, we shall see later that the full time-dependent Maxwell's equations only allow information to propagate at the speed of light (*i.e.*, they do not violate Relativity). How can these two statements be reconciled? The crucial point is that the scalar potential cannot be measured directly, it can only be inferred from the electric field. In the time-dependent case, there are two parts to the electric field: that part which comes from the scalar potential, and that part which comes from the vector potential [see Equation (4.28)]. So, if the scalar potential responds immediately to some distance rearrangement of charge density then it does not necessarily follow that the electric field also has an immediate response. What actually happens is that the change in the part of the electric field which comes from the scalar potential is balanced by an equal and opposite change in the part which comes from the vector potential, so that the overall electric field remains unchanged. This state of affairs persists at least until sufficient time has elapsed for a light signal to travel from the distant charges to the region in question. Thus, Relativity is not violated, since it is the electric field, and not the scalar potential, which carries physically accessible information.

It is clear that the apparent action at a distance nature of Equation (4.32) is highly misleading. This suggests, very strongly, that the Coulomb gauge is not the optimum gauge in the time-dependent case. A more sensible choice is the so-called *Lorenz gauge*:

$$\nabla\cdot\mathbf{A} = -\epsilon_0\mu_0\,\frac{\partial\phi}{\partial t}. \tag{4.33}$$

It can be shown, by analogy with earlier arguments (see Section 3.12), that it is always possible to make a gauge transformation such that the above equation is satisfied at a given instance in time. Substituting the Lorenz gauge condition into Equation (4.30), we obtain

$$\epsilon_0\mu_0\,\frac{\partial^2\phi}{\partial t^2} - \nabla^2\phi = \frac{\rho}{\epsilon_0}. \tag{4.34}$$

It turns out that this is a three-dimensional wave equation in which information propagates at the speed of light. But, more of this later. Note that the magnetically induced part of the electric field (*i.e.*, $-\partial \mathbf{A}/\partial t$) is not purely solenoidal in the Lorenz gauge. This is a slight disadvantage of the Lorenz gauge with respect to the Coulomb gauge. However, this disadvantage is more than offset by other advantages which will become apparent presently. Incidentally, the fact that the part of the electric field which we ascribe to magnetic induction changes when we change the gauge suggests that the separation of the field into magnetically induced and charge induced components is not unique in the general time-varying case (*i.e.*, it is a convention).

4.5 THE DISPLACEMENT CURRENT

Michael Faraday revolutionized Physics in 1830 by showing that electricity and magnetism were interrelated phenomena. He achieved this breakthrough by careful experimentation. Between 1864 and 1873, James Clerk Maxwell achieved a similar breakthrough by pure thought. Of course, this was only possible because he was able to take the experimental results of Coulomb, Ampère, Faraday, *etc.*, as his starting point. Prior to 1864, the laws of electromagnetism were written in integral form. Thus, Gauss's law was (in SI units) *the flux of the electric field through a closed surface equals the total enclosed charge, divided by* ϵ_0. The no-magnetic-monopole law was *the flux of the magnetic field through any closed surface is zero*. Faraday's law was *the electromotive force generated around a closed loop equals minus the rate of change of the magnetic flux through the loop*. Finally, Ampère's circuital law was *the line integral of the magnetic field around a closed loop equals the total current flowing through the loop, multiplied by* μ_0. Maxwell's first great achievement was to realize that these laws could be expressed as a set of first-order partial differential equations. Of course, he wrote his equations out in component form, because modern vector notation did not come into vogue until about the time of the First World War. In modern notation, Maxwell first wrote:

$$\nabla \cdot \mathbf{E} = \frac{\rho}{\epsilon_0}, \tag{4.35}$$

$$\nabla \cdot \mathbf{B} = 0, \tag{4.36}$$

$$\nabla \times \mathbf{E} = -\frac{\partial \mathbf{B}}{\partial t}, \tag{4.37}$$

$$\nabla \times \mathbf{B} = \mu_0 \, \mathbf{j}. \tag{4.38}$$

Maxwell's second great achievement was to realize that these equations are wrong.

We can see that there is something slightly unusual about Equations (4.35)–(4.38): *i.e.*, they are very asymmetric with respect to electric and magnetic fields. After all, time-varying magnetic fields can induce electric fields, but electric fields apparently cannot affect magnetic fields in any way. However, there is a far more serious problem associated with the above equations, which we alluded to earlier. Consider the integral form of the last Maxwell equation (*i.e.*, Ampère's circuital law)

$$\oint_C \mathbf{B} \cdot d\mathbf{l} = \mu_0 \int_S \mathbf{j} \cdot d\mathbf{S}. \tag{4.39}$$

This says that the line integral of the magnetic field around a closed loop C is equal to μ_0 times the flux of the current density through the loop. The problem is that the flux of the current density through a loop is not, in general, a well-defined quantity. In order for the flux to be well-defined, the integral of $\mathbf{j} \cdot d\mathbf{S}$ over some surface S attached to a loop C must depend on C, but not on the details of S. This is only the case if

$$\nabla \cdot \mathbf{j} = 0. \tag{4.40}$$

Unfortunately, the above condition is only satisfied for non-time-varying fields.

Why do we say that, in general, $\nabla \cdot \mathbf{j} \neq 0$? Well, consider the flux of $\mathbf{j}$ out of some closed surface S enclosing a volume V. This is clearly equivalent to the rate at which charge flows out of S. However, if charge is a conserved quantity (and we certainly believe that it is) then the rate at which charge flows out of S must equal the rate of decrease of the charge contained in volume V. Thus,

$$\oint_S \mathbf{j} \cdot d\mathbf{S} = -\frac{\partial}{\partial t}\int_V \rho \, dV. \tag{4.41}$$

Making use of Gauss' theorem, this yields

$$\nabla \cdot \mathbf{j} = -\frac{\partial \rho}{\partial t}. \tag{4.42}$$

Thus, $\nabla \cdot \mathbf{j} = 0$ is only true in a steady-state (*i.e.*, when $\partial/\partial t \equiv 0$).

The problem with Ampère's circuital law is well illustrated by the following very famous example. Consider a long straight wire interrupted by a parallel plate capacitor. Suppose that C is some loop which circles the wire. In the time-independent situation, the capacitor acts like a break in the wire, so no current flows, and no magnetic field is generated.

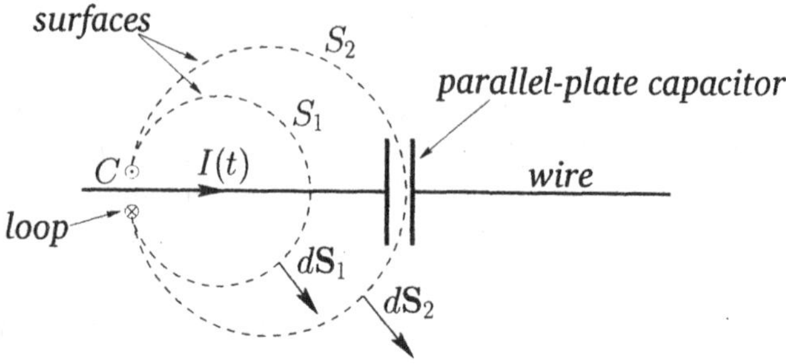

Figure 4.2: *Application of Ampère's circuital law to a charging, or discharging, capacitor.*

There is clearly no problem with Ampère's law in this case. However, in the time-dependent situation, a transient current flows in the wire as the capacitor charges up, or charges down, and so a transient magnetic field is generated. Thus, the line integral of the magnetic field around C is (transiently) non-zero. According to Ampère's circuital law, the flux of the current density through any surface attached to C should also be (transiently) non-zero. Let us consider two such surfaces. The first surface, S_1, intersects the wire—see Figure 4.2. This surface causes us no problem, since the flux of $\mathbf{j}$ though the surface is clearly non-zero (because it intersects a current-carrying wire). The second surface, S_2, passes between the plates of the capacitor, and, therefore, does not intersect the wire at all. Clearly, the flux of the current density through this surface is zero. The current density fluxes through surfaces S_1 and S_2 are obviously different. However, both surfaces are attached to the same loop C, so the fluxes should be the same, according to Ampère's law (4.39). It would appear that Ampère's circuital law is about to disintegrate. Note, however, that although the surface S_2 does not intersect any electric current, it does pass through a region of strong changing electric field as it threads between the plates of the charging (or discharging) capacitor. Perhaps, if we add a term involving $\partial\mathbf{E}/\partial t$ to the right-hand side of Equation (4.38) then we can somehow fix up Ampère's circuital law? This is, essentially, how Maxwell reasoned more than one hundred years ago.

Let us try out this scheme. Suppose that we write

$$\nabla \times \mathbf{B} = \mu_0 \mathbf{j} + \lambda \frac{\partial \mathbf{E}}{\partial t}, \tag{4.43}$$

instead of Equation (4.38). Here, λ is some constant. Does this resolve our problem? We want the flux of the right-hand side of the above equation through some loop C to be well-defined; *i.e.*, it should only depend on C, and not the particular surface S (which spans C) upon which it is evaluated. This is another way of saying that we want the divergence of the right-hand side to be zero. In fact, we can see that this is necessary for self-consistency, since the divergence of the left-hand side is automatically zero. So, taking the divergence of Equation (4.43), we obtain

$$0 = \mu_0 \nabla \cdot \mathbf{j} + \lambda \frac{\partial \nabla \cdot \mathbf{E}}{\partial t}. \tag{4.44}$$

But, we know that

$$\nabla \cdot \mathbf{E} = \frac{\rho}{\epsilon_0}, \tag{4.45}$$

so combining the previous two equations we arrive at

$$\mu_0 \nabla \cdot \mathbf{j} + \frac{\lambda}{\epsilon_0} \frac{\partial \rho}{\partial t} = 0. \tag{4.46}$$

Now, our charge conservation law (4.42) can be written

$$\nabla \cdot \mathbf{j} + \frac{\partial \rho}{\partial t} = 0. \tag{4.47}$$

The previous two equations are in agreement provided $\lambda = \epsilon_0 \mu_0$. So, if we modify the final Maxwell equation such that it reads

$$\nabla \times \mathbf{B} = \mu_0 (\mathbf{j} + \mathbf{j}_d), \tag{4.48}$$

where

$$\mathbf{j}_d = \epsilon_0 \frac{\partial \mathbf{E}}{\partial t}, \tag{4.49}$$

then we find that the divergence of the right-hand side is zero as a consequence of charge conservation. The extra term, $\mathbf{j}_d$, is called the *displacement current* density (this name was invented by Maxwell). In summary, we have shown that although the flux of the real current density through a loop is *not* well-defined, if we form the sum of the real current density and the displacement current density then the flux of this new quantity through a loop *is* well-defined.

Of course, the displacement current is not a current at all. It is, in fact, associated with the induction of magnetic fields by time-varying electric fields. Maxwell came up with this rather curious name because many of his ideas regarding electric and magnetic fields were completely wrong. For instance, Maxwell believed in the ether (a tenuous invisible medium permeating all space), and he thought that electric and magnetic fields were some sort of stresses in this medium. He also thought that the displacement current was associated with displacements of the ether (hence, the name). The reason that these misconceptions did not invalidate his equations is quite simple. Maxwell based his equations on the results of experiments, and he added in his extra term so as to make these equations mathematically self-consistent. Both of these steps are valid irrespective of the existence or non-existence of the ether.

Now, the field equations (4.35)–(4.38) are derived directly from the results of famous nineteenth-century experiments. So, if a new term involving the time derivative of the electric field needs to be added to one of these equations, for the sake of mathematical consistency, why is there is no corresponding nineteenth-century experimental result which demonstrates this fact? Well, it turns out that the new term describes an effect which is far too small to have been observed in the nineteenth century. Let us demonstrate this.

First, we shall show that it is comparatively easy to detect the induction of an electric field by a changing magnetic field in a desktop laboratory experiment. The Earth's magnetic field is about 1 gauss (that is, 10^{-4} tesla). Magnetic fields generated by electromagnets (which will fit on a laboratory desktop) are typically about one hundred times bigger than this. Let us, therefore, consider a hypothetical experiment in which a 100 gauss magnetic field is switched on suddenly. Suppose that the field ramps up in one tenth of a second. What electromotive force is generated in a 10 centimeter square loop of wire located in this field? Faraday's law is written

$$V = -\frac{\partial}{\partial t}\oint \mathbf{B}\cdot d\mathbf{S} \sim \frac{B\,A}{t}, \tag{4.50}$$

where $B = 0.01$ tesla is the field-strength, $A = 0.01$ m^2 is the area of the loop, and $t = 0.1$ seconds is the ramp time. It follows that $V \sim 1$ millivolt. Well, one millivolt is easily detectable. In fact, most hand-held laboratory voltmeters are calibrated in millivolts. It is, thus, clear that we would have no difficulty whatsoever detecting the magnetic induction of electric fields in a nineteenth-century-style laboratory experiment.

Let us now consider the electric induction of magnetic fields. Suppose that our electric field is generated by a parallel plate capacitor of spacing one centimeter which is charged up to 100 volts. This gives a field of 10^4 volts per meter. Suppose, further, that the capacitor is discharged in one tenth of a second. The law of electric induction is obtained by integrating Equation (4.48), and neglecting the first term on the right-hand side. Thus,

$$\oint \mathbf{B} \cdot d\mathbf{l} = \epsilon_0 \mu_0 \frac{\partial}{\partial t} \int \mathbf{E} \cdot d\mathbf{S}. \tag{4.51}$$

Let us consider a loop 10 centimeters square. What is the magnetic field generated around this loop (which we could try to measure with a Hall probe)? Very approximately, we find that

$$l\,B \sim \epsilon_0 \mu_0 \frac{E\,l^2}{t}, \tag{4.52}$$

where $l = 0.1$ meters is the dimensions of the loop, B is the magnetic field-strength, $E = 10^4$ volts per meter is the electric field, and $t = 0.1$ seconds is the decay time of the field. We obtain $B \sim 10^{-9}$ gauss. Modern technology is unable to detect such a small magnetic field, so we cannot really blame nineteenth-century physicists for not discovering electric induction experimentally.

Note, however, that the displacement current *is* detectable in some modern experiments. Suppose that we take an FM radio signal, amplify it so that its peak voltage is one hundred volts, and then apply it to the parallel plate capacitor in the previous hypothetical experiment. What size of magnetic field would this generate? Well, a typical FM signal oscillates at 10^9 Hz, so t in the previous example changes from 0.1 seconds to 10^{-9} seconds. Thus, the induced magnetic field is about 10^{-1} gauss. This is certainly detectable by modern technology. Hence, we conclude that if the electric field is oscillating sufficiently rapidly then electric induction of magnetic fields is an observable effect. In fact, there is a virtually infallible rule for deciding whether or not the displacement current can be neglected in Equation (4.48). If *electromagnetic radiation* is important then the displacement current must be included. On the other hand, if electromagnetic radiation is unimportant then the displacement current can be safely neglected. Clearly, Maxwell's inclusion of the displacement current in Equation (4.48) was a vital step in his later realization that his equations allowed propagating wave-like solutions. These solutions are, of course, electromagnetic waves. But, more of this later.

We are now in a position to write out the full set of Maxwell equations:

$$\nabla \cdot \mathbf{E} = \frac{\rho}{\epsilon_0}, \tag{4.53}$$

$$\nabla \cdot \mathbf{B} = 0, \tag{4.54}$$

$$\nabla \times \mathbf{E} = -\frac{\partial \mathbf{B}}{\partial t}, \tag{4.55}$$

$$\nabla \times \mathbf{B} = \mu_0 \, \mathbf{j} + \epsilon_0 \mu_0 \, \frac{\partial \mathbf{E}}{\partial t}. \tag{4.56}$$

These four partial differential equations constitute a *complete* description of the behavior of electric and magnetic fields. The first equation describes how electric fields are induced by electric charges. The second equation says that there is no such thing as a magnetic monopole. The third equation describes the induction of electric fields by changing magnetic fields, and the fourth equation describes the generation of magnetic fields by electric currents, and the induction of magnetic fields by changing electric fields. Note that, with the inclusion of the displacement current, these equations treat electric and magnetic fields on an equal footing: *i.e.*, electric fields can induce magnetic fields, and *vice versa*. Equations (4.53)–(4.56) succinctly sum up the experimental results of Coulomb, Ampère, and Faraday. They are called *Maxwell's equations* because James Clerk Maxwell was the first to write them down (in component form). Maxwell also modified them so as to make them mathematically self-consistent.

As an example of a calculation involving the displacement current, let us find the current and displacement current densities associated with the decaying charge distribution

$$\rho(r, t) = \frac{\rho_0 \exp(-t/\tau)}{r^2 + a^2}, \tag{4.57}$$

where r is a spherical polar coordinate, ρ_0 is a constant, and τ and a are positive constants. Now, according to charge conservation,

$$\nabla \cdot \mathbf{j} = -\frac{\partial \rho}{\partial t}. \tag{4.58}$$

By symmetry, we expect $\mathbf{j} = \mathbf{j}(r, t)$. Hence, it follows that $\mathbf{j} = j_r(r, t) \, \mathbf{e}_r$ [since only a radial current has a non-zero divergence when $\mathbf{j} = \mathbf{j}(r)$]. Hence, the above equation yields

$$\frac{1}{r^2} \frac{\partial (r^2 \, j_r)}{\partial r} = -\frac{\partial \rho}{\partial t} = \frac{\rho_0 \exp(-t/\tau)}{\tau \, (r^2 + a^2)}. \tag{4.59}$$

This expression can be integrated, subject to the sensible boundary condition $j_r(0) = 0$, to give

$$j_r(r) = \frac{\rho_0}{\tau} \, e^{-t/\tau} \left[\frac{r - a \tan^{-1}(r/a)}{r^2} \right]. \tag{4.60}$$

Now, the electric field generated by the decaying charge distribution satisfies

$$\nabla \cdot \mathbf{E} = \frac{\rho}{\epsilon_0}. \tag{4.61}$$

Since $\partial\rho/\partial t = -\rho/\tau$, it can be seen, from a comparison of Equations (4.58) and (4.61), that

$$\mathbf{E} = \frac{\tau}{\epsilon_0}\,\mathbf{j}. \tag{4.62}$$

However, the displacement current density is given by

$$\mathbf{j}_d = \epsilon_0\,\frac{\partial \mathbf{E}}{\partial t} = -\mathbf{j}, \tag{4.63}$$

since $\partial\mathbf{j}/\partial t = -\mathbf{j}/\tau$. Hence, we conclude that the displacement current density cancels out the true current density, so that $\mathbf{j} + \mathbf{j}_d = \mathbf{0}$. This is just as well, since $\nabla \times \mathbf{B} = \mu_0\,(\mathbf{j} + \mathbf{j}_d)$. But, if $\mathbf{B} = \mathbf{B}(r, t)$, by symmetry, then $\nabla \times \mathbf{B}$ has no radial component—see Equation (2.169). Thus, if the current and displacement current are constrained, by symmetry, to be radial, then they must sum to zero, else the fourth Maxwell equation cannot be satisfied. In fact, no magnetic field is generated in this particular example, which also implies that there is no induced electric field.

4.6 POTENTIAL FORMULATION

We have seen that Equations (4.54) and (4.55) are automatically satisfied if we write the electric and magnetic fields in terms of potentials: *i.e.*,

$$\mathbf{E} = -\nabla\phi - \frac{\partial \mathbf{A}}{\partial t}, \tag{4.64}$$

$$\mathbf{B} = \nabla \times \mathbf{A}. \tag{4.65}$$

This prescription is not unique, but we can make it unique by adopting the following conventions:

$$\phi(\mathbf{r}) \to 0 \quad \text{as} \quad |\mathbf{r}| \to \infty, \tag{4.66}$$

$$\nabla \cdot \mathbf{A} = -\epsilon_0\mu_0\,\frac{\partial \phi}{\partial t}. \tag{4.67}$$

The above equations can be combined with Equation (4.53) to give

$$\epsilon_0\mu_0\,\frac{\partial^2\phi}{\partial t^2} - \nabla^2\phi = \frac{\rho}{\epsilon_0}. \tag{4.68}$$

Let us now consider Equation (4.56). Substitution of Equations (4.64) and (4.65) into this formula yields

$$\nabla \times \nabla \times \mathbf{A} \equiv \nabla(\nabla \cdot \mathbf{A}) - \nabla^2 \mathbf{A} = \mu_0 \mathbf{j} - \epsilon_0 \mu_0 \frac{\partial \nabla \phi}{\partial t} - \epsilon_0 \mu_0 \frac{\partial^2 \mathbf{A}}{\partial t^2}, \quad (4.69)$$

or

$$\epsilon_0 \mu_0 \frac{\partial^2 \mathbf{A}}{\partial t^2} - \nabla^2 \mathbf{A} = \mu_0 \mathbf{j} - \nabla \left(\nabla \cdot \mathbf{A} + \epsilon_0 \mu_0 \frac{\partial \phi}{\partial t} \right). \quad (4.70)$$

We can now see quite clearly where the Lorenz gauge condition (4.33) comes from. The above equation is, in general, very complicated, since it involves both the vector and scalar potentials. But, if we adopt the Lorenz gauge then the last term on the right-hand side becomes zero, and the equation simplifies considerably, and ends up only involving the vector potential. Thus, we find that Maxwell's equations reduce to the following equations:

$$\epsilon_0 \mu_0 \frac{\partial^2 \phi}{\partial t^2} - \nabla^2 \phi = \frac{\rho}{\epsilon_0}, \quad (4.71)$$

$$\epsilon_0 \mu_0 \frac{\partial^2 \mathbf{A}}{\partial t^2} - \nabla^2 \mathbf{A} = \mu_0 \mathbf{j}. \quad (4.72)$$

Of course, this is the same (scalar) equation written four times over. In steady-state (*i.e.*, $\partial/\partial t = 0$), the equation in question reduces to Poisson's equation, which we know how to solve. With the $\partial^2/\partial t^2$ terms included, it becomes a slightly more complicated equation (in fact, a driven three-dimensional wave equation).

4.7 ELECTROMAGNETIC WAVES

This is an appropriate point at which to demonstrate that Maxwell's equations possess wave-like solutions which can propagate through a vacuum. Let us start from Maxwell's equations in free space (*i.e.*, with no charges and no currents):

$$\nabla \cdot \mathbf{E} = 0, \quad (4.73)$$

$$\nabla \cdot \mathbf{B} = 0, \quad (4.74)$$

$$\nabla \times \mathbf{E} = -\frac{\partial \mathbf{B}}{\partial t}, \quad (4.75)$$

$$\nabla \times \mathbf{B} = \epsilon_0 \mu_0 \frac{\partial \mathbf{E}}{\partial t}. \quad (4.76)$$

Note that these equations exhibit a nice symmetry between the electric and magnetic fields.

There is an easy way to show that the above equations possess wave-like solutions, and a hard way. The easy way is to assume that the solutions are going to be wave-like beforehand. Specifically, let us search for plane-wave solutions of the form:

$$\mathbf{E}(\mathbf{r},t) = \mathbf{E}_0 \cos(\mathbf{k}\cdot\mathbf{r} - \omega\, t), \tag{4.77}$$

$$\mathbf{B}(\mathbf{r},t) = \mathbf{B}_0 \cos(\mathbf{k}\cdot\mathbf{r} - \omega\, t - \phi). \tag{4.78}$$

Here, $\mathbf{E}_0$ and $\mathbf{B}_0$ are constant vectors, $\mathbf{k}$ is called the wave-vector, and ω is the angular frequency. The frequency in hertz, f, is related to the angular frequency via $\omega = 2\pi\, f$. This frequency is conventionally defined to be positive. The quantity ϕ is a phase difference between the electric and magnetic fields. Actually, it is more convenient to write

$$\mathbf{E} = \mathbf{E}_0\, e^{\,i\,(\mathbf{k}\cdot\mathbf{r}-\omega\, t)}, \tag{4.79}$$

$$\mathbf{B} = \mathbf{B}_0\, e^{\,i\,(\mathbf{k}\cdot\mathbf{r}-\omega\, t)}, \tag{4.80}$$

where, by convention, the physical solution is the *real part* of the above equations. The phase difference ϕ is absorbed into the constant vector $\mathbf{B}_0$ by allowing it to become complex. Thus, $\mathbf{B}_0 \rightarrow \mathbf{B}_0\, e^{-i\,\phi}$. In general, the vector $\mathbf{E}_0$ is also complex.

Now, a wave maximum of the electric field satisfies

$$\mathbf{k}\cdot\mathbf{r} = \omega\, t + n\, 2\pi, \tag{4.81}$$

where n is an integer. The solution to this equation is a set of equally spaced parallel planes (one plane for each possible value of n), whose normals are parallel to the wave-vector $\mathbf{k}$, and which propagate in the direction of $\mathbf{k}$ with phase-velocity

$$v = \frac{\omega}{k}. \tag{4.82}$$

The spacing between adjacent planes (*i.e.*, the wavelength) is given by

$$\lambda = \frac{2\pi}{k} \tag{4.83}$$

—see Figure 4.3.

Consider a general plane-wave vector field

$$\mathbf{A} = \mathbf{A}_0\, e^{\,i\,(\mathbf{k}\cdot\mathbf{r}-\omega\, t)}. \tag{4.84}$$

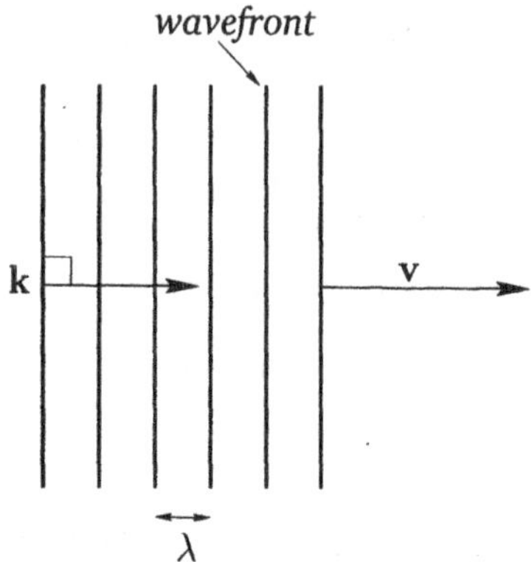

Figure 4.3: *Wavefronts associated with a plane-wave.*

What is the divergence of **A**? This is easy to evaluate. We have

$$\nabla\cdot\mathbf{A} = \frac{\partial A_x}{\partial x} + \frac{\partial A_y}{\partial y} + \frac{\partial A_z}{\partial z} = (A_{0x}\,\mathrm{i}\,k_x + A_{0y}\,\mathrm{i}\,k_y + A_{0z}\,\mathrm{i}\,k_z)\,\mathrm{e}^{\,\mathrm{i}\,(\mathbf{k}\cdot\mathbf{r}-\omega t)}$$
$$= \mathrm{i}\,\mathbf{k}\cdot\mathbf{A}. \tag{4.85}$$

How about the curl of **A**? This is slightly more difficult. We have

$$(\nabla\times\mathbf{A})_x = \frac{\partial A_z}{\partial y} - \frac{\partial A_y}{\partial z} = (\mathrm{i}\,k_y A_z - \mathrm{i}\,k_z A_y)$$
$$= \mathrm{i}\,(\mathbf{k}\times\mathbf{A})_x, \tag{4.86}$$

which easily generalizes to

$$\nabla\times\mathbf{A} = \mathrm{i}\,\mathbf{k}\times\mathbf{A}. \tag{4.87}$$

Hence, we can see that vector field operations on a plane-wave simplify to replacing the ∇ operator with $\mathrm{i}\,\mathbf{k}$.

The first Maxwell equation (4.73) reduces to

$$\mathrm{i}\,\mathbf{k}\cdot\mathbf{E}_0 = 0, \tag{4.88}$$

using the assumed electric and magnetic fields, (4.79) and (4.80), and Equation (4.85). Thus, the electric field is *perpendicular* to the direction of propagation of the wave. Likewise, the second Maxwell equation gives

$$\mathrm{i}\,\mathbf{k}\cdot\mathbf{B}_0 = 0, \tag{4.89}$$

implying that the magnetic field is also *perpendicular* to the direction of propagation. Clearly, the wave-like solutions of Maxwell's equation are a type of *transverse wave*. The third Maxwell equation yields

$$\mathrm{i}\,\mathbf{k} \times \mathbf{E}_0 = \mathrm{i}\,\omega\,\mathbf{B}_0, \tag{4.90}$$

where use has been made of Equation (4.87). Dotting this equation with $\mathbf{E}_0$ gives

$$\mathbf{E}_0 \cdot \mathbf{B}_0 = \frac{\mathbf{E}_0 \cdot \mathbf{k} \times \mathbf{E}_0}{\omega} = 0. \tag{4.91}$$

Thus, the electric and magnetic fields are mutually perpendicular. Dotting equation (4.90) with $\mathbf{B}_0$ yields

$$\mathbf{B}_0 \cdot \mathbf{k} \times \mathbf{E}_0 = \omega\,B_0^2 > 0. \tag{4.92}$$

Thus, the vectors $\mathbf{E}_0$, $\mathbf{B}_0$, and $\mathbf{k}$ are *mutually perpendicular*, and form a right-handed set. The final Maxwell equation gives

$$\mathrm{i}\,\mathbf{k} \times \mathbf{B}_0 = -\mathrm{i}\,\epsilon_0\mu_0\,\omega\,\mathbf{E}_0. \tag{4.93}$$

Combining this with Equation (4.90) yields

$$\mathbf{k} \times (\mathbf{k} \times \mathbf{E}_0) = (\mathbf{k} \cdot \mathbf{E}_0)\,\mathbf{k} - k^2\,\mathbf{E}_0 = -k^2\,\mathbf{E}_0 = -\epsilon_0\mu_0\,\omega^2\,\mathbf{E}_0, \tag{4.94}$$

or

$$k^2 = \epsilon_0\mu_0\,\omega^2, \tag{4.95}$$

where use has been made of Equation (4.88). However, we know, from Equation (4.82), that the phase-velocity c is related to the magnitude of the wave-vector and the angular wave frequency via $c = \omega/k$. Thus, we obtain

$$c = \frac{1}{\sqrt{\epsilon_0\mu_0}}. \tag{4.96}$$

So, we have found transverse plane-wave solutions of the free-space Maxwell equations propagating at some phase-velocity c, which is given by a combination of ϵ_0 and μ_0, and is thus the *same* for all frequencies and wavelengths. The constants ϵ_0 and μ_0 are easily measurable. The former is related to the force acting between stationary electric charges, and the latter to the force acting between steady electric currents. Both of these constants were fairly well-known in Maxwell's time. Maxwell, incidentally, was the first person to look for wave-like solutions of his equations, and, thus, to derive Equation (4.96). The modern values of ϵ_0 and μ_0 are

$$\epsilon_0 = 8.8542 \times 10^{-12}\,\mathrm{C^2\,N^{-1}\,m^{-2}}, \tag{4.97}$$

$$\mu_0 = 4\pi \times 10^{-7}\,\mathrm{N\,A^{-2}}. \tag{4.98}$$

Let us use these values to find the phase-velocity of "electromagnetic waves." We obtain

$$c = \frac{1}{\sqrt{\epsilon_0 \mu_0}} = 2.998 \times 10^8\,\mathrm{m\,s^{-1}}. \tag{4.99}$$

Of course, we immediately recognize this as the velocity of light. Maxwell also made this connection back in the 1870s. He conjectured that light, whose nature had previously been unknown, was a form of electromagnetic radiation. This was a remarkable prediction. After all, Maxwell's equations were derived from the results of benchtop laboratory experiments involving charges, batteries, coils, and currents, which apparently had nothing whatsoever to do with light.

Maxwell was able to make another remarkable prediction. The wavelength of light was well-known in the late nineteenth century from studies of diffraction through slits, *etc.* Visible light actually occupies a surprisingly narrow wavelength range. The shortest wavelength blue light which is visible has $\lambda = 0.4$ microns (one micron is 10^{-6} meters). The longest wavelength red light which is visible has $\lambda = 0.76$ microns. However, there is nothing in our analysis which suggests that this particular range of wavelengths is special. Electromagnetic waves can have *any* wavelength. Maxwell concluded that visible light was a small part of a vast spectrum of previously undiscovered types of electromagnetic radiation. Since Maxwell's time, virtually all of the non-visible parts of the electromagnetic spectrum have been observed.

Table 4.1 gives a brief guide to the electromagnetic spectrum. Electromagnetic waves are of particular importance to us because they are our main source of information regarding the Universe around us. Radio waves and microwaves (which are comparatively hard to scatter) have provided much of our knowledge about the center of our own galaxy. This is completely unobservable in visible light, which is strongly scattered by interstellar gas and dust lying in the galactic plane. For the same reason, the spiral arms of our galaxy can only be mapped out using radio waves. Infrared radiation is useful for detecting protostars, which are not yet hot enough to emit visible radiation. Of course, visible radiation is still the mainstay of Astronomy. Satellite-based ultraviolet observations

Radiation type	Wavelength range (m)
Gamma Rays	$< 10^{-11}$
X-Rays	10^{-11}–10^{-9}
Ultraviolet	10^{-9}–10^{-7}
Visible	10^{-7}–10^{-6}
Infrared	10^{-6}–10^{-4}
Microwave	10^{-4}–10^{-1}
TV-FM	10^{-1}–10^{1}
Radio	$> 10^{1}$

Table 4.1 *The electromagnetic spectrum.*

have yielded invaluable insights into the structure and distribution of distant galaxies. Finally, X-ray and γ-ray Astronomy usually concentrates on exotic objects, such as pulsars and supernova remnants.

Equations (4.88), (4.90), and the relation $c = \omega/k$, imply that

$$B_0 = \frac{E_0}{c}. \tag{4.100}$$

Thus, the magnetic field associated with an electromagnetic wave is smaller in magnitude than the electric field by a factor c. Consider a free charge interacting with an electromagnetic wave. The force exerted on the charge is given by the Lorentz formula

$$\mathbf{f} = q\,(\mathbf{E} + \mathbf{v} \times \mathbf{B}). \tag{4.101}$$

The ratio of the electric and magnetic forces is

$$\frac{f_{\text{magnetic}}}{f_{\text{electric}}} \sim \frac{v\,B_0}{E_0} \sim \frac{v}{c}. \tag{4.102}$$

So, unless the charge is moving close to the velocity of light (*i.e.*, unless the charge is relativistic), the electric force greatly exceeds the magnetic force. Clearly, in most terrestrial situations, electromagnetic waves are an essentially *electrical* phenomenon (as far as their interaction with matter goes). For this reason, electromagnetic waves are usually characterized by their wave-vector **k** (which specifies the direction of propagation and the wavelength) and the plane of polarization (*i.e.*, the plane of oscillation) of the associated electric field. For a given wave-vector **k**, the electric field can have any direction in the plane normal to $\mathbf{k}$. However, there are only two *independent* directions in a plane (*i.e.*, we can only define two linearly independent vectors in a plane). This implies that there are only two independent polarizations of an electromagnetic wave, once its direction of propagation is specified.

But, how do electromagnetic waves propagate through a vacuum? After all, most types of wave require a medium before they can propagate (*e.g.*, sound waves require air). The answer to this question is evident from Equations (4.75) and (4.76). According to these equations, the time variation of the electric component of the wave induces the magnetic component, and the time variation of the magnetic component induces the electric. In other words, electromagnetic waves are *self-sustaining*, and therefore require no medium through which to propagate.

Let us now search for the wave-like solutions of Maxwell's equations in free-space the hard way. Suppose that we take the curl of the fourth Maxwell equation, (4.76). We obtain

$$\nabla \times \nabla \times \mathbf{B} = \nabla(\nabla \cdot \mathbf{B}) - \nabla^2\mathbf{B} = -\nabla^2\mathbf{B} = \epsilon_0\mu_0 \frac{\partial\, \nabla \times \mathbf{E}}{\partial t}. \tag{4.103}$$

Here, we have made use of the fact that $\nabla \cdot \mathbf{B} = 0$. The third Maxwell equation, (4.75), yields

$$\left(\nabla^2 - \frac{1}{c^2}\frac{\partial^2}{\partial t^2}\right)\mathbf{B} = \mathbf{0}, \tag{4.104}$$

where use has been made of Equation (4.99). A similar equation can be obtained for the electric field by taking the curl of Equation (4.75):

$$\left(\nabla^2 - \frac{1}{c^2}\frac{\partial^2}{\partial t^2}\right)\mathbf{E} = \mathbf{0}. \tag{4.105}$$

We have found that electric and magnetic fields both satisfy equations of the form

$$\left(\nabla^2 - \frac{1}{c^2}\frac{\partial^2}{\partial t^2}\right)\mathbf{A} = \mathbf{0} \tag{4.106}$$

in free space. As is easily verified, the most general solution to this equation is

$$A_x = F_x(\mathbf{n} \cdot \mathbf{r} - c\,t), \tag{4.107}$$

$$A_y = F_y(\mathbf{n} \cdot \mathbf{r} - c\,t), \tag{4.108}$$

$$A_z = F_z(\mathbf{n} \cdot \mathbf{r} - c\,t), \tag{4.109}$$

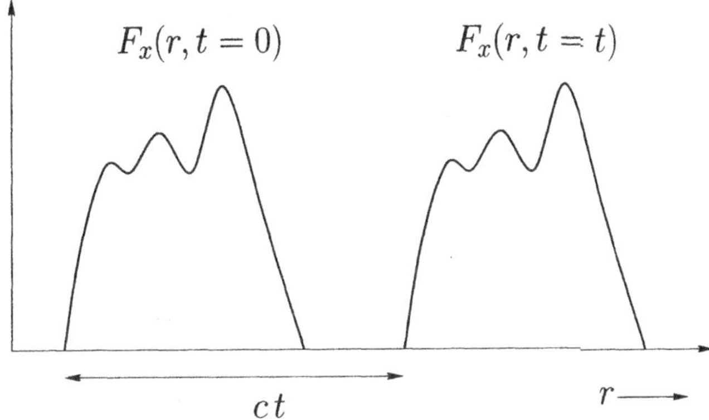

Figure 4.4: *An arbitrary wave-pulse.*

where $\mathbf{n}$ is a unit vector, and $F_x(\phi)$, $F_y(\phi)$, and $F_z(\phi)$ are arbitrary one-dimensional scalar functions. Looking along the direction of $\mathbf{n}$, so that $\mathbf{n} \cdot \mathbf{r} = r$, we find that

$$A_x = F_x(r - c\,t), \tag{4.110}$$

$$A_y = F_y(r - c\,t), \tag{4.111}$$

$$A_z = F_z(r - c\,t). \tag{4.112}$$

The x-component of this solution is shown schematically in Figure 4.4. It clearly propagates in r with velocity c. If we look along a direction which is perpendicular to $\mathbf{n}$ then $\mathbf{n} \cdot \mathbf{r} = 0$, and there is no propagation. Thus, the components of $\mathbf{A}$ are *arbitrarily shaped pulses* which propagate, *without changing shape*, along the direction of $\mathbf{n}$ with velocity c. These pulses can be related to the sinusoidal plane-wave solutions which we found earlier by Fourier transformation: *e.g.*,

$$F_x(r - c\,t) = \frac{1}{\sqrt{2\pi}} \int_{-\infty}^{\infty} \bar{F}_x(k)\, e^{\,i\,k\,(r-c\,t)}\, dk, \tag{4.113}$$

where

$$\bar{F}_x(k) = \frac{1}{\sqrt{2\pi}} \int_{-\infty}^{\infty} F_x(x)\, e^{-i\,k\,x}\, dx. \tag{4.114}$$

Thus, any arbitrary shaped pulse propagating in the direction of $\mathbf{n}$ with velocity c can be broken down into a superposition of sinusoidal oscillations of different wavevectors, $k\,\mathbf{n}$, propagating in the same direction with the same velocity.

4.8 GREEN'S FUNCTIONS

The solution of the steady-state Maxwell equations essentially boils down to solving Poisson's equation

$$\nabla^2 u = v, \tag{4.115}$$

where $v(\mathbf{r})$ is denoted the source function. The potential $u(\mathbf{r})$ satisfies the boundary condition

$$u(\mathbf{r}, t) \to 0 \qquad \text{as } |\mathbf{r}| \to \infty, \tag{4.116}$$

provided that the source function is reasonably localized. The solutions to Poisson's equation are superposable (because the equation is linear). This property is exploited in the Green's function method of solving this equation. The Green's function $G(\mathbf{r}, \mathbf{r}')$ is the potential generated by a unit amplitude point source, located at $\mathbf{r}'$, which satisfies the appropriate boundary conditions. Thus,

$$\nabla^2 G(\mathbf{r}, \mathbf{r}') = \delta(\mathbf{r} - \mathbf{r}'). \tag{4.117}$$

Any source function $v(\mathbf{r})$ can be represented as a weighted sum of point sources

$$v(\mathbf{r}) = \int \delta(\mathbf{r} - \mathbf{r}')\, v(\mathbf{r}')\, d^3\mathbf{r}'. \tag{4.118}$$

It follows from superposability that the potential generated by the source $v(\mathbf{r})$ can be written as the similarly weighted sum of point source driven potentials (*i.e.*, Green's functions)

$$u(\mathbf{r}) = \int G(\mathbf{r}, \mathbf{r}')\, v(\mathbf{r}')\, d^3\mathbf{r}'. \tag{4.119}$$

We found earlier that the Green's function for Poisson's equation is

$$G(\mathbf{r}, \mathbf{r}') = -\frac{1}{4\pi} \frac{1}{|\mathbf{r} - \mathbf{r}'|}. \tag{4.120}$$

It follows that the general solution to Equation (4.115) is written

$$u(\mathbf{r}) = -\frac{1}{4\pi} \int \frac{v(\mathbf{r}')}{|\mathbf{r} - \mathbf{r}'|}\, d^3\mathbf{r}'. \tag{4.121}$$

Note that the point source driven potential (4.120) is perfectly sensible. It is spherically symmetric about the source, and falls off smoothly with increasing distance from the source.

The solution of the time-dependent Maxwell equations essentially boils down to solving the three-dimensional wave equation

$$\left(\nabla^2 - \frac{1}{c^2}\frac{\partial^2}{\partial t^2}\right) u = v, \tag{4.122}$$

where $v(\mathbf{r}, t)$ is a time-varying source function. The potential $u(\mathbf{r}, t)$ satisfies the boundary conditions

$$u(\mathbf{r}, t) \to 0 \qquad \text{as } |\mathbf{r}| \to \infty \text{ and } |t| \to \infty. \tag{4.123}$$

The solutions to Equation (4.122) are superposable (since the equation is linear), so a Green's function method of solution is again appropriate. The Green's function $G(\mathbf{r}, \mathbf{r}'; t, t')$ is the potential generated by a point *impulse* located at position $\mathbf{r}'$ and applied at time t'. Thus,

$$\left(\nabla^2 - \frac{1}{c^2}\frac{\partial^2}{\partial t^2}\right) G(\mathbf{r}, \mathbf{r}'; t, t') = \delta(\mathbf{r} - \mathbf{r}')\,\delta(t - t'). \tag{4.124}$$

Of course, the Green's function must satisfy the correct boundary conditions. A general source $v(\mathbf{r}, t)$ can be built up from a weighted sum of point impulses

$$v(\mathbf{r}, t) = \iint \delta(\mathbf{r} - \mathbf{r}')\,\delta(t - t')\,v(\mathbf{r}', t')\,d^3\mathbf{r}'\,dt'. \tag{4.125}$$

It follows that the potential generated by $v(\mathbf{r}, t)$ can be written as the similarly weighted sum of point impulse driven potentials

$$u(\mathbf{r}, t) = \iint G(\mathbf{r}, \mathbf{r}'; t, t')\,v(\mathbf{r}', t')\,d^3\mathbf{r}'\,dt'. \tag{4.126}$$

So, how do we find the Green's function?

Consider

$$G(\mathbf{r}, \mathbf{r}'; t, t') = \frac{F(t - t' - |\mathbf{r} - \mathbf{r}'|/c)}{|\mathbf{r} - \mathbf{r}'|}, \tag{4.127}$$

where $F(\phi)$ is a general scalar function. Let us try to prove the following theorem:

$$\left(\nabla^2 - \frac{1}{c^2}\frac{\partial^2}{\partial t^2}\right) G = -4\pi F(t - t')\,\delta(\mathbf{r} - \mathbf{r}'). \tag{4.128}$$

At a general point, $\mathbf{r} \neq \mathbf{r}'$, the above expression reduces to

$$\left(\nabla^2 - \frac{1}{c^2}\frac{\partial^2}{\partial t^2}\right) G = 0. \tag{4.129}$$

Hence, we basically have to show that G is a valid solution of the free-space wave equation. Now, we can easily demonstrate that

$$\frac{\partial|\mathbf{r}-\mathbf{r}'|}{\partial x} = \frac{x - x'}{|\mathbf{r}-\mathbf{r}'|}. \tag{4.130}$$

It follows by simple differentiation that

$$\frac{\partial^2 G}{\partial x^2} = \left(\frac{3(x-x')^2 - |\mathbf{r}-\mathbf{r}'|^2}{|\mathbf{r}-\mathbf{r}'|^5}\right) F + \left(\frac{3(x-x')^2 - |\mathbf{r}-\mathbf{r}'|^2}{|\mathbf{r}-\mathbf{r}'|^4}\right)\frac{F'}{c} + \frac{(x-x')^2}{|\mathbf{r}-\mathbf{r}'|^3}\frac{F''}{c^2}, \tag{4.131}$$

where $F'(\phi) = dF(\phi)/d\phi$, *etc.* We can derive analogous equations for $\partial^2 G/\partial y^2$ and $\partial^2 G/\partial z^2$. Thus,

$$\nabla^2 G = \frac{\partial^2 G}{\partial x^2} + \frac{\partial^2 G}{\partial y^2} + \frac{\partial^2 G}{\partial z^2} = \frac{F''}{|\mathbf{r}-\mathbf{r}'|\,c^2} = \frac{1}{c^2}\frac{\partial^2 G}{\partial t^2}, \tag{4.132}$$

giving

$$\left(\nabla^2 - \frac{1}{c^2}\frac{\partial^2}{\partial t^2}\right) G = 0, \tag{4.133}$$

which is the desired result. Consider, now, the region around $\mathbf{r} = \mathbf{r}'$. It is clear that the dominant term on the right-hand side of Equation (4.131) as $|\mathbf{r}-\mathbf{r}'| \to 0$ is the first one, which is essentially $F\,\partial^2(|\mathbf{r}-\mathbf{r}'|^{-1})/\partial x^2$. It is also clear that $(1/c^2)(\partial^2 G/\partial t^2)$ is negligible compared to this term. Thus, as $|\mathbf{r}-\mathbf{r}'| \to 0$ we find that

$$\left(\nabla^2 - \frac{1}{c^2}\frac{\partial^2}{\partial t^2}\right) G \to F(t-t')\,\nabla^2\left(\frac{1}{|\mathbf{r}-\mathbf{r}'|}\right). \tag{4.134}$$

However, according to Equations (4.117) and (4.120),

$$\nabla^2\left(\frac{1}{|\mathbf{r}-\mathbf{r}'|}\right) = -4\pi\,\delta(\mathbf{r}-\mathbf{r}'). \tag{4.135}$$

We conclude that

$$\left(\nabla^2 - \frac{1}{c^2}\frac{\partial^2}{\partial t^2}\right) G = -4\pi\, F(t - t')\, \delta(\mathbf{r} - \mathbf{r}'), \tag{4.136}$$

which is the desired result.

Let us now make the special choice

$$F(\phi) = -\frac{\delta(\phi)}{4\pi}. \tag{4.137}$$

It follows from Equation (4.136) that

$$\left(\nabla^2 - \frac{1}{c^2}\frac{\partial^2}{\partial t^2}\right) G = \delta(\mathbf{r} - \mathbf{r}')\, \delta(t - t'). \tag{4.138}$$

Thus,

$$G(\mathbf{r}, \mathbf{r}'; t, t') = -\frac{1}{4\pi}\frac{\delta(t - t' - |\mathbf{r} - \mathbf{r}'|/c)}{|\mathbf{r} - \mathbf{r}'|} \tag{4.139}$$

is the Green's function for the driven wave equation (4.122).

The time-dependent Green's function (4.139) is the same as the steady-state Green's function (4.120), apart from the delta-function appearing in the former. What does this delta-function do? Well, consider an observer at point $\mathbf{r}$. Because of the delta-function, our observer only measures a non-zero potential at one particular time

$$t = t' + \frac{|\mathbf{r} - \mathbf{r}'|}{c}. \tag{4.140}$$

It is clear that this is the time the impulse was applied at position $\mathbf{r}'$ (*i.e.*, t') *plus* the time taken for a light signal to travel between points $\mathbf{r}'$ and $\mathbf{r}$. At time $t > t'$, the locus of all the points at which the potential is non-zero is

$$|\mathbf{r} - \mathbf{r}'| = c\,(t - t'). \tag{4.141}$$

In other words, it is a sphere centered on $\mathbf{r}'$ whose radius is the distance traveled by light in the time interval since the impulse was applied at position $\mathbf{r}'$. Thus, the Green's function (4.139) describes a *spherical wave* which emanates from position $\mathbf{r}'$ at time t', and propagates at the speed of light. The amplitude of the wave is inversely proportional to the distance from the source.

4.9 RETARDED POTENTIALS

We are now in a position to solve Maxwell's equations. Recall that the steady-state Maxwell equations reduce to

$$\nabla^2\phi = -\frac{\rho}{\epsilon_0}, \tag{4.142}$$

$$\nabla^2\mathbf{A} = -\mu_0\,\mathbf{j}. \tag{4.143}$$

The solutions to these equations are easily found using the Green's function for Poisson's equation, (4.120):

$$\phi(\mathbf{r}) = \frac{1}{4\pi\epsilon_0}\int \frac{\rho(\mathbf{r}')}{|\mathbf{r}-\mathbf{r}'|}\,d^3\mathbf{r}' \tag{4.144}$$

$$\mathbf{A}(\mathbf{r}) = \frac{\mu_0}{4\pi}\int \frac{\mathbf{j}(\mathbf{r}')}{|\mathbf{r}-\mathbf{r}'|}\,d^3\mathbf{r}'. \tag{4.145}$$

The time-dependent Maxwell equations reduce to

$$\left(\nabla^2 - \frac{1}{c^2}\frac{\partial^2}{\partial t^2}\right)\phi = -\frac{\rho}{\epsilon_0}, \tag{4.146}$$

$$\left(\nabla^2 - \frac{1}{c^2}\frac{\partial^2}{\partial t^2}\right)\mathbf{A} = -\mu_0\,\mathbf{j}. \tag{4.147}$$

We can solve these equations using the time-dependent Green's function, (4.139). From Equation (4.126), we find that

$$\phi(\mathbf{r},t) = \frac{1}{4\pi\epsilon_0}\iint \frac{\delta(t-t'-|\mathbf{r}-\mathbf{r}'|/c)\,\rho(\mathbf{r}',t')}{|\mathbf{r}-\mathbf{r}'|}\,d^3\mathbf{r}'\,dt', \tag{4.148}$$

with a similar equation for **A**. Using the well-known property of delta-functions, these equations yield

$$\phi(\mathbf{r},t) = \frac{1}{4\pi\epsilon_0}\int \frac{\rho(\mathbf{r}',t-|\mathbf{r}-\mathbf{r}'|/c)}{|\mathbf{r}-\mathbf{r}'|}\,d^3\mathbf{r}' \tag{4.149}$$

$$\mathbf{A}(\mathbf{r},t) = \frac{\mu_0}{4\pi}\int \frac{\mathbf{j}(\mathbf{r}',t-|\mathbf{r}-\mathbf{r}'|/c)}{|\mathbf{r}-\mathbf{r}'|}\,d^3\mathbf{r}'. \tag{4.150}$$

These are the general solutions to Maxwell's equations. Note that the time-dependent solutions, (4.149) and (4.150), are the same as the steady-state solutions, (4.144) and (4.145), apart from the weird way in which time appears in the former. According to Equations (4.149) and (4.150), if we want to work out the potentials at position $\mathbf{r}$ and time t

then we have to perform integrals of the charge density and current density over all space (just like in the steady-state situation). However, when we calculate the contribution of charges and currents at position $\mathbf{r}'$ to these integrals we do not use the values at time t, instead we use the values at some earlier time $t - |\mathbf{r} - \mathbf{r}'|/c$. What is this earlier time? It is simply the latest time at which a light signal emitted from position $\mathbf{r}'$ would be received at position $\mathbf{r}$ before time t. This is called the *retarded time*. Likewise, the potentials (4.149) and (4.150) are called *retarded potentials*. It is often useful to adopt the following notation

$$A(\mathbf{r}', t - |\mathbf{r} - \mathbf{r}'|/c) \equiv [A(\mathbf{r}', t)]. \tag{4.151}$$

The square brackets denote retardation (*i.e.*, using the retarded time instead of the real time). Using this notation Equations (4.149) and (4.150), become

$$\phi(\mathbf{r}) = \frac{1}{4\pi\epsilon_0} \int \frac{[\rho(\mathbf{r}')]}{|\mathbf{r} - \mathbf{r}'|}\, d^3\mathbf{r}', \tag{4.152}$$

$$\mathbf{A}(\mathbf{r}) = \frac{\mu_0}{4\pi} \int \frac{[\mathbf{j}(\mathbf{r}')]}{|\mathbf{r} - \mathbf{r}'|}\, d^3\mathbf{r}'. \tag{4.153}$$

The time dependence in the above equations is taken as read.

We are now in a position to understand electromagnetism at its most fundamental level. A charge distribution $\rho(\mathbf{r}, t)$ can be thought of theoretically as being built up out of a collection, or series, of charges which instantaneously come into existence, at some point $\mathbf{r}'$ and some time t', and then disappear again. Mathematically, this is written

$$\rho(\mathbf{r}, t) = \iint \delta(\mathbf{r} - \mathbf{r}')\delta(t - t')\, \rho(\mathbf{r}', t')\, d^3\mathbf{r}' dt'. \tag{4.154}$$

Likewise, we can think of a current distribution $\mathbf{j}(\mathbf{r}, t)$ as built up out of a collection, or series, of currents which instantaneously appear and then disappear:

$$\mathbf{j}(\mathbf{r}, t) = \iint \delta(\mathbf{r} - \mathbf{r}')\delta(t - t')\, \mathbf{j}(\mathbf{r}', t')\, d^3\mathbf{r}' dt'. \tag{4.155}$$

Each of these ephemeral charges and currents excites a spherical wave in the appropriate potential. Thus, the charge density at $\mathbf{r}'$ and t' sends out a wave in the scalar potential:

$$\phi(\mathbf{r}, t) = \frac{\rho(\mathbf{r}', t')}{4\pi\epsilon_0} \frac{\delta(t - t' - |\mathbf{r} - \mathbf{r}'|/c)}{|\mathbf{r} - \mathbf{r}'|}. \tag{4.156}$$

Likewise, the current density at $\mathbf{r}'$ and t' sends out a wave in the vector potential:

$$\mathbf{A}(\mathbf{r},t) = \frac{\mu_0\,\mathbf{j}(\mathbf{r}',t')}{4\pi}\,\frac{\delta(t-t'-|\mathbf{r}-\mathbf{r}'|/c)}{|\mathbf{r}-\mathbf{r}'|}. \tag{4.157}$$

These waves can be thought of as messengers which inform other charges and currents about the charges and currents present at position $\mathbf{r}'$ and time t'. However, these messengers travel at a finite speed: *i.e.*, the speed of light. So, by the time they reach other charges and currents their message is a little out of date. Every charge and every current in the Universe emits these spherical waves. The resultant scalar and vector potential fields are given by Equations (4.152) and (4.153). Of course, we can turn these fields into electric and magnetic fields using Equations (4.64) and (4.65). We can then evaluate the force exerted on charges using the Lorentz formula. We can see that we have now escaped from the apparent action at a distance nature of Coulomb's law and the Biot-Savart law. Electromagnetic information is carried by spherical waves in the vector and scalar potentials, and, therefore, travels at the velocity of light. Thus, if we change the position of a charge then a distant charge can only respond after a time delay sufficient for a spherical wave to propagate from the former to the latter charge.

Consider a thought experiment in which a charge q appears at position $\mathbf{r}_0$ at time t_1, persists for a while, and then disappears at time t_2. What is the electric field generated by such a charge? Using Equation (4.150), we find that

$$\begin{aligned}\phi(\mathbf{r},t) &= \frac{q}{4\pi\epsilon_0}\frac{1}{|\mathbf{r}-\mathbf{r}_0|} && \text{for } t_1 \le t-|\mathbf{r}-\mathbf{r}_0|/c \le t_2\\ &= 0 && \text{otherwise.}\end{aligned} \tag{4.158}$$

Now, $\mathbf{E} = -\nabla\phi$ (since there are no currents, and therefore no vector potential is generated), so

$$\begin{aligned}\mathbf{E}(\mathbf{r},t) &= \frac{q}{4\pi\epsilon_0}\frac{\mathbf{r}-\mathbf{r}_0}{|\mathbf{r}-\mathbf{r}_0|^3} && \text{for } t_1 \le t-|\mathbf{r}-\mathbf{r}_0|/c \le t_2\\ &= \mathbf{0} && \text{otherwise.}\end{aligned} \tag{4.159}$$

This solution is shown pictorially in Figure 4.5. We can see that the charge effectively emits a Coulomb electric field which propagates radially away from the charge at the speed of light. Likewise, it is easy to show that a current-carrying wire effectively emits an Ampèrian magnetic field at the speed of light.

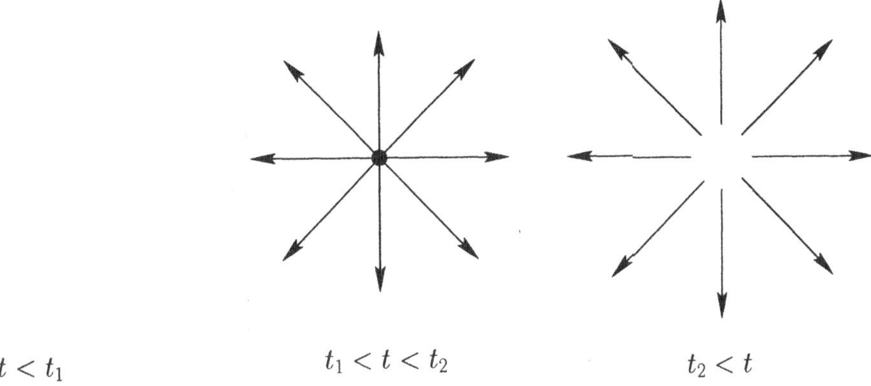

Figure 4.5: *Electric field due to a charge which appears at* $t = t_1$ *and disappears at* $t = t_2$.

We can now appreciate the essential difference between time-dependent electromagnetism and the action-at-a-distance laws of Coulomb and Biot-Savart. In the latter theories, the field-lines act rather like rigid wires attached to charges (or circulating around currents). If the charges (or currents) move then so do the field-lines, leading inevitably to unphysical action at a distance-type behavior. In the time-dependent theory, charges act rather like water sprinklers: *i.e.*, they spray out the Coulomb field in all directions at the speed of light. Similarly, current-carrying wires throw out magnetic field loops at the speed of light. If we move a charge (or current) then field-lines emitted beforehand are not affected, so the field at a distant charge (or current) only responds to the change in position after a time delay sufficient for the field to propagate between the two charges (or currents) at the speed of light.

As we mentioned previously, it is not entirely obvious that electric and magnetic fields have a real existence in Coulomb's law and the Biot-Savart law. After all, the only measurable quantities are the forces acting between charges and currents. We can certainly describe the force on a given charge or current, due to the other charges and currents in the Universe, in terms of the local electric and magnetic fields, but we have no way of knowing whether these fields persist when the charge or current is not present (*i.e.*, we could argue that electric and magnetic fields are just a convenient way of calculating forces, but, in reality, the forces are transmitted directly between charges and currents by some form of magic). On the other hand, it is patently obvious that electric and

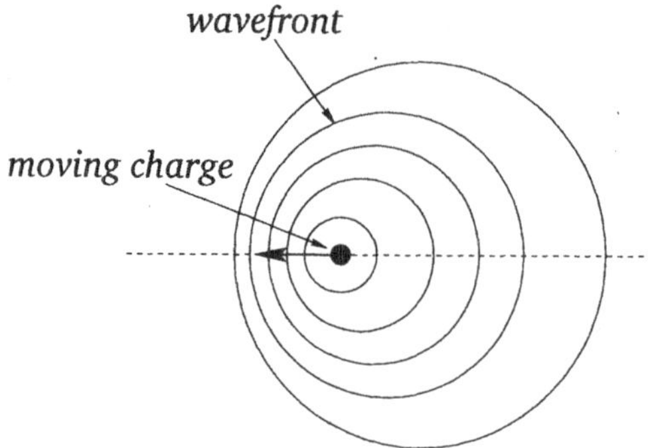

Figure 4.6: *Spherical wavefronts emitted by a moving charge.*

magnetic fields have a real existence in the time-dependent theory of electromagnetism. For instance, consider the following thought experiment. Suppose that a charge q_1 comes into existence for a period of time, emits a Coulomb field, and then disappears. Suppose that a distant charge q_2 interacts with this field, but is sufficiently far from the first charge that by the time the field arrives the first charge has already disappeared. The force exerted on the second charge is only ascribable to the electric field—it cannot be ascribed to the first charge, because this charge no longer exists by the time the force is exerted. In this example, the electric field clearly transmits energy and momentum between the two charges. Anything which possesses energy and momentum is "real" in a physical sense. Later on in this book, we shall demonstrate that electric and magnetic fields conserve energy and momentum.

Let us now consider a moving charge. Such a charge is continually emitting spherical waves in the scalar potential, and the resulting wavefront pattern is sketched in Figure 4.6. Clearly, the wavefronts are more closely spaced in front of the charge than they are behind it, suggesting that the electric field in front is stronger than the field behind. In a medium, such as water or air, where waves travel at a finite speed, c (say), it is possible to get a very interesting effect if the wave source travels at some velocity v which *exceeds* the wave speed. This is illustrated in Figure 4.7. The locus of the outermost wavefront is now a cone instead of a sphere. The wave intensity on the cone is extremely large. In fact, this is a *shockwave* The half-angle θ of the shockwave cone is simply $\sin^{-1}(c/v)$. In water, shockwaves are produced by fast-moving boats. We call these *bow waves*. In air, shockwaves are produced by speeding bullets and supersonic jets. In the latter case, they are called *sonic booms*. Is there

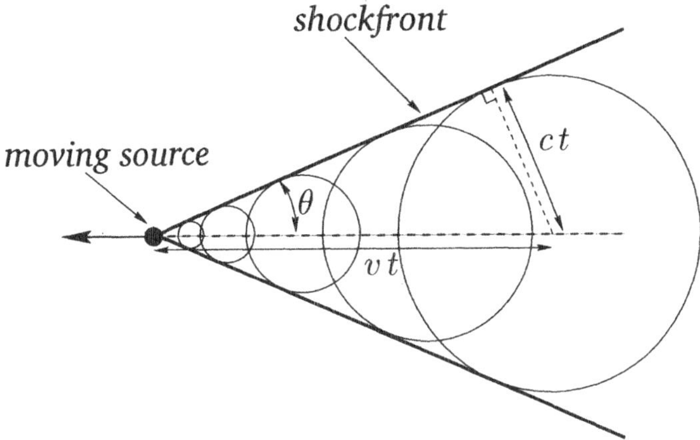

Figure 4.7: *A shockwave.*

any such thing as an electromagnetic shockwave? At first sight, it would appear not. After all, electromagnetic waves travel at the speed of light, and no wave source (*i.e.*, electrically charged particle) can travel faster than this velocity. This is a rather disappointing conclusion. However, when an electromagnetic wave travels through a transparent dielectric medium a remarkable thing happens. The oscillating electric field of the wave induces a slight separation of the positive and negative charges in the atoms which make up the medium. We call separated positive and negative charges an electric dipole. Of course, the atomic dipoles oscillate in sympathy with the field which induces them. However, an oscillating electric dipole radiates electromagnetic waves. Amazingly, when we add the original wave to these induced waves, it is exactly as if the original wave propagates through the medium in question at a velocity which is *slower* than the velocity of light in vacuum. Suppose, now, that we shoot a charged particle through the medium faster than the slowed down velocity of electromagnetic waves. This is possible since the waves are traveling slower than the velocity of light in vacuum. In practice, the particle has to be traveling pretty close to the velocity of light in vacuum (*i.e.*, it has to be relativistic), but modern particle accelerators produce copious amounts of such particles. We can now get an electromagnetic shockwave. We expect such a wave to generate an intense cone of radiation, similar to the bow wave produced by a fast ship. In fact, this type of radiation has been observed. It is called *Cherenkov radiation*, and it is very useful in High-Energy Physics. Cherenkov radiation is typically produced by surrounding a particle accelerator with perspex blocks.

Relativistic charged particles emanating from the accelerator pass through the perspex traveling faster than the local velocity of light, and therefore emit Cherenkov radiation. We know the velocity of light (c_*, say) in perspex (this can be worked out from the refractive index), so if we can measure the half angle θ of the Cherenkov radiation cone emitted by each particle then we can evaluate the particle speed v via the geometric relation $\sin\theta = c_*/v$.

4.10 ADVANCED POTENTIALS?

We have defined the retarded time

$$t_r = t - |\mathbf{r} - \mathbf{r}'|/c \tag{4.160}$$

as the latest time at which a light signal emitted from position $\mathbf{r}'$ would reach position $\mathbf{r}$ before time t. We have also shown that the solution to Maxwell's equations can be written in terms of retarded potentials:

$$\phi(\mathbf{r}, t) = \frac{1}{4\pi\epsilon_0}\int \frac{\rho(\mathbf{r}', t_r)}{|\mathbf{r} - \mathbf{r}'|}\, d^3\mathbf{r}', \tag{4.161}$$

etc. But, is this the most general solution? Suppose that we define the *advanced time.*

$$t_a = t + |\mathbf{r} - \mathbf{r}'|/c. \tag{4.162}$$

This is the time a light signal emitted at time t from position $\mathbf{r}$ would reach position $\mathbf{r}'$. It turns out that we can also write a solution to Maxwell's equations in terms of *advanced potentials*:

$$\phi(\mathbf{r}, t) = \frac{1}{4\pi\epsilon_0}\int \frac{\rho(\mathbf{r}', t_a)}{|\mathbf{r} - \mathbf{r}'|}\, d^3\mathbf{r}', \tag{4.163}$$

etc. In fact, mathematically speaking, this is just as good a solution to Maxwell's equation as the one involving retarded potentials. Consider the Green's function corresponding to our retarded potential solution:

$$\phi(\mathbf{r}, t) = \frac{\rho(\mathbf{r}', t')}{4\pi\epsilon_0}\frac{\delta(t - t' - |\mathbf{r} - \mathbf{r}'|/c)}{|\mathbf{r} - \mathbf{r}'|}, \tag{4.164}$$

with a similar equation for the vector potential. This says that the charge density present at position $\mathbf{r}'$ and time t' emits a spherical wave in the scalar potential *which propagates forward in time*. The Green's function corresponding to our advanced potential solution is

$$\phi(\mathbf{r},t) = \frac{\rho(\mathbf{r}',t')}{4\pi\epsilon_0}\frac{\delta(t-t'+|\mathbf{r}-\mathbf{r}'|/c)}{|\mathbf{r}-\mathbf{r}'|}. \tag{4.165}$$

This says that the charge density present at position $\mathbf{r}'$ and time t' emits a spherical wave in the scalar potential *which propagates backward in time*. Obviously, the advanced solution is usually rejected, on physical grounds, because it violates causality (*i.e.*, it allows effects to exist prior to causes).

Now, the wave equation for the scalar potential,

$$\left(\nabla^2 - \frac{1}{c^2}\frac{\partial^2}{\partial t^2}\right)\phi = -\frac{\rho}{\epsilon_0}, \tag{4.166}$$

is manifestly symmetric in time (*i.e.*, it is invariant under the transformation $t \to -t$). Thus, mathematically speaking, backward-traveling waves are just as good a solution to this equation as forward-traveling waves. The equation is also symmetric in space (*i.e.*, it is invariant under the transformation $\mathbf{r} \to -\mathbf{r}$). So, why do we adopt the Green's function (4.164) which is symmetric in space (*i.e.*, invariant under $\mathbf{r} \to -\mathbf{r}$) but asymmetric in time (*i.e.*, not invariant under $t \to -t$)? Would it not be more consistent to adopt the completely symmetric Green's function

$$\phi(\mathbf{r},t) = \frac{\rho(\mathbf{r}',t')}{4\pi\epsilon_0}\frac{1}{2}\left(\frac{\delta(t-t'-|\mathbf{r}-\mathbf{r}'|/c)}{|\mathbf{r}-\mathbf{r}'|} + \frac{\delta(t-t'+|\mathbf{r}-\mathbf{r}'|/c)}{|\mathbf{r}-\mathbf{r}'|}\right)? \tag{4.167}$$

According to this Green's function, a given charge emits half of its waves running forward in time (*i.e.*, retarded waves), and the other half running backward in time (*i.e.*, advanced waves). This sounds completely crazy! However, in the 1940s Richard P. Feynman and John A. Wheeler pointed out that under certain circumstances this prescription gives the right answer. Consider a charge interacting with "the rest of the Universe," where the "rest of the Universe" denotes all of the distant charges in the Universe, and is, by implication, a very long way from our original charge. Suppose that the "rest of the Universe" is a perfect reflector of advanced waves and a perfect absorber of retarded waves. The waves emitted by the charge can be written schematically as

$$F = \frac{1}{2}(\text{retarded}) + \frac{1}{2}(\text{advanced}). \tag{4.168}$$

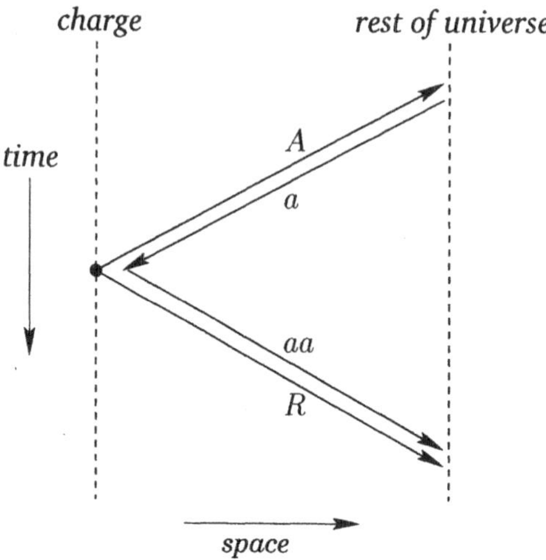

Figure 4.8: *A space-time diagram illustrating the Feynman-Wheeler solution.*

Likewise, the response of the rest of the universe is written

$$R = \frac{1}{2}(\text{retarded}) - \frac{1}{2}(\text{advanced}). \tag{4.169}$$

This is illustrated in the space-time diagram shown in Figure 4.8. Here, A and R denote the advanced and retarded waves emitted by the charge, respectively. The advanced wave travels to "the rest of the Universe" and is reflected: *i.e.*, the distant charges oscillate in response to the advanced wave and emit a retarded wave a, as shown. The retarded wave a is a spherical wave which converges on the original charge, passes through the charge, and then diverges again. The divergent wave is denoted aa. Note that a looks like a negative advanced wave emitted by the charge, whereas aa looks like a positive retarded wave. This is essentially what Equation (4.169) says. The retarded waves R and aa are absorbed by "the rest of the Universe."

If we add the waves emitted by the charge to the response of "the rest of the Universe" we obtain

$$F' = F + R = (\text{retarded}). \tag{4.170}$$

Thus, charges *appear* to emit only retarded waves, which agrees with our everyday experience. Clearly, we have side-stepped the problem of adopting a time-asymmetric Green's function by adopting time-asymmetric boundary conditions to the Universe: *i.e.*, the distant charges in the Universe absorb retarded waves and reflect advanced waves. This is possible because the absorption takes place at the end of the Universe, and the reflection takes place at the beginning of the Universe. It is quite plausible that the state of the Universe (and, hence, its interaction with electromagnetic waves) is completely different at these two epochs. It should be pointed out that the Feynman-Wheeler model runs into trouble when an attempt is made to combine Electromagnetism with Quantum Mechanics. These difficulties have yet to be resolved, so the present status of this model is that it is "an interesting idea," but it is still not fully accepted into the canon of Physics.

4.11 RETARDED FIELDS

We have found the solution to Maxwell's equations in terms of retarded potentials. Let us now construct the associated retarded electric and magnetic fields using

$$\mathbf{E} = -\nabla\phi - \frac{\partial \mathbf{A}}{\partial t}, \tag{4.171}$$

$$\mathbf{B} = \nabla \times \mathbf{A}. \tag{4.172}$$

It is helpful to write

$$\mathbf{R} = \mathbf{r} - \mathbf{r}', \tag{4.173}$$

where $R = |\mathbf{r} - \mathbf{r}'|$. The retarded time becomes $t_r = t - R/c$, and a general retarded quantity is written $[F(\mathbf{r}', t)] \equiv F(\mathbf{r}', t_r)$. Thus, we can write the retarded potential solutions of Maxwell's equations in the especially compact form:

$$\phi(\mathbf{r}, t) = \frac{1}{4\pi\epsilon_0} \int \frac{[\rho]}{R}\, d^3\mathbf{r}', \tag{4.174}$$

$$\mathbf{A}(\mathbf{r}, t) = \frac{\mu_0}{4\pi} \int \frac{[\mathbf{j}]}{R}\, d^3\mathbf{r}'. \tag{4.175}$$

It is easily seen that

$$\begin{aligned} \nabla\phi &= \frac{1}{4\pi\epsilon_0} \int \left([\rho]\nabla(R^{-1}) + \frac{[\partial\rho/\partial t]}{R} \nabla t_r \right) d^3\mathbf{r}' \\ &= -\frac{1}{4\pi\epsilon_0} \int \left(\frac{[\rho]}{R^3}\mathbf{R} + \frac{[\partial\rho/\partial t]}{c\,R^2}\mathbf{R} \right) d^3\mathbf{r}', \end{aligned} \tag{4.176}$$

where use has been made of

$$\nabla R = \frac{\mathbf{R}}{R}, \quad \nabla(R^{-1}) = -\frac{\mathbf{R}}{R^3}, \quad \nabla t_r = -\frac{\mathbf{R}}{c\,R}. \tag{4.177}$$

Likewise,

$$\begin{aligned}\nabla \times \mathbf{A} &= \frac{\mu_0}{4\pi}\int\left(\nabla(R^{-1}) \times [\mathbf{j}] + \frac{\nabla t_r \times [\partial \mathbf{j}/\partial t]}{R}\right) d^3\mathbf{r}' \\ &= -\frac{\mu_0}{4\pi}\int\left(\frac{\mathbf{R} \times [\mathbf{j}]}{R^3} + \frac{\mathbf{R} \times [\partial \mathbf{j}/\partial t]}{c\,R^2}\right) d^3\mathbf{r}'.\end{aligned} \tag{4.178}$$

Equations (4.171), (4.172), (4.176), and (4.178) can be combined to give

$$\mathbf{E} = \frac{1}{4\pi\epsilon_0}\int\left([\rho]\,\frac{\mathbf{R}}{R^3} + \left[\frac{\partial \rho}{\partial t}\right]\frac{\mathbf{R}}{c\,R^2} - \frac{[\partial \mathbf{j}/\partial t]}{c^2\,R}\right) d^3\mathbf{r}', \tag{4.179}$$

which is the time-dependent generalization of Coulomb's law, and

$$\mathbf{B} = \frac{\mu_0}{4\pi}\int\left(\frac{[\mathbf{j}] \times \mathbf{R}}{R^3} + \frac{[\partial \mathbf{j}/\partial t] \times \mathbf{R}}{c\,R^2}\right) d^3\mathbf{r}', \tag{4.180}$$

which is the time-dependent generalization of the Biot-Savart law.

Suppose that the typical variation time-scale of our charges and currents is t_0. Let us define $R_0 = c\,t_0$, which is the distance a light ray travels in time t_0. We can evaluate Equations (4.179) and (4.180) in two asymptotic limits: the *near field* region $R \ll R_0$, and the *far field* region $R \gg R_0$. In the near field region,

$$\frac{|t - t_r|}{t_0} = \frac{R}{R_0} \ll 1, \tag{4.181}$$

so the difference between retarded time and standard time is relatively small. This allows us to expand retarded quantities in a Taylor series. Thus,

$$[\rho] \simeq \rho + \frac{\partial \rho}{\partial t}(t_r - t) + \frac{1}{2}\frac{\partial^2 \rho}{\partial t^2}(t_r - t)^2 + \cdots, \tag{4.182}$$

giving

$$[\rho] \simeq \rho - \frac{\partial \rho}{\partial t}\frac{R}{c} + \frac{1}{2}\frac{\partial^2 \rho}{\partial t^2}\frac{R^2}{c^2} + \cdots. \tag{4.183}$$

Expansion of the retarded quantities in the near field region yields

$$\mathbf{E}(\mathbf{r},t) \simeq \frac{1}{4\pi\epsilon_0}\int\left(\frac{\rho\,\mathbf{R}}{R^3} - \frac{1}{2}\frac{\partial^2\rho}{\partial t^2}\frac{\mathbf{R}}{c^2 R} - \frac{\partial\mathbf{j}/\partial t}{c^2 R} + \cdots\right)d^3\mathbf{r}', \tag{4.184}$$

$$\mathbf{B}(\mathbf{r},t) \simeq \frac{\mu_0}{4\pi}\int\left(\frac{\mathbf{j}\times\mathbf{R}}{R^3} - \frac{1}{2}\frac{(\partial^2\mathbf{j}/\partial t^2)\times\mathbf{R}}{c^2 R} + \cdots\right)d^3\mathbf{r}'. \tag{4.185}$$

In Equation (4.184), the first term on the right-hand side corresponds to Coulomb's law, the second term is the lowest order correction to Coulomb's law due to retardation effects, and the third term corresponds to Faraday induction. In Equation (4.185), the first term on the right-hand side is the Biot-Savart law, and the second term is the lowest order correction to the Biot-Savart law due to retardation effects. Note that the retardation corrections are only of order $(R/R_0)^2$. We might suppose, from looking at Equations (4.179) and (4.180), that the corrections should be of order R/R_0. However, all of the order R/R_0 terms canceled out in the previous expansion. Suppose, then, that we have an electric circuit sitting on a laboratory benchtop. Let the currents in the circuit change on a typical time-scale of one tenth of a second. In this time, light can travel about 3×10^7 meters, so $R_0 \sim 30{,}000$ kilometers. The length-scale of the experiment is about one meter, so $R = 1$ meter. Thus, the retardation corrections are of relative order $(3\times 10^7)^{-2} \sim 10^{-15}$. It is clear that we are fairly safe just using Coulomb's law, Faraday's law, and the Biot-Savart law to analyze the fields generated by this type of circuit.

In the far field region, $R \gg R_0$, Equations (4.179) and (4.180) are dominated by the terms which vary like R^{-1}, so that

$$\mathbf{E}(\mathbf{r},t) \simeq -\frac{1}{4\pi\epsilon_0}\int\frac{[\partial\mathbf{j}_\perp/\partial t]}{c^2 R}\,d^3\mathbf{r}', \tag{4.186}$$

$$\mathbf{B}(\mathbf{r},t) \simeq \frac{\mu_0}{4\pi}\int\frac{[\partial\mathbf{j}_\perp/\partial t]\times\mathbf{R}}{c\,R^2}\,d^3\mathbf{r}', \tag{4.187}$$

where

$$\mathbf{j}_\perp = \mathbf{j} - \frac{(\mathbf{j}\cdot\mathbf{R})}{R^2}\,\mathbf{R}. \tag{4.188}$$

Here, use has been made of $[\partial\rho/\partial t] = -[\nabla\cdot\mathbf{j}]$ and $[\nabla\cdot\mathbf{j}] \simeq -[\partial\mathbf{j}/\partial t]\cdot\mathbf{R}/cR$. Suppose that our charges and currents are localized to some finite region of space in the vicinity of the origin, and that the extent of the current-and-charge-containing region is much less than $|\mathbf{r}|$. It follows that retarded quantities can be written

$$[\rho(\mathbf{r}', t)] \simeq \rho(\mathbf{r}', t - r/c), \tag{4.189}$$

etc. Thus, the electric field reduces to

$$\mathbf{E}(\mathbf{r}, t) \simeq -\frac{1}{4\pi\epsilon_0} \frac{\left[\int \partial \mathbf{j}_\perp/\partial t \, d^3\mathbf{r}'\right]}{c^2 \, r}, \tag{4.190}$$

whereas the magnetic field is given by

$$\mathbf{B}(\mathbf{r}, t) \simeq \frac{1}{4\pi\epsilon_0} \frac{\left[\int \partial \mathbf{j}_\perp/\partial t \, d^3\mathbf{r}'\right] \times \mathbf{r}}{c^3 \, r^2}. \tag{4.191}$$

Here, $[\cdots]$ merely denotes evaluation at the retarded time $t - r/c$. Note that

$$\frac{E}{B} = c, \tag{4.192}$$

and

$$\mathbf{E} \cdot \mathbf{B} = 0. \tag{4.193}$$

This configuration of electric and magnetic fields is characteristic of an *electromagnetic wave* (see Section 4.7). Thus, Equations (4.190) and (4.191) describe an electromagnetic wave propagating *radially* away from the charge-and-current-containing region. Note that the wave is driven by time-varying electric currents. Now, charges moving with a constant velocity constitute a steady current, so a non-steady current is associated with *accelerating charges*. We conclude that accelerating electric charges emit electromagnetic waves. The wave fields, (4.190) and (4.191), fall off like the inverse of the distance from the wave source. This behavior should be contrasted with that of Coulomb or Biot-Savart fields, which fall off like the inverse square of the distance from the source. It is the fact that wave fields attenuate fairly gently with increasing distance from the source which makes Astronomy possible. If wave fields obeyed an inverse square law then no appreciable radiation would reach us from the rest of the Universe.

In conclusion, electric and magnetic fields look simple in the near field region (they are just Coulomb fields, *etc.*) and also in the far field region (they are just electromagnetic waves). Only in the intermediate region, $R \sim R_0$, do things start to get really complicated (so we generally avoid looking in this region!).

4.12 MAXWELL'S EQUATIONS

This marks the end of our theoretical investigation of Maxwell's equations. Let us now summarize what we have learned so far. The field equations which govern electric and magnetic fields are written:

$$\nabla \cdot \mathbf{E} = \frac{\rho}{\epsilon_0}, \tag{4.194}$$

$$\nabla \cdot \mathbf{B} = 0, \tag{4.195}$$

$$\nabla \times \mathbf{E} = -\frac{\partial \mathbf{B}}{\partial t}, \tag{4.196}$$

$$\nabla \times \mathbf{B} = \mu_0 \mathbf{j} + \frac{1}{c^2}\frac{\partial \mathbf{E}}{\partial t}. \tag{4.197}$$

These equations can be integrated to give

$$\oint_S \mathbf{E} \cdot d\mathbf{S} = \frac{1}{\epsilon_0}\int_V \rho \, dV, \tag{4.198}$$

$$\oint_S \mathbf{B} \cdot d\mathbf{S} = 0, \tag{4.199}$$

$$\oint_C \mathbf{E} \cdot d\mathbf{l} = -\frac{\partial}{\partial t}\int_S \mathbf{B} \cdot d\mathbf{S}, \tag{4.200}$$

$$\oint_C \mathbf{B} \cdot d\mathbf{l} = \mu_0 \int_{S'} \mathbf{j} \cdot d\mathbf{S} + \frac{1}{c^2}\frac{\partial}{\partial t}\int_{S'} \mathbf{E} \cdot d\mathbf{S}. \tag{4.201}$$

Here, S is a surface enclosing a volume V, and S' a surface attached to a closed curve C.

Equations (4.195) and (4.196) are automatically satisfied by writing

$$\mathbf{E} = -\nabla\phi - \frac{\partial \mathbf{A}}{\partial t}, \tag{4.202}$$

$$\mathbf{B} = \nabla \times \mathbf{A}. \tag{4.203}$$

This prescription is not unique (there are many choices of ϕ and $\mathbf{A}$ which generate the same fields), but we can make it unique by adopting the following conventions:

$$\phi(\mathbf{r}, t) \to 0 \qquad \text{as } |\mathbf{r}| \to \infty, \tag{4.204}$$

and

$$\frac{1}{c^2}\frac{\partial \phi}{\partial t} + \nabla \cdot \mathbf{A} = 0. \tag{4.205}$$

The latter convention is known as the *Lorenz gauge condition*. Equations (4.194) and (4.197) reduce to

$$\left(\nabla^2 - \frac{1}{c^2}\frac{\partial^2}{\partial t^2}\right)\phi = -\frac{\rho}{\epsilon_0}, \tag{4.206}$$

$$\left(\nabla^2 - \frac{1}{c^2}\frac{\partial^2}{\partial t^2}\right)\mathbf{A} = -\mu_0\,\mathbf{j}. \tag{4.207}$$

These are driven wave equations of the general form

$$\left(\nabla^2 - \frac{1}{c^2}\frac{\partial^2}{\partial t^2}\right)u = v. \tag{4.208}$$

The Green's function for this equation which satisfies sensible boundary conditions, and is consistent with causality, is

$$G(\mathbf{r},\mathbf{r}';t,t') = -\frac{1}{4\pi}\frac{\delta(t-t'-|\mathbf{r}-\mathbf{r}'|/c)}{|\mathbf{r}-\mathbf{r}'|}. \tag{4.209}$$

Thus, the solutions to Equations (4.206) and (4.207) are

$$\phi(\mathbf{r},t) = \frac{1}{4\pi\epsilon_0}\int\frac{[\rho]}{R}\,d^3\mathbf{r}', \tag{4.210}$$

$$\mathbf{A}(\mathbf{r},t) = \frac{\mu_0}{4\pi}\int\frac{[\mathbf{j}]}{R}\,d^3\mathbf{r}', \tag{4.211}$$

where $R = |\mathbf{r}-\mathbf{r}'|$, and $[A] \equiv A(\mathbf{r}',t-R/c)$. These solutions can be combined with Equations (4.202) and (4.203) to give

$$\mathbf{E}(\mathbf{r},t) = \frac{1}{4\pi\epsilon_0}\int\left([\rho]\,\frac{\mathbf{R}}{R^3} + \left[\frac{\partial\rho}{\partial t}\right]\frac{\mathbf{R}}{c\,R^2} - \frac{[\partial\mathbf{j}/\partial t]}{c^2\,R}\right)d^3\mathbf{r}', \tag{4.212}$$

$$\mathbf{B}(\mathbf{r},t) = \frac{\mu_0}{4\pi}\int\left(\frac{[\mathbf{j}]\times\mathbf{R}}{R^3} + \frac{[\partial\mathbf{j}/\partial t]\times\mathbf{R}}{c\,R^2}\right)d^3\mathbf{r}'. \tag{4.213}$$

Equations (4.194)–(4.213) constitute the complete theory of classical electromagnetism. We can express the same information in terms of field equations [Equations (4.194)–(4.197)], integrated field equations [Equations (4.198)–(4.201)], retarded electromagnetic potentials [Equations (4.210) and (4.211)], and retarded electromagnetic fields [Equations (4.212) and (4.213)]. Let us now consider the applications of this theory.

4.13 EXERCISES

4.1. Consider a particle accelerator in which charged particles are constrained to move in a circle in the x-y plane by a z-directed magnetic field. If the magnetic field-strength is gradually increased then the particles are accelerated by the induced electric field. This type of accelerator is called a *betatron*. It is preferable to keep the radius of the particle orbit constant during the acceleration. Show that this is possible provided that the magnetic field distribution is such that the average field over the area of the orbit is twice the field at the circumference. Assume that the field is symmetric about the center of the orbit, and that the particles are subrelativistic.

4.2. A charged particle executes a circular orbit in the plane perpendicular to a uniform magnetic field of strength B. If the magnitude of the field is very gradually increased then the induced electric field accelerates the charge. Demonstrate that in a single rotation

$$\frac{\Delta K}{K} \simeq \frac{\Delta B}{B},$$

where K is the particle's kinetic energy. Hence, deduce that the ratio K/B is approximately constant during the field ramp. By considering the circulating charge as a circular current loop, show that the charge's effective magnetic moment is

$$m = \frac{K}{B},$$

and is, thus, approximately conserved during the field ramp. Does the radius of the orbit increase or decrease as the magnetic field-strength increases?

4.3. Demonstrate that

$$\int_V \mathbf{j}\, d^3\mathbf{r} = \int_S \mathbf{r}\, \mathbf{j} \cdot d\mathbf{S} - \int_V \mathbf{r}\, \nabla \cdot \mathbf{j}\, d^3\mathbf{r},$$

where S is a surface enclosing some volume V. Hence, deduce that for a distribution of charges and currents localized to some finite region of space

$$\int \mathbf{j}\, d^3\mathbf{r} = \frac{d\mathbf{p}}{dt},$$

where the integral is over the whole volume of the distribution, and

$$\mathbf{p} = \int \mathbf{r}\, \rho\, d^3\mathbf{r}$$

is the distribution's electric dipole moment.

4.4. Given that Ampère's circuital law implies the Biot-Savart law, show that Faraday's law implies that

$$\mathbf{E}(\mathbf{r},t) = -\frac{1}{4\pi}\int \frac{[\partial\mathbf{B}(\mathbf{r}',t)/\partial t] \times (\mathbf{r}-\mathbf{r}')}{|\mathbf{r}-\mathbf{r}'|^3}\, d^3\mathbf{r}'.$$

Consider a thin iron ring of major radius a and minor cross-sectional area A. A uniform circulating magnetic field $B(t)$ is produced inside the iron by current flowing in a wire wound toroidally onto the ring. Show that the electric field induced on the major axis of the ring is

$$\mathbf{E}(z,t) = -\frac{a^2\,A}{2\,(a^2+z^2)^{3/2}}\left(\frac{dB}{dt}\right)\mathbf{e}_z,$$

where z is measured from the plane of the ring, in a right-handed sense with respect to the circulating magnetic field. Demonstrate that

$$\int_{-\infty}^{\infty} E_z(z,t)\,dz = -A\,\frac{dB}{dt}.$$

Derive this result directly from Faraday's law.

4.5. A Rogowski coil consists of a thin wire wound uniformly onto a non-magnetic ring-shaped former of major radius a and constant cross-sectional area A. Suppose that there are N turns around the ring. If a time-dependent current $I(t)$ passes *anywhere* through the ring show that the voltage induced in the wire is

$$V = \frac{\mu_0}{2\pi}\,\frac{N\,A}{a}\,\frac{dI}{dt}.$$

4.6. In a certain region of space the charge density takes the form

$$\rho = \rho_0\,e^{-\lambda\,r},$$

where r is a spherical polar coordinate, ρ_0 a spatial constant, and λ a positive constant. Find the electric field generated by this charge distribution. Suppose that

$$\rho_0 = \rho_{00}\,e^{-\gamma\,t},$$

where ρ_{00} is a spatial and temporal constant, and γ a positive constant. What is the current density associated with the time-varying charge density? What is the displacement current generated by the changing electric field? Find the magnetic field generated by these current distributions.

4.7. An alternating current $I = I_0\cos(\omega\,t)$ flows down a long straight wire of negligible thickness, and back along a thin co-axial conducting cylindrical shell of radius R.

(a) In which direction does the induced electric field $\mathbf{E}$ point (radial, circumferential, or longitudinal)?

(b) Find $\mathbf{E}$ as a function of r (perpendicular distance from the wire).

(c) Find the displacement current density $\mathbf{j}_d$.

(d) Integrate $\mathbf{j}_d$ to obtain the total displacement current

$$I_d = \int \mathbf{j}_d \cdot d\mathbf{S}.$$

(e) What is the ratio of I_d to I? If the outer cylinder were 2 mm in diameter, how high would the frequency ω have to be (in Hz) for I_d to be 1% of I?

4.8. Consider the one-dimensional free-space electromagnetic wave equation

$$\frac{\partial^2 E}{\partial z^2} = \epsilon_0 \mu_0 \frac{\partial^2 E}{\partial t^2},$$

where E is the electric field-strength in the x-direction. Show that a change of variables,

$$\alpha = t + \sqrt{\epsilon_0 \mu_0}\, z,$$

$$\beta = t - \sqrt{\epsilon_0 \mu_0}\, z,$$

causes the equation to assume a form which can easily be integrated. Hence, deduce that

$$E(z, t) = F(\alpha) + G(\beta),$$

where F and G are arbitrary functions. Interpret this solution.

4.9. Given the free-space electromagnetic wave electric field

$$\mathbf{E} = E_0 \cos[k\,(z - c\,t)]\,\mathbf{e}_x + E_0 \sin[k\,(z + c\,t)]\,\mathbf{e}_y,$$

where k is real, find the corresponding magnetic field $\mathbf{B}$.

4.10. Given a plane electromagnetic wave propagating in free-space along the positive z-direction, and polarized such that

$$\mathbf{E} = \mathbf{E}_0 \sin\left[k\,(z - c\,t)\right],$$

where k is real, and $\mathbf{E}_0$ is a constant vector, show that it is possible to set the scalar potential ϕ to zero. Find a possible vector potential $\mathbf{A}$ which satisfies the Lorenz gauge.

4.11. The electric field of a plane electromagnetic wave propagating in the z-direction is written

$$\mathbf{E} = A\, e^{i\,(k\,z - \omega\,t)}\,\mathbf{e}_x + B\, e^{i\,(k\,z - \omega\,t)}\,\mathbf{e}_y,$$

where A and B are complex numbers, and k and ω are positive real numbers. Of course, the physical electric field is the real part of the above expression. Suppose that

$$A = |E|/\sqrt{2}, \qquad B = A\, e^{i\,\phi},$$

where $|E|$ is the maximum magnitude of the electric field vector, and ϕ is real. Demonstrate that, in general, the tip of the electric field vector traces out an ellipse in the x-y plane whose major axis is tilted at 45° with respect to the x-axis. Show that the major/minor radii of the ellipse are $|E|\,|\cos\phi/2|$ and $|E|\,|\sin\phi/2|$, and that the electric field vector rotates clockwise (looking down the z-axis) when $0° \leq \phi \leq 180°$, and counterclockwise otherwise. Demonstrate that for the special cases when $\phi = 0°, 180°$, the electric field vector traces out a *straight-line*, passing through the origin, in the x-y plane. These cases correspond to so-called *linearly polarized* waves. Show that for the special cases when $\phi = 90°, 270°$ the electric field vector traces out a *circle* in the x-y plane, rotating clockwise and counterclockwise, respectively. These cases correspond to left-hand and right-hand *circularly polarized* waves, respectively. (The handedness is determined with respect to the direction of wave propagation and the sense of rotation of the electric field in the plane perpendicular to this direction.) Note that in the general case the electric field vector traces out an *ellipse* in the x-y plane, rotating clockwise and counterclockwise, respectively, depending on whether or not $0° \leq \phi \leq 180°$. These cases correspond to left-hand and right-hand *elliptically polarized* waves, respectively.

4.12. A magnetic monopole of monopole charge q_m placed at the origin generates a radial magnetic field of the form

$$B_r(r) = \frac{\mu_0\, c\, q_m}{4\pi\, r^2}.$$

Using this result, and assuming that the number of magnetic monopoles in the Universe is a conserved quantity (like the number of electric charges), show that when Maxwell's equations are generalized to take magnetic monopoles into account they take the form

$$\nabla\cdot\mathbf{E} = \frac{\rho}{\epsilon_0},$$
$$\nabla\cdot\mathbf{H} = \frac{\rho_m}{\epsilon_0},$$
$$\nabla\times\mathbf{E} = -\,\mu_0 c\,\mathbf{j}_m - \frac{\partial\mathbf{H}}{\partial\tau},$$
$$\nabla\times\mathbf{H} = \mu_0 c\,\mathbf{j} + \frac{\partial\mathbf{E}}{\partial\tau},$$

where $\mathbf{H} = c\,\mathbf{B}$ and $\tau = c\,t$. Here, ρ_m is the number density of monopoles, and $\mathbf{j}_m$ is the monopole current density.

4.13. Consider a distribution of charges and currents which is localized in a region of linear extent a in the vicinity of the origin. Demonstrate that the lowest order electric and magnetic field generated by such a distribution in the far field region, $|\mathbf{r}| \gg a$, is

$$\mathbf{E}(\mathbf{r}, t) \simeq -\frac{1}{4\pi\epsilon_0\, c^2}\,\frac{\mathbf{e}_e \times ([\ddot{\mathbf{p}}] \times \mathbf{e}_r)}{r},$$

$$\mathbf{B}(\mathbf{r}, t) \simeq \frac{1}{4\pi\epsilon_0\, c^3}\,\frac{[\ddot{\mathbf{p}}] \times \mathbf{e}_r}{r},$$

where $\mathbf{e}_r = \mathbf{r}/r$, $\mathbf{p} = \int \mathbf{r}\,\rho\, d^3\mathbf{r}$ is the electric dipole moment of the distribution, and $\dot{}$ denotes a time derivative. Here, $[\cdots]$ implies evaluation at the retarded time $t - r/c$. Suppose that $\mathbf{p} = p_0 \cos(\omega\, t)\,\mathbf{e}_z$. Show that the far field electric and magnetic fields take the form

$$\mathbf{E}(\mathbf{r}, t) = -\frac{\omega^2\, p_0}{4\pi\epsilon_0\, c^2}\,\cos[\omega\,(t - r/c)]\,\frac{\sin\theta}{r}\,\mathbf{e}_\theta,$$

$$\mathbf{B}(\mathbf{r}, t) = -\frac{\omega^2\, p_0}{4\pi\epsilon_0\, c^3}\,\cos[\omega\,(t - r/c)]\,\frac{\sin\theta}{r}\,\mathbf{e}_\phi.$$

Here, θ and ϕ are spherical polar coordinates.

Chapter 5 ELECTROSTATIC CALCULATIONS

5.1 INTRODUCTION

In this chapter, we shall make a detailed investigation of the electric fields generated by stationary charge distributions using Maxwell's equations. In particular, we shall examine the interaction of electrostatic fields with ohmic conductors.

5.2 ELECTROSTATIC ENERGY

Consider a collection of N static point charges q_i located at position vectors $\mathbf{r}_i$, respectively (where i runs from 1 to N). What is the electrostatic energy stored in such a collection? In other words, how much work would we have to do in order to assemble the charges, starting from an initial state in which they are all at rest and very widely separated?

Well, we know that a static electric field is conservative, and can consequently be written in terms of a scalar potential:

$$\mathbf{E} = -\nabla\phi. \tag{5.1}$$

We also know that the electric force on a charge q located at position $\mathbf{r}$ is written

$$\mathbf{f} = q\,\mathbf{E}(\mathbf{r}). \tag{5.2}$$

The work *we* would have to do against electrical forces in order to *slowly* move the charge from point P to point Q is simply

$$W = -\int_P^Q \mathbf{f}\cdot d\mathbf{l} = -q\int_P^Q \mathbf{E}\cdot d\mathbf{l} = q\int_P^Q \nabla\phi\cdot d\mathbf{l} = q\,[\phi(Q) - \phi(P)]. \tag{5.3}$$

The negative sign in the above expression comes about because we would have to exert a force $-\mathbf{f}$ on the charge, in order to counteract the force exerted by the electric field. Recall, finally, that the scalar potential field generated by a point charge q located at position $\mathbf{r}'$ is

$$\phi(\mathbf{r}) = \frac{1}{4\pi\epsilon_0}\frac{q}{|\mathbf{r}-\mathbf{r}'|}. \tag{5.4}$$

Let us build up our collection of charges one by one. It takes no work to bring the first charge from infinity, since there is no electric field to fight against. Let us clamp this charge in position at $\mathbf{r}_1$. In order to bring the second charge into position at $\mathbf{r}_2$, we have to do work against the electric field generated by the first charge. According to Equations (5.3) and Equations (5.4), this work is given by

$$W_2 = \frac{1}{4\pi\epsilon_0}\frac{q_2\, q_1}{|\mathbf{r}_2-\mathbf{r}_1|}. \tag{5.5}$$

Let us now bring the third charge into position. Since electric fields and scalar potentials are superposable, the work done whilst moving the third charge from infinity to $\mathbf{r}_3$ is simply the sum of the works done against the electric fields generated by charges 1 and 2 taken in isolation:

$$W_3 = \frac{1}{4\pi\epsilon_0}\left(\frac{q_3\, q_1}{|\mathbf{r}_3-\mathbf{r}_1|}+\frac{q_3\, q_2}{|\mathbf{r}_3-\mathbf{r}_2|}\right). \tag{5.6}$$

Thus, the total work done in assembling the three charges is given by

$$W = \frac{1}{4\pi\epsilon_0}\left(\frac{q_2\, q_1}{|\mathbf{r}_2-\mathbf{r}_1|}+\frac{q_3\, q_1}{|\mathbf{r}_3-\mathbf{r}_1|}+\frac{q_3\, q_2}{|\mathbf{r}_3-\mathbf{r}_2|}\right). \tag{5.7}$$

This result can easily be generalized to N charges:

$$W = \frac{1}{4\pi\epsilon_0}\sum_{i=1}^{N}\sum_{j<i}^{N}\frac{q_i\, q_j}{|\mathbf{r}_i-\mathbf{r}_j|}. \tag{5.8}$$

The restriction that j must be less than i makes the above summation rather messy. If we were to sum without restriction (other than $j \neq i$) then each pair of charges would be counted twice. It is convenient to do just this, and then to divide the result by two. Thus, we obtain

$$W = \frac{1}{2}\frac{1}{4\pi\epsilon_0}\sum_{i=1}^{N}\sum_{\substack{j=1\\ j\neq i}}^{N}\frac{q_i\, q_j}{|\mathbf{r}_i-\mathbf{r}_j|}. \tag{5.9}$$

This is the *potential energy* (*i.e.*, the difference between the total energy and the kinetic energy) of a collection of charges. We can think of this quantity as the work required to bring stationary charges from infinity and assemble them in the required formation. Alternatively, it is the kinetic energy which would be released if the collection were dissolved, and the charges returned to infinity. But where is this potential energy stored? Let us investigate further.

Equation (5.9) can be written

$$W = \frac{1}{2}\sum_{i=1}^{N} q_i\,\phi_i, \tag{5.10}$$

where

$$\phi_i = \frac{1}{4\pi\epsilon_0}\sum_{\substack{j=1\\ j\neq i}}^{N}\frac{q_j}{|\mathbf{r}_i - \mathbf{r}_j|} \tag{5.11}$$

is the scalar potential experienced by the ith charge due to the other charges in the distribution.

Let us now consider the potential energy of a continuous charge distribution. It is tempting to write

$$W = \frac{1}{2}\int \rho\,\phi\, d^3\mathbf{r}, \tag{5.12}$$

by analogy with Equations (5.10) and (5.11), where

$$\phi(\mathbf{r}) = \frac{1}{4\pi\epsilon_0}\int \frac{\rho(\mathbf{r}')}{|\mathbf{r} - \mathbf{r}'|}\, d^3\mathbf{r}' \tag{5.13}$$

is the familiar scalar potential generated by a continuous charge distribution of charge density $\rho(\mathbf{r})$. Let us try this out. We know from Maxwell's equations that

$$\rho = \epsilon_0\,\nabla\cdot\mathbf{E}, \tag{5.14}$$

so Equation (5.12) can be written

$$W = \frac{\epsilon_0}{2}\int \phi\,\nabla\cdot\mathbf{E}\, d^3\mathbf{r}. \tag{5.15}$$

Now, vector field theory yields the standard result

$$\nabla\cdot(\mathbf{E}\,\phi) = \phi\,\nabla\cdot\mathbf{E} + \mathbf{E}\cdot\nabla\phi. \tag{5.16}$$

However, $\nabla\phi = -\mathbf{E}$, so we obtain

$$W = \frac{\epsilon_0}{2}\left[\int \nabla\cdot(\mathbf{E}\,\phi)\,d^3\mathbf{r} + \int E^2\,d^3\mathbf{r}\right]. \tag{5.17}$$

Application of Gauss' theorem gives

$$W = \frac{\epsilon_0}{2}\left(\oint_S \phi\,\mathbf{E}\cdot d\mathbf{S} + \int_V E^2\,dV\right), \tag{5.18}$$

where V is some volume which encloses all of the charges, and S is its bounding surface. Let us assume that V is a sphere, centered on the origin, and let us take the limit in which the radius r of this sphere goes to infinity. We know that, in general, the electric field at large distances from a bounded charge distribution looks like the field of a point charge, and, therefore, falls off like $1/r^2$. Likewise, the potential falls off like $1/r$—see Exercise 3.7. However, the surface area of the sphere increases like r^2. Hence, it is clear that, in the limit as $r \to \infty$, the surface integral in Equation (5.18) falls off like $1/r$, and is consequently zero. Thus, Equation (5.18) reduces to

$$W = \frac{\epsilon_0}{2}\int E^2\,d^3\mathbf{r}, \tag{5.19}$$

where the integral is over all space. This is a very interesting result. It tells us that the potential energy of a continuous charge distribution is stored in the *electric field* generated by the distribution. Of course, we now have to assume that an electric field possesses an *energy density*

$$U = \frac{\epsilon_0}{2}E^2. \tag{5.20}$$

We can easily check that Equation (5.19) is correct. Suppose that we have a charge Q which is uniformly distributed within a sphere of radius a centered on the origin. Let us imagine building up this charge distribution from a succession of thin spherical layers of infinitesimal thickness. At each stage, we gather a small amount of charge dq from infinity, and spread it over the surface of the sphere in a thin layer extending from r to $r + dr$. We continue this process until the final radius of the sphere is a. If $q(r)$ is the sphere's charge when it has attained radius r, then the work done in bringing a charge dq to its surface is

$$dW = \frac{1}{4\pi\epsilon_0}\frac{q(r)\,dq}{r}. \tag{5.21}$$

This follows from Equation (5.5), since the electric field generated outside a spherical charge distribution is the same as that of a point charge

$q(r)$ located at its geometric center ($r = 0$)—see Section 3.4. If the constant charge density of the sphere is ρ then

$$q(r) = \frac{4\pi}{3}\, r^3\, \rho, \tag{5.22}$$

and

$$dq = 4\pi\, r^2\, \rho\, dr. \tag{5.23}$$

Thus, Equation (5.21) becomes

$$dW = \frac{4\pi}{3\epsilon_0}\, \rho^2\, r^4\, dr. \tag{5.24}$$

The total work needed to build up the sphere from nothing to radius a is plainly

$$W = \frac{4\pi}{3\epsilon_0}\, \rho^2 \int_0^a r^4\, dr = \frac{4\pi}{15\epsilon_0}\, \rho^2\, a^5. \tag{5.25}$$

This can also be written in terms of the total charge $Q = (4\pi/3)\, a^3\, \rho$ as

$$W = \frac{3}{5}\frac{Q^2}{4\pi\epsilon_0\, a}. \tag{5.26}$$

Now that we have evaluated the potential energy of a spherical charge distribution by the direct method, let us work it out using Equation (5.19). We shall assume that the electric field is both radial and spherically symmetric, so that $\mathbf{E} = E_r(r)\, \mathbf{e}_r$. Application of Gauss' law,

$$\oint_S \mathbf{E} \cdot d\mathbf{S} = \frac{1}{\epsilon_0} \int_V \rho\, dV, \tag{5.27}$$

where V is a sphere of radius r, centered on the origin, gives

$$E_r(r) = \frac{Q}{4\pi\epsilon_0}\frac{r}{a^3} \tag{5.28}$$

for $r < a$, and

$$E_r(r) = \frac{Q}{4\pi\epsilon_0\, r^2} \tag{5.29}$$

for $r \geq a$. Equations (5.19), (5.28), and (5.29) yield

$$W = \frac{Q^2}{8\pi\epsilon_0}\left(\frac{1}{a^6}\int_0^a r^4\, dr + \int_a^\infty \frac{dr}{r^2}\right), \tag{5.30}$$

which reduces to

$$W = \frac{Q^2}{8\pi\epsilon_0\, a}\left(\frac{1}{5}+1\right) = \frac{3}{5}\frac{Q^2}{4\pi\epsilon_0\, a}. \tag{5.31}$$

Thus, Equation (5.19) gives the correct answer.

The reason that we have checked Equation (5.19) so carefully is that, on close inspection, it is found to be inconsistent with Equation (5.10), from which it was supposedly derived! For instance, the energy given by Equation (5.19) is manifestly positive definite, whereas the energy given by Equation (5.10) can be negative (it is certainly negative for a collection of two point charges of opposite sign). The inconsistency was introduced into our analysis when we replaced Equation (5.11) with Equation (5.13). In Equation (5.11), the self-interaction of the ith charge with its own electric field is specifically excluded, whereas it is included in Equation (5.13). Thus, the potential energies (5.10) and (5.19) are different because in the former we start from ready-made point charges, whereas in the latter we build up the whole charge distribution from scratch. Hence, if we were to work out the potential energy of a point charge distribution using Equation (5.19) then we would obtain the energy (5.10) *plus* the energy required to assemble the point charges. What is the energy required to assemble a point charge? In fact, it is *infinite*. To see this, let us suppose, for the sake of argument, that our point charges actually consist of charge uniformly distributed in small spheres of radius b. According to Equation (5.26), the energy required to assemble the ith point charge is

$$W_i = \frac{3}{5}\frac{q_i^2}{4\pi\epsilon_0\, b}. \tag{5.32}$$

We can think of this as the self-energy of the ith charge. Thus, we can write

$$W = \frac{\epsilon_0}{2}\int E^2\, d^3\mathbf{r} = \frac{1}{2}\sum_{i=1}^{N} q_i\, \phi_i + \sum_{i=1}^{N} W_i \tag{5.33}$$

which enables us to reconcile Equations (5.10) and (5.19). Unfortunately, if our point charges really are point charges then $b \to 0$, and the self-energy of each charge becomes infinite. Thus, the potential energies predicted by Equations (5.10) and (5.19) differ by an infinite amount. What does this all mean? We have to conclude that the idea of locating electrostatic potential energy in the electric field is inconsistent with the existence of point charges. One way out of this difficulty would be to say that elementary charges, such as electrons, are not points objects, but instead have finite spatial extents. Regrettably, there is no experimental evidence to back up this assertion. Alternatively, we could say that our classical theory of electromagnetism breaks down on very small length-scales due to quantum effects. Unfortunately, the quantum mechanical version of electromagnetism (which is called Quantum Electrodynamics) suffers from the same infinities in the self-energies of charged particles as the classical version. There is a prescription, called *renormalization*, for steering round these infinities, and getting finite answers which agree with experimental data to extraordinary accuracy. However, nobody really understands why this prescription works. Indeed, the problem of the infinite self-energies of elementary charged particles is still an unresolved issue in Physics.

5.3 OHM'S LAW

A *conductor* is a medium which contains free electric charges (usually electrons) which drift in the presence of an applied electric field, giving rise to an electric current flowing in the same direction as the field. The well-known relationship between the current and the voltage in a typical conductor is given by *Ohm's law*: *i.e.*,

$$V = I\,R, \tag{5.34}$$

where V is the voltage drop across a conductor of electrical resistance R through which a current I flows. Incidentally, the unit of electrical resistance is the ohm (Ω), which is equivalent to a volt per ampere.

Let us generalize Ohm's law so that it is expressed in terms of $\mathbf{E}$ and $\mathbf{j}$, rather than V and I. Consider a length l of a conductor of uniform cross-sectional area A through which a current I flows. In general, we expect the electrical resistance of the conductor to be proportional to its length, l, and inversely proportional to its cross-sectional area, A (*i.e.*, we expect that it is harder to push an electrical current down a long rather than a short wire, and easier to push a current down a wide rather than a narrow conducting channel). Thus, we can write

$$R = \eta\,\frac{l}{A}. \tag{5.35}$$

Here, the constant η is called the *resistivity* of the conducting medium, and is measured in units of ohm-meters. Hence, Ohm's law becomes

$$V = \eta \frac{l}{A} I. \tag{5.36}$$

However, $I/A = j_z$ (supposing that the conductor is aligned along the z-axis) and $V/l = E_z$, so the above equation reduces to

$$E_z = \eta \, j_z. \tag{5.37}$$

There is nothing special about the z-axis (in an isotropic conducting medium), so the previous formula immediately generalizes to

$$\mathbf{E} = \eta \, \mathbf{j}. \tag{5.38}$$

This is the vector form of Ohm's law.

It is fairly easy to account for the above equation physically. Consider a metal which has n free electrons per unit volume. Of course, the metal also has a fixed lattice of metal ions whose charge per unit volume is equal and opposite to that of the free electrons, rendering the medium electrically neutral. In the presence of an electric field $\mathbf{E}$, a given free electron accelerates (from rest at $t = 0$) such that its drift velocity is written $\mathbf{v} = -(e/m_e)\, t\, \mathbf{E}$, where $-e$ is the electron charge, and m_e the electron mass. Suppose that, on average, a drifting electron collides with a metal ion once every τ seconds. Given that a metal ion is much more massive than an electron, we expect a free electron to lose all of the momentum it had previously acquired from the electric field during such a collision. It follows that the mean drift velocity of the free electrons is $\bar{\mathbf{v}} = -(e\,\tau/2\,m_e)\,\mathbf{E}$. Hence, the mean current density is $\mathbf{j} = (n\,e^2\,\tau/2\,m_e)\,\mathbf{E}$. Thus, the resistivity can be written

$$\eta = \frac{n\,e^2\,\tau}{2\,m_e}. \tag{5.39}$$

We conclude that the resistivity of a typical conducting medium is determined by the number density of free electrons, as well as the mean collision rate of these electrons with the fixed ions. The fact that $\mathbf{j} \propto \mathbf{E}$ leads immediately to the relation (5.35) between resistivity and resistance.

A free charge q which moves through a voltage drop V acquires an energy $q\,V$ from the electric field. In a conducting medium, this energy is dissipated as *heat* (the conversion to heat takes place each time a free charge collides with a fixed ion). This type of heating is called *ohmic heating*. Suppose that N charges per unit time pass through a conductor. The current flowing is obviously $I = N q$. The total energy gained by the charges, which appears as heat inside the conductor, is

$$P = N q V = I V \tag{5.40}$$

per unit time. Thus, the heating power is

$$P = I V = I^2 R = \frac{V^2}{R}. \tag{5.41}$$

Equations (5.40) and (5.41) generalize to

$$P = \mathbf{j} \cdot \mathbf{E} = \eta j^2, \tag{5.42}$$

where P is now the power dissipated per unit volume inside the conducting medium.

5.4 CONDUCTORS

Most (but not all) electrical conductors obey Ohm's law. Such conductors are termed *ohmic*. Suppose that we apply an electric field to an ohmic conductor. What is going to happen? According to Equation (5.38), the electric field drives currents. These currents redistribute the charge inside the conductor until the original electric field is canceled out. At this point, the currents stop flowing. It might be objected that the currents could keep flowing in closed loops. According to Ohm's law, this would require a non-zero emf, $\oint \mathbf{E} \cdot d\mathbf{l}$, acting around each loop (unless the conductor is a *superconductor*, with $\eta = 0$). However, we know that in a steady-state

$$\oint_C \mathbf{E} \cdot d\mathbf{l} = 0 \tag{5.43}$$

around any closed loop C. This proves that a steady-state emf acting around a closed loop inside a conductor is impossible. The only other alternative is

$$\mathbf{j} = \mathbf{E} = \mathbf{0} \tag{5.44}$$

everywhere inside the conductor. It immediately follows from the Maxwell equation $\nabla \cdot \mathbf{E} = \rho/\epsilon_0$ that

$$\rho = 0. \tag{5.45}$$

So, there are no electric charges in the interior of a conductor. But, how can a conductor cancel out an applied electric field if it contains no charges? The answer is that all of the charges reside on the *surface* of the conductor. In reality, the charges lie within one or two atomic layers of the surface (see any textbook on solid-state physics). The difference in scalar potential between two points P and Q is simply

$$\phi(Q) - \phi(P) = \int_P^Q \nabla\phi \cdot d\mathbf{l} = -\int_P^Q \mathbf{E} \cdot d\mathbf{l}. \tag{5.46}$$

However, if P and Q both lie inside the same conductor then it is clear from Equations (5.44) and (5.46) that the potential difference between P and Q is zero. This is true no matter where P and Q are situated inside the conductor, so we conclude that the scalar potential must be *uniform* inside a conductor. A corollary of this is that the surface of a conductor is an equipotential (*i.e.*, ϕ = constant) surface.

So, the electric field inside a conductor is zero. We can demonstrate that the field within an empty cavity lying inside a conductor is zero as well, provided that there are no charges within the cavity. Let us, first of all, apply Gauss' law to a surface S which surrounds the cavity, but lies wholly within the conducting medium—see Figure 5.1. Since the electric field is zero inside a conductor, it follows that zero net charge is enclosed by S. This does not preclude the possibility that there are equal amounts of positive and negative charges distributed on the inner surface of the conductor. However, we can easily rule out this possibility using the steady-state relation

$$\oint_C \mathbf{E} \cdot d\mathbf{l} = 0, \tag{5.47}$$

for any closed loop C. If there are any electric field-lines inside the cavity then they must run from the positive to the negative surface charges. Consider a closed loop C which straddles the cavity and the conductor, such as the one shown in Figure 5.1. In the presence of field-lines, it is clear that the line integral of **E** along that portion of the loop which lies inside the cavity is non-zero. However, the line integral of **E** along that portion of the loop which runs through the conducting medium is obviously zero (since $\mathbf{E} = \mathbf{0}$ inside a conductor). Thus, the line integral of the field around the closed loop C is non-zero, which clearly contradicts Equation (5.47). In fact, this equation implies that the line integral of the electric field along any path which runs through the cavity, from one point on the interior surface of the conductor to

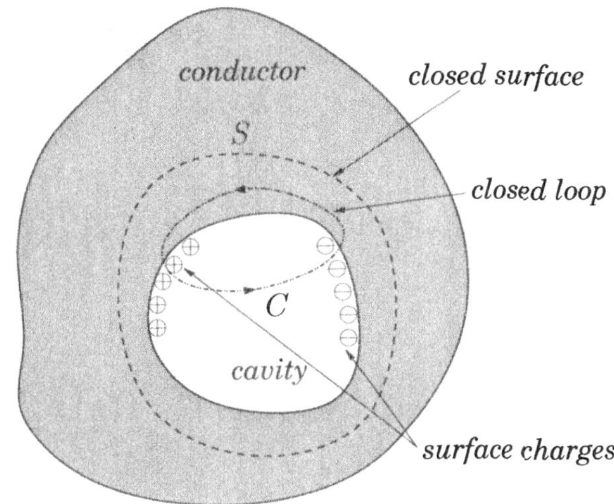

Figure 5.1: *An empty cavity inside a conductor.*

another, is zero. This can only be the case if the electric field itself is zero everywhere inside the cavity. (The above argument is not entirely rigorous. In particular, it is not clear that it fails if there are charges inside the cavity. We shall discuss an improved argument later on in this chapter.)

We have shown that if a charge-free cavity is completely enclosed by a conductor then no stationary distribution of charges outside the conductor can ever produce any electric fields inside the cavity. It follows that we can shield a sensitive piece of electrical equipment from stray external electric fields by placing it inside a metal can. In fact, a wire mesh cage will do, as long as the mesh spacing is not too wide. Such a cage is known as a *Faraday cage*.

Let us consider some small region on the surface of a conductor. Suppose that the local surface charge density is σ, and that the electric field just outside the conductor is $\mathbf{E}$. Note that this field must be directed *normal* to the surface of the conductor. Any parallel component would be shorted out by surface currents. Another way of saying this is that the surface of a conductor is an equipotential. We know that $\nabla\phi$ is always perpendicular to an equipotential, so $\mathbf{E} = -\nabla\phi$ must be locally perpendicular to a conducting surface. Let us use Gauss' law,

$$\oint_S \mathbf{E}\cdot d\mathbf{S} = \frac{1}{\epsilon_0}\int_V \rho\, dV, \tag{5.48}$$

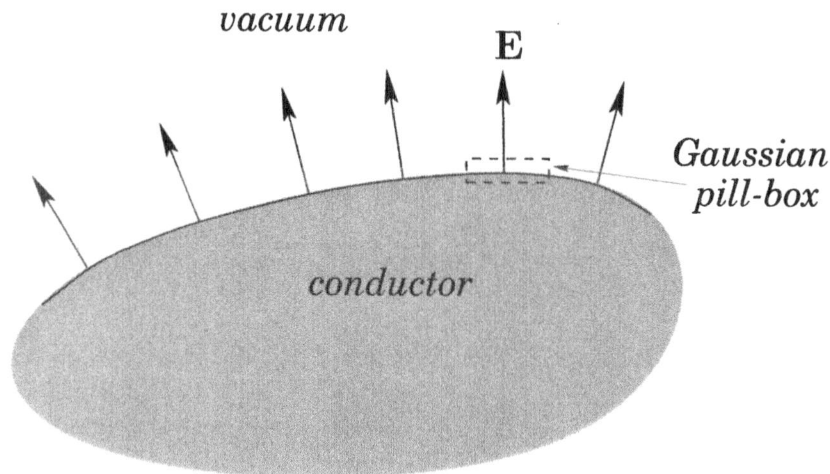

Figure 5.2: *The surface of a conductor.*

where V is a so-called *Gaussian pill-box*—see Figure 5.2. This is a pill-box-shaped volume whose two ends are aligned parallel to the surface of the conductor, with the surface running between them, and whose sides are perpendicular to the surface. It is clear that **E** is parallel to the sides of the box, so the sides make no contribution to the surface integral. The end of the box which lies inside the conductor also makes no contribution, since $\mathbf{E} = \mathbf{0}$ inside a conductor. Thus, the only non-zero contribution to the surface integral comes from the end lying in free space. This contribution is simply $E_\perp A$, where $E_\perp$ denotes an outward pointing (from the conductor) normal electric field, and A is the cross-sectional area of the box. The charge enclosed by the box is simply σA, from the definition of a surface charge density. Thus, Gauss' law yields

$$E_\perp = \frac{\sigma}{\epsilon_0} \tag{5.49}$$

as the relationship between the normal electric field immediately outside a conductor and the surface charge density.

Let us look at the electric field generated by a sheet charge distribution a little more carefully. Suppose that the charge per unit area is σ. By symmetry, we expect the field generated below the sheet to be the mirror image of that above the sheet (at least, locally). Thus, if we integrate Gauss' law over a pill-box of cross-sectional area A, as shown in Figure 5.3, then the two ends both contribute $E_{sheet} A$ to the surface

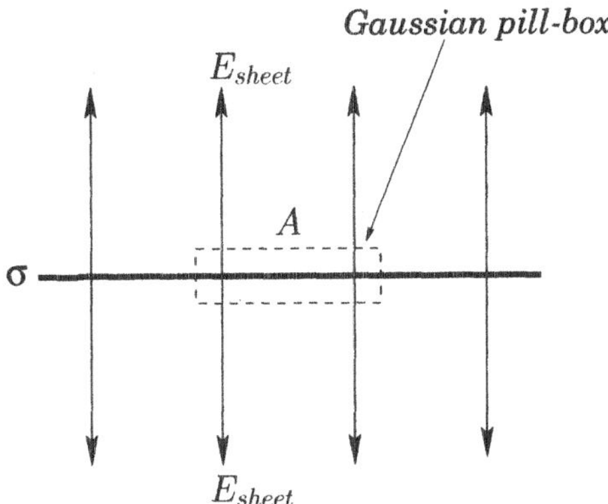

Figure 5.3: *The electric field of a sheet charge.*

integral, where E_{sheet} is the normal electric field generated above and below the sheet. The charge enclosed by the pill-box is just $\sigma\,A$. Thus, Gauss' law yields a symmetric electric field

$$E_{sheet} = \begin{cases} +\sigma/(2\,\epsilon_0) & \text{above} \\ -\sigma/(2\,\epsilon_0) & \text{below} \end{cases}. \tag{5.50}$$

So, how do we get the asymmetric electric field of a conducting surface, which is zero immediately below the surface (*i.e.*, inside the conductor) and non-zero immediately above it? Clearly, we have to add in an external field (*i.e.*, a field which is not generated locally by the sheet charge). The requisite field is

$$E_{ext} = \frac{\sigma}{2\,\epsilon_0} \tag{5.51}$$

both above and below the charge sheet. The total field is the sum of the field generated locally by the charge sheet and the external field. Thus, we obtain

$$E_{total} = \begin{cases} +\sigma/\epsilon_0 & \text{above} \\ 0 & \text{below} \end{cases}, \tag{5.52}$$

which is in agreement with Equation (5.49).

Now, the external field exerts a force on the charge sheet. Of course, the field generated locally by the sheet itself cannot exert a local force (*i.e.*, the charge sheet cannot exert a force on itself). Thus, the force per unit area acting on the surface of a conductor always acts outward, and is given by

$$p = \sigma\, E_{ext} = \frac{\sigma^2}{2\,\epsilon_0}. \tag{5.53}$$

We conclude that there is an *electrostatic pressure* acting on any charged conductor. This effect can be observed by charging up soap bubbles: the additional electrostatic pressure eventually causes them to burst. The electrostatic pressure can also be written

$$p = \frac{\epsilon_0}{2}\, E_\perp^{\,2}, \tag{5.54}$$

where $E_\perp$ is the field-strength immediately above the surface of the conductor. Note that, according to the above formula, the electrostatic pressure is equivalent to the energy density of the electric field immediately outside the conductor. This is not a coincidence. Suppose that the conductor expands normally by an average distance dx, due to the electrostatic pressure. The electric field is excluded from the region into which the conductor expands. The volume of this region is $dV = A\, dx$, where A is the surface area of the conductor. Thus, the energy of the electric field decreases by an amount $dE = U\, dV = (\epsilon_0/2)\, E_\perp^{\,2}\, dV$, where U is the energy density of the field. This decrease in energy can be ascribed to the work which the field does on the conductor in order to make it expand. This work is $dW = p\, A\, dx$, where p is the force per unit area that the field exerts on the conductor. Thus, $dE = dW$, from energy conservation, giving

$$p = \frac{\epsilon_0}{2}\, E_\perp^{\,2}. \tag{5.55}$$

Incidentally, this technique for calculating a force, given an expression for the energy of a system as a function of some adjustable parameter, is called *the principle of virtual work*.

We have seen that an electric field is excluded from the inside of a conductor, but not from the outside, giving rise to a net *outward* force. We can account for this fact by saying that the field exerts a *negative* pressure $(\epsilon_0/2)\, E_\perp^{\,2}$ on the conductor. Now, we know that if we evacuate a closed metal can then the pressure difference between the inside and the outside eventually causes it to *implode*. Likewise, if we place the can in a strong electric field then the pressure difference between the inside and the outside will eventually cause it to *explode*. How big a field do we need before the electrostatic pressure difference is the same as that obtained by evacuating the can? In other words, what electric field exerts a negative pressure of one atmosphere (*i.e.*, 10^5 newtons per meter squared) on the can? The answer is a field of strength $E \sim 10^8$ volts per meter. Fortunately, this is a rather large electric field, so there is no danger of your car exploding when you turn on the radio!

5.5 BOUNDARY CONDITIONS ON THE ELECTRIC FIELD

What are the general boundary conditions satisfied by the electric field at the interface between two different media: *e.g.*, the interface between a vacuum and a conductor? Consider an interface P between two media 1 and 2. Let us, first of all, apply Gauss' law,

$$\oint_S \mathbf{E} \cdot d\mathbf{S} = \frac{1}{\epsilon_0} \int_V \rho \, dV, \tag{5.56}$$

to a Gaussian pill-box S of cross-sectional area A whose two ends are locally parallel to the interface—see Figure 5.4. The ends of the box can be made arbitrarily close together. In this limit, the flux of the electric field out of the sides of the box is obviously negligible, and the only contribution to the flux comes from the two ends. In fact,

$$\oint_S \mathbf{E} \cdot d\mathbf{S} = (E_{\perp 1} - E_{\perp 2})\, A, \tag{5.57}$$

where $E_{\perp 1}$ is the perpendicular (to the interface) electric field in medium 1 at the interface, *etc.* The charge enclosed by the pill-box is simply σA, where σ is the sheet charge density on the interface. Note that any volume distribution of charge gives rise to a negligible contribution to the right-hand side of Equation (5.56), in the limit where the two ends of the pill-box are very closely spaced. Thus, Gauss' law yields

$$E_{\perp 1} - E_{\perp 2} = \frac{\sigma}{\epsilon_0} \tag{5.58}$$

at the interface: *i.e.*, the presence of a charge sheet on an interface causes a discontinuity in the perpendicular component of the electric field. What about the parallel electric field? Let us apply Faraday's law to a rectangular loop C whose long sides, length l, run parallel to the interface,

$$\oint_C \mathbf{E} \cdot d\mathbf{l} = -\frac{\partial}{\partial t} \int_S \mathbf{B} \cdot d\mathbf{S} \tag{5.59}$$

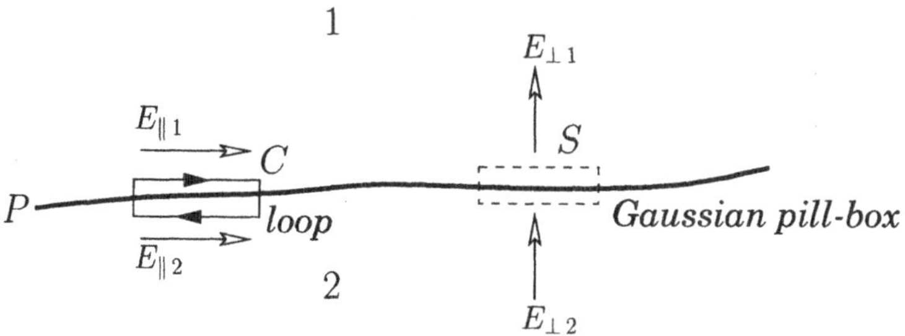

Figure 5.4: *Boundary conditions on the electric field.*

—see Figure 5.4. The length of the short sides is assumed to be arbitrarily small. Hence, the dominant contribution to the loop integral comes from the long sides:

$$\oint_C \mathbf{E} \cdot d\mathbf{l} = (E_{\parallel 1} - E_{\parallel 2})\, l, \tag{5.60}$$

where $E_{\parallel 1}$ is the parallel (to the interface) electric field in medium 1 at the interface, *etc.* The flux of the magnetic field through the loop is approximately $B_{\perp} A$, where $B_{\perp}$ is the component of the magnetic field which is normal to the loop, and A the area of the loop. But, $A \to 0$ as the short sides of the loop are shrunk to zero. So, unless the magnetic field becomes infinite at the interface (and we shall assume that it does not), the flux also tends to zero. Thus,

$$E_{\parallel 1} - E_{\parallel 2} = 0: \tag{5.61}$$

i.e., there can be no discontinuity in the parallel component of the electric field across an interface.

5.6 CAPACITORS

It is clear that we can store electrical charge on the surface of a conductor. However, electric fields will be generated immediately above this surface. Now, the conductor can only successfully store charge if it is electrically insulated from its surroundings. Of course, air is a very good insulator. Unfortunately, air ceases to be an insulator when the electric field-strength through it exceeds some critical value which is about

$E_{crit} \sim 10^6$ volts per meter. This phenomenon, which is called *breakdown*, is associated with the formation of sparks. The most well-known example of the breakdown of air is during a lightning strike. Thus, a good charge storing device is one which holds a relatively large amount of charge, but only generates relatively small external electric fields (so as to avoid breakdown). Such a device is called a *capacitor*.

Consider two thin, parallel, conducting plates of cross-sectional area A which are separated by a *small* distance d (*i.e.*, $d \ll \sqrt{A}$). Suppose that each plate carries an equal and opposite charge $\pm Q$ (where $Q > 0$). We expect this charge to spread evenly over the plates to give an effective sheet charge density $\pm\sigma = Q/A$ on each plate. Suppose that the upper plate carries a positive charge and that the lower carries a negative charge. According to Equation (5.50), the field generated by the upper plate is normal to the plate and of magnitude

$$E_{upper} = \begin{cases} +\sigma/(2\epsilon_0) & \text{above} \\ -\sigma/(2\epsilon_0) & \text{below} \end{cases}. \tag{5.62}$$

Likewise, the field generated by the lower plate is

$$E_{lower} = \begin{cases} -\sigma/(2\epsilon_0) & \text{above} \\ +\sigma/(2\epsilon_0) & \text{below} \end{cases}. \tag{5.63}$$

Note that we are neglecting any "leakage" of the field at the edges of the plates. This is reasonable provided that the plates are relatively closely spaced. The total field is the sum of the two fields generated by the upper and lower plates. Thus, the net field is normal to the plates, and of magnitude

$$E_\perp = \begin{cases} \sigma/\epsilon_0 & \text{between} \\ 0 & \text{otherwise} \end{cases} \tag{5.64}$$

—see Figure 5.5. Since the electric field is uniform, the potential difference between the plates is simply

$$V = E_\perp\, d = \frac{\sigma\, d}{\epsilon_0}. \tag{5.65}$$

Now, it is conventional to measure the capacity of a conductor, or set of conductors, to store charge, but generate small external electric fields, in terms of a parameter called the *capacitance*. This parameter is usually denoted C. The capacitance of a charge-storing device is simply

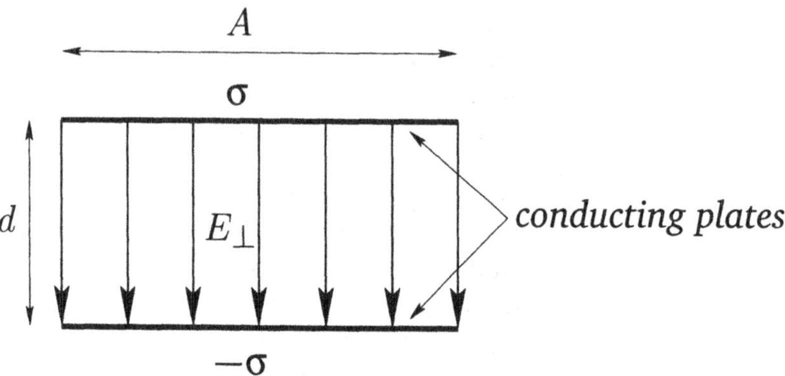

Figure 5.5: *The electric field of a parallel plate capacitor.*

the ratio of the charge stored to the potential difference generated by this charge: *i.e.*,

$$C = \frac{Q}{V}. \tag{5.66}$$

Clearly, a good charge-storing device has a high capacitance. Incidentally, capacitance is measured in farads (F), which are equivalent to coulombs per volt. This is a rather unwieldy unit, since capacitors in electrical circuits typically have capacitances which are only about one millionth of a farad. For a parallel plate capacitor, we have

$$C = \frac{\sigma A}{V} = \frac{\epsilon_0 A}{d}. \tag{5.67}$$

Note that the capacitance only depends on *geometric* quantities, such as the area and spacing of the plates. This is a consequence of the superposability of electric fields. If we double the charge on a set of conductors then we double the electric fields generated around them, and we, therefore, double the potential difference between the conductors. Thus, the potential difference between the conductors is always directly proportional to the charge on the conductors: the constant of proportionality (the inverse of the capacitance) can only depend on geometry.

Suppose that the charge $\pm Q$ on each plate of a parallel plate capacitor is built up gradually by transferring small amounts of charge from one plate to another. If the instantaneous charge on the plates is $\pm q$, and an infinitesimal amount of positive charge dq is transferred from the negatively charged to the positively charge plate, then the work done is $dW = V\,dq = q\,dq/C$, where V is the instantaneous voltage difference between the plates. Note that the voltage difference is such that it opposes any increase in the charge on either plate. The total work done in charging the capacitor is

$$W = \frac{1}{C}\int_0^Q q\,dq = \frac{Q^2}{2C} = \frac{1}{2}\,C\,V^2, \tag{5.68}$$

where use has been made of Equation (5.66). The energy stored in the capacitor is the same as the work required to charge up the capacitor. Thus, the stored energy is

$$W = \frac{1}{2}\,C\,V^2. \tag{5.69}$$

This is a general result which holds for all types of capacitor.

The energy of a charged parallel plate capacitor is actually stored in the electric field between the plates. This field is of approximately constant magnitude $E_\perp = V/d$, and occupies a region of volume $A\,d$. Thus, given the energy density of an electric field, $U = (\epsilon_0/2)\,E^2$, the energy stored in the electric field is

$$W = \frac{\epsilon_0}{2}\frac{V^2}{d^2}\,A\,d = \frac{1}{2}\,C\,V^2, \tag{5.70}$$

where use has been made of Equation (5.67). Note that Equations (5.68) and (5.70) agree with one another. The fact that the energy of a capacitor is stored in its electric field is also a general result.

The idea, which we discussed earlier, that an electric field exerts a negative pressure $(\epsilon_0/2)\,E_\perp^{\,2}$ on conductors immediately suggests that the two plates in a parallel plate capacitor *attract* one another with a mutual force

$$F = \frac{\epsilon_0}{2}\,E_\perp^{\,2}\,A = \frac{1}{2}\frac{C\,V^2}{d}. \tag{5.71}$$

It is not actually necessary to have two oppositely charged conductors in order to make a capacitor. Consider an isolated conducting sphere of radius a which carries an electric charge Q. The spherically symmetric radial electric field generated outside the sphere is given by

$$E_r(r > a) = \frac{Q}{4\pi\epsilon_0\,r^2}. \tag{5.72}$$

It follows that the potential difference between the sphere and infinity—or, more realistically, some large, relatively distant reservoir of charge

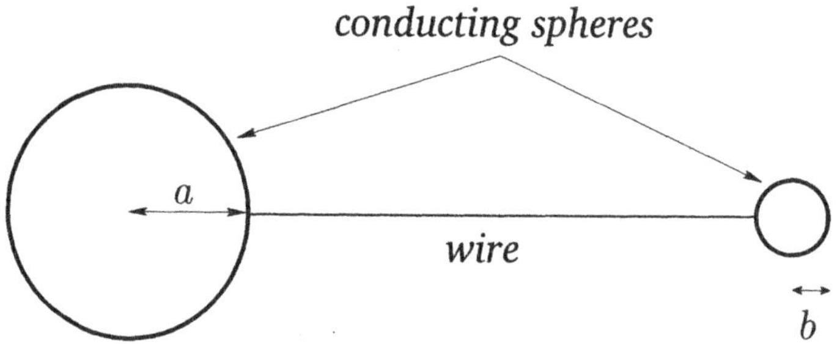

Figure 5.6: *Two conducting spheres connected by a wire.*

such as the Earth—is

$$V = \frac{Q}{4\pi\epsilon_0\, a}. \tag{5.73}$$

Thus, the capacitance of the sphere is

$$C = \frac{Q}{V} = 4\pi\epsilon_0\, a. \tag{5.74}$$

The energy of a spherical capacitor when it carries a charge Q is again given by $(1/2)\, C\, V^2$. It can easily be demonstrated that this is really the energy contained in the electric field surrounding the capacitor.

Suppose that we have two spheres of radii a and b, respectively, which are connected by a *long* electric wire—see Figure 5.6. The wire allows charge to move back and forth between the spheres until they reach the same potential (with respect to infinity). Let Q_a be the charge on the first sphere, and Q_b the charge on the second sphere. Of course, the total charge $Q_0 = Q_a + Q_b$ carried by the two spheres is a conserved quantity. It follows from Equation (5.73) that

$$\frac{Q_a}{Q_0} = \frac{a}{a+b}, \tag{5.75}$$

$$\frac{Q_b}{Q_0} = \frac{b}{a+b}. \tag{5.76}$$

Note that if one sphere is much smaller than the other one, *e.g.*, $b \ll a$, then the large sphere grabs most of the charge: *i.e.*,

$$\frac{Q_a}{Q_b} \simeq \frac{a}{b} \gg 1. \tag{5.77}$$

The ratio of the electric fields generated just above the surfaces of the two spheres follows from Equations (5.72) and (5.77):

$$\frac{E_b}{E_a} \simeq \frac{a}{b}. \tag{5.78}$$

Note that if $b \ll a$ then the field just above the smaller sphere is far larger than that above the larger sphere. Equation (5.78) is a simple example of a far more general rule: *i.e.*, the electric field above some point on the surface of a conductor is inversely proportional to the local radius of curvature of the surface.

It is clear that if we wish to store significant amounts of charge on a conductor then the surface of the conductor must be made as *smooth* as possible. Any sharp spikes on the surface will inevitably have comparatively small radii of curvature. Intense local electric fields are thus generated around such spikes. These fields can easily exceed the critical field for the breakdown of air, leading to sparking and the eventual loss of the charge on the conductor. Sparking can also be very destructive, because the associated electric currents flow through very localized regions, giving rise to intense ohmic heating.

As a final example, consider two coaxial conducting cylinders of radii a and b, where $a < b$. Suppose that the charge per unit length carried by the outer and inner cylinders is $+\lambda$ and $-\lambda$, respectively. We can safely assume that $\mathbf{E} = E_r(r)\,\mathbf{e}_r$, by symmetry (adopting standard cylindrical polar coordinates). Let us apply Gauss' law to a cylindrical surface of radius r, coaxial with the conductors, and of length l. For $a < r < b$, we find that

$$2\pi\, r\, l\, E_r(r) = \frac{\lambda\, l}{\epsilon_0}, \tag{5.79}$$

so that

$$E_r = \frac{\lambda}{2\pi\epsilon_0\, r} \tag{5.80}$$

for $a < r < b$. It is fairly obvious that $E_r = 0$ if r is not in the range a to b. The potential difference between the inner and outer cylinders is

$$V = -\int_{\text{outer}}^{\text{inner}} \mathbf{E}\cdot d\mathbf{l} = \int_{\text{inner}}^{\text{outer}} \mathbf{E}\cdot d\mathbf{l}$$

$$= \int_a^b E_r\, dr = \frac{\lambda}{2\pi\epsilon_0}\int_a^b \frac{dr}{r}, \tag{5.81}$$

so

$$V = \frac{\lambda}{2\pi\epsilon_0} \ln \frac{b}{a}. \tag{5.82}$$

Thus, the capacitance per unit length of the two cylinders is

$$C = \frac{\lambda}{V} = \frac{2\pi\epsilon_0}{\ln b/a}. \tag{5.83}$$

5.7 POISSON'S EQUATION

Now, we know that in a steady-state we can write

$$\mathbf{E} = -\nabla\phi, \tag{5.84}$$

with the scalar potential satisfying Poisson's equation:

$$\nabla^2\phi = -\frac{\rho}{\epsilon_0}. \tag{5.85}$$

We even know the general solution to this equation:

$$\phi(\mathbf{r}) = \frac{1}{4\pi\epsilon_0} \int \frac{\rho(\mathbf{r}')}{|\mathbf{r} - \mathbf{r}'|} \, d^3\mathbf{r}'. \tag{5.86}$$

So, what else is there to say about Poisson's equation? Well, consider a positive (say) point charge in the vicinity of an uncharged, insulated, conducting sphere. The charge attracts negative charges to the near side of the sphere, and repels positive charges to the far side. The surface charge distribution induced on the sphere is such that the surface is maintained at a constant electrical potential. We now have a problem. We cannot use formula (5.86) to work out the potential $\phi(\mathbf{r})$ around the sphere, since we do not know beforehand how the charges induced on its conducting surface are distributed. The only things which we know about the surface are that it is an equipotential, and carries zero net charge. Clearly, the solution (5.86) to Poisson's equation is completely useless in the presence of conducting surfaces. Let us now try to develop some techniques for solving Poisson's equation which allow us to solve real problems (which invariably involve conductors).

5.8 THE UNIQUENESS THEOREM

We have already seen the great value of the uniqueness theorem for Poisson's equation (or Laplace's equation) in our discussion of the Helmholtz theorem (see Section 3.11). Let us now examine the uniqueness theorem in detail.

Consider a volume V bounded by some surface S—see Figure 5.7. Suppose that we are given the charge density ρ throughout V, and the value of the scalar potential ϕ_S on S. Is this sufficient information to uniquely specify the scalar potential throughout V? Suppose, for the sake of argument, that the solution is not unique. Let there be two different potentials ϕ_1 and ϕ_2 which satisfy

$$\nabla^2\phi_1 = -\frac{\rho}{\epsilon_0}, \tag{5.87}$$

$$\nabla^2\phi_2 = -\frac{\rho}{\epsilon_0} \tag{5.88}$$

throughout V, and

$$\phi_1 = \phi_S, \tag{5.89}$$

$$\phi_2 = \phi_S \tag{5.90}$$

on S. We can form the difference between these two potentials:

$$\phi_3 = \phi_1 - \phi_2. \tag{5.91}$$

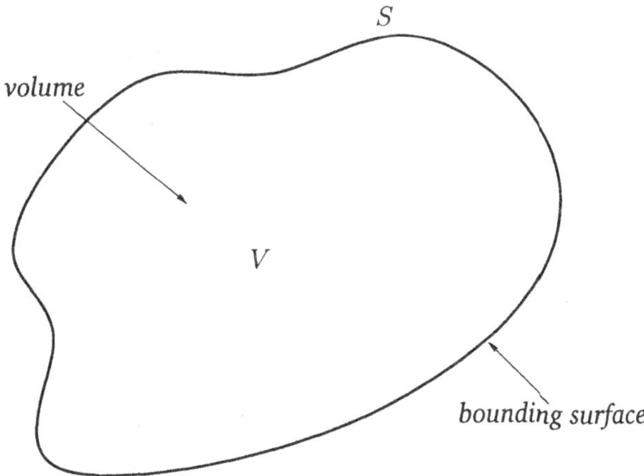

Figure 5.7: *The first uniqueness theorem.*

The potential ϕ_3 clearly satisfies

$$\nabla^2\phi_3 = 0 \tag{5.92}$$

throughout V, and

$$\phi_3 = 0 \tag{5.93}$$

on S.

Now, according to vector field theory,

$$\nabla \cdot (\phi_3 \, \nabla\phi_3) \equiv (\nabla\phi_3)^2 + \phi_3\nabla^2\phi_3. \tag{5.94}$$

Thus, using Gauss' theorem,

$$\int_V \left[(\nabla\phi_3)^2 + \phi_3\nabla^2\phi_3\right] dV = \oint_S \phi_3\nabla\phi_3 \cdot d\mathbf{S}. \tag{5.95}$$

But, $\nabla^2\phi_3 = 0$ throughout V, and $\phi_3 = 0$ on S, so the above equation reduces to

$$\int_V (\nabla\phi_3)^2 \, dV = 0. \tag{5.96}$$

Note that $(\nabla\phi_3)^2$ is a *positive definite* quantity. The only way in which the volume integral of a positive definite quantity can be zero is if that quantity itself is zero throughout the volume. This is not necessarily the case for a non-positive definite quantity: we could have positive and negative contributions from various regions inside the volume which cancel one another out. Thus, since $(\nabla\phi_3)^2$ is positive definite, it follows that

$$\phi_3 = \text{constant} \tag{5.97}$$

throughout V. However, we know that $\phi_3 = 0$ on S, so we get

$$\phi_3 = 0 \tag{5.98}$$

throughout V. In other words,

$$\phi_1 = \phi_2 \tag{5.99}$$

throughout V and on S. Our initial assumption that ϕ_1 and ϕ_2 are two different solutions of Poisson's equation, satisfying the same boundary conditions, turns out to be incorrect. Hence, the solution is unique.

The fact that the solutions to Poisson's equation are unique is very useful. It means that if we find a solution to this equation—no matter

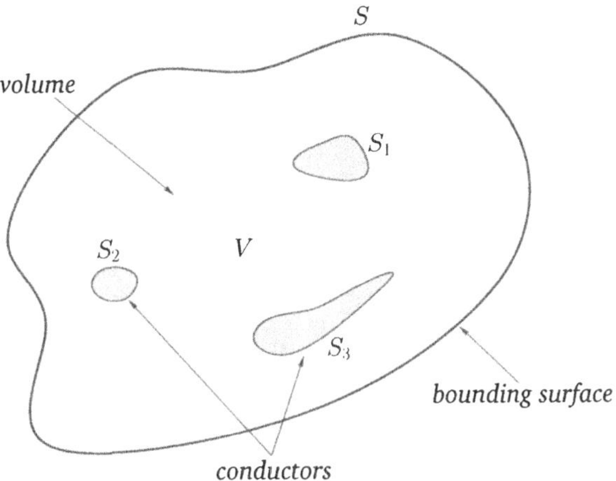

Figure 5.8: *The second uniqueness theorem.*

how contrived the derivation—then this is the only possible solution. One immediate use of the uniqueness theorem is to prove that the electric field inside an empty cavity situated within a conductor is zero. Recall that our previous proof of this was rather involved, and was also not particularly rigorous (see Section 5.4). Now, we know that the interior surface of the conductor is at some constant potential ϕ_0, say. So, we have $\phi = \phi_0$ on the boundary of the cavity, and $\nabla^2\phi = 0$ inside the cavity (since it contains no charges). One rather obvious solution to this problem is $\phi = \phi_0$ throughout the cavity. Since the solutions to Poisson's equation are unique, this is the *only* solution. Thus,

$$\mathbf{E} = -\nabla\phi = -\nabla\phi_0 = \mathbf{0} \tag{5.100}$$

inside the cavity.

Suppose that some volume V contains a number of conductors—see Figure 5.8. We know that the surface of each conductor is an equipotential, but, in general, we do not know the potential of a given conductor (unless we are specifically told that the conductor is earthed, *etc.*). However, if the conductors are insulated then it is plausible that we might know the charge on each conductor. Suppose that there are N conductors, each carrying a known charge Q_i ($i = 1$ to N), and suppose that the region V containing these conductors is filled by a known charge density ρ, and bounded by some surface S which is either infinity or an enclosing conductor. Is this sufficient information to uniquely specify the electric field throughout V?

Well, suppose that it is not sufficient information, so that there are two different fields $\mathbf{E}_1$ and $\mathbf{E}_2$ which satisfy

$$\nabla \cdot \mathbf{E}_1 = \frac{\rho}{\epsilon_0}, \tag{5.101}$$

$$\nabla \cdot \mathbf{E}_2 = \frac{\rho}{\epsilon_0} \tag{5.102}$$

throughout V, with

$$\oint_{S_i} \mathbf{E}_1 \cdot d\mathbf{S}_i = \frac{Q_i}{\epsilon_0}, \tag{5.103}$$

$$\oint_{S_i} \mathbf{E}_2 \cdot d\mathbf{S}_i = \frac{Q_i}{\epsilon_0} \tag{5.104}$$

on the surface of the ith conductor, and, finally,

$$\oint_S \mathbf{E}_1 \cdot d\mathbf{S} = \frac{Q_{\text{total}}}{\epsilon_0}, \tag{5.105}$$

$$\oint_S \mathbf{E}_2 \cdot d\mathbf{S} = \frac{Q_{\text{total}}}{\epsilon_0} \tag{5.106}$$

over the bounding surface, where

$$Q_{\text{total}} = \sum_{i=1}^{N} Q_i + \int_V \rho \, dV \tag{5.107}$$

is the total charge contained in volume V.

Let us form the difference field

$$\mathbf{E}_3 = \mathbf{E}_1 - \mathbf{E}_2. \tag{5.108}$$

It is clear that

$$\nabla \cdot \mathbf{E}_3 = 0 \tag{5.109}$$

throughout V, and

$$\oint_{S_i} \mathbf{E}_3 \cdot d\mathbf{S}_i = 0 \tag{5.110}$$

for all i, with

$$\oint_S \mathbf{E}_3 \cdot d\mathbf{S} = 0. \tag{5.111}$$

Now, we know that each conductor is at a constant potential, so if

$$\mathbf{E}_3 = -\nabla\phi_3, \tag{5.112}$$

then ϕ_3 is a constant on the surface of each conductor. Furthermore, if the outer surface S is infinity then $\phi_1 = \phi_2 = \phi_3 = 0$ on this surface. On the other hand, if the outer surface is an enclosing conductor then ϕ_3 is a constant on it. Either way, ϕ_3 is constant on S.

Consider the vector identity

$$\nabla\cdot(\phi_3\,\mathbf{E}_3) \equiv \phi_3\,\nabla\cdot\mathbf{E}_3 + \mathbf{E}_3\cdot\nabla\phi_3. \tag{5.113}$$

We have $\nabla\cdot\mathbf{E}_3 = 0$ throughout V, and $\nabla\phi_3 = -\mathbf{E}_3$, so the above identity reduces to

$$\nabla\cdot(\phi_3\,\mathbf{E}_3) = -E_3^2 \tag{5.114}$$

throughout V. Integrating over V, and making use of Gauss' theorem, yields

$$\int_V E_3^2\,dV = \sum_{i=1}^{N}\oint_{S_i}\phi_3\,\mathbf{E}_3\cdot d\mathbf{S}_i - \oint_S \phi_3\,\mathbf{E}_3\cdot d\mathbf{S}. \tag{5.115}$$

However, ϕ_3 is a constant on the surfaces S_i and S. So, making use of Equations (5.110) and (5.111), we obtain

$$\int_V E_3^2\,dV = 0. \tag{5.116}$$

Of course, E_3^2 is a positive definite quantity, so the above relation implies that

$$\mathbf{E}_3 = \mathbf{0} \tag{5.117}$$

throughout V: *i.e.*, the fields $\mathbf{E}_1$ and $\mathbf{E}_2$ are identical throughout V. Hence, the solution is unique.

For a general electrostatic problem involving charges and conductors, it is clear that if we are given either the potential at the surface of each conductor or the charge carried by each conductor (plus the charge density throughout the volume, *etc.*) then we can uniquely determine the electric field. There are many other uniqueness theorems which generalize this result still further: *e.g.*, we could be given the potentials on the surfaces of some of the conductors, and the charges on the surfaces of the others, and the solution would still be unique.

At this point, it is worth noting that there are also uniqueness theorems associated with magnetostatics. For instance, if the current density, **j**, is specified throughout some volume V, and either the magnetic field, **B**, or the vector potential, **A**, is specified on the bounding surface S, then the magnetic field is uniquely determined throughout V and on S. The proof of this proposition proceeds along the usual lines. Suppose that the magnetic field is not uniquely determined. In other words, suppose there are two different magnetic fields, $\mathbf{B}_1$ and $\mathbf{B}_2$, satisfying

$$\nabla \times \mathbf{B}_1 = \mu_0 \mathbf{j}, \tag{5.118}$$

$$\nabla \times \mathbf{B}_2 = \mu_0 \mathbf{j}, \tag{5.119}$$

throughout V. Suppose, further, that either $\mathbf{B}_1 = \mathbf{B}_2 = \mathbf{B}_S$ or $\mathbf{A}_1 = \mathbf{A}_2 = \mathbf{A}_S$ on S. Forming the difference field, $\mathbf{B}_3 = \mathbf{B}_1 - \mathbf{B}_2$, we have

$$\nabla \times \mathbf{B}_3 = \mathbf{0} \tag{5.120}$$

throughout V, and either $\mathbf{B}_3 = \mathbf{0}$ or $\mathbf{A}_3 = \mathbf{0}$ on S. Now, according to vector field theory,

$$\int_V \left[(\nabla \times \mathbf{U})^2 - \mathbf{U} \cdot \nabla \times \nabla \times \mathbf{U}\right] dV \equiv \oint_S \mathbf{U} \times (\nabla \times \mathbf{U}) \cdot d\mathbf{S}. \tag{5.121}$$

Setting $\mathbf{U} = \mathbf{A}_3$, and using $\mathbf{B}_3 = \nabla \times \mathbf{A}_3$ and Equation (5.120), we obtain

$$\int_V B_3^2 \, dV = \oint_S \mathbf{A}_3 \times \mathbf{B}_3 \cdot d\mathbf{S}. \tag{5.122}$$

However, we know that either $\mathbf{B}_3$ or $\mathbf{A}_3$ is zero on S. Hence, we get

$$\int_V B_3^2 \, dV = 0. \tag{5.123}$$

Since B_3^2 is positive definite, the only way in which the above equation can be satisfied is if B_3 is zero throughout V. Hence, $\mathbf{B}_1 = \mathbf{B}_2$ throughout V, and the solution is unique.

5.9 ONE-DIMENSIONAL SOLUTIONS OF POISSON'S EQUATION

So, how do we actually solve Poisson's equation,

$$\frac{\partial^2 \phi}{\partial x^2} + \frac{\partial^2 \phi}{\partial y^2} + \frac{\partial^2 \phi}{\partial z^2} = -\frac{\rho(x, y, z)}{\epsilon_0}, \tag{5.124}$$

in practice? In general, the answer is that we use a computer. However, there are a few situations, possessing a high degree of symmetry, where it is possible to find analytic solutions. Let us discuss some of these situations.

Suppose, first of all, that there is no variation of quantities in (say) the y- and z-directions. In this case, Poisson's equation reduces to an ordinary differential equation in x, the solution of which is relatively straightforward. Consider, for instance, a *vacuum diode*, in which electrons are emitted from a hot cathode and accelerated toward an anode, which is held at a large positive potential V with respect to the cathode. We can think of this as an essentially one-dimensional problem. Suppose that the cathode is at $x = 0$ and the anode at $x = d$. Poisson's equation takes the form

$$\frac{d^2\phi}{dx^2} = -\frac{\rho(x)}{\epsilon_0}, \tag{5.125}$$

where $\phi(x)$ satisfies the boundary conditions $\phi(0) = 0$ and $\phi(d) = V$. By energy conservation, an electron emitted from rest at the cathode has an x-velocity $v(x)$ which satisfies

$$\frac{1}{2}\, m_e\, v^2(x) - e\,\phi(x) = 0. \tag{5.126}$$

Here, m_e and $-e$ are the mass and charge of an electron, respectively. Finally, in a steady-state, the electric current I (between the anode and cathode) is independent of x (otherwise, charge will continually build up at some points). In fact,

$$I = -\rho(x)\, v(x)\, A, \tag{5.127}$$

where A is the cross-sectional area of the diode. The previous three equations can be combined to give

$$\frac{d^2\phi}{dx^2} = \frac{I}{\epsilon_0\, A}\left(\frac{m_e}{2\,e}\right)^{1/2}\phi^{-1/2}. \tag{5.128}$$

The solution of the above equation which satisfies the boundary conditions is

$$\phi(x) = V\left(\frac{x}{d}\right)^{4/3}, \tag{5.129}$$

with

$$I = \frac{4}{9}\frac{\epsilon_0\, A}{d^2}\left(\frac{2\,e}{m_e}\right)^{1/2} V^{3/2}. \tag{5.130}$$

This relationship between the current and the voltage in a vacuum diode is called the *Child-Langmuir law*.

Let us now consider the solution of Poisson's equation in more than one dimension.

5.10 THE METHOD OF IMAGES

Suppose that we have a point charge q held a distance d from an infinite, grounded, conducting plate—see Figure 5.9. Let the plate lie in the x-y plane, and suppose that the point charge is located at coordinates $(0, 0, d)$. What is the scalar potential generated in the region above the plate? This is not a simple question, because the point charge induces surface charges on the plate, and we do not know beforehand how these charges are distributed.

Well, what do we know in this problem? We know that the conducting plate is an equipotential surface. In fact, the potential of the plate is zero, since it is grounded. We also know that the potential at infinity is zero (this is our usual boundary condition for the scalar potential). Thus, we need to solve Poisson's equation in the region $z > 0$, with a single point charge q at position (0, 0, d), subject to the boundary conditions

$$\phi(x, y, 0) = 0, \tag{5.131}$$

and

$$\phi(x, y, z) \to 0 \qquad \text{as } x^2 + y^2 + z^2 \to \infty. \tag{5.132}$$

Let us forget about the real problem, for a moment, and concentrate on a slightly different one. We refer to this as the *analog problem*—see Figure 5.9. In the analog problem, we have a charge q located at $(0, 0, d)$ and a charge $-q$ located at (0, 0, -d), with no conductors present. We can easily find the scalar potential for this problem, since we know where all the charges are located. We get

$$\phi_{\text{analog}}(x, y, z) = \frac{1}{4\pi\epsilon_0}\left\{\frac{q}{\sqrt{x^2 + y^2 + (z - d)^2}} - \frac{q}{\sqrt{x^2 + y^2 + (z + d)^2}}\right\}. \tag{5.133}$$

Note, however, that

$$\phi_{\text{analog}}(x, y, 0) = 0, \tag{5.134}$$

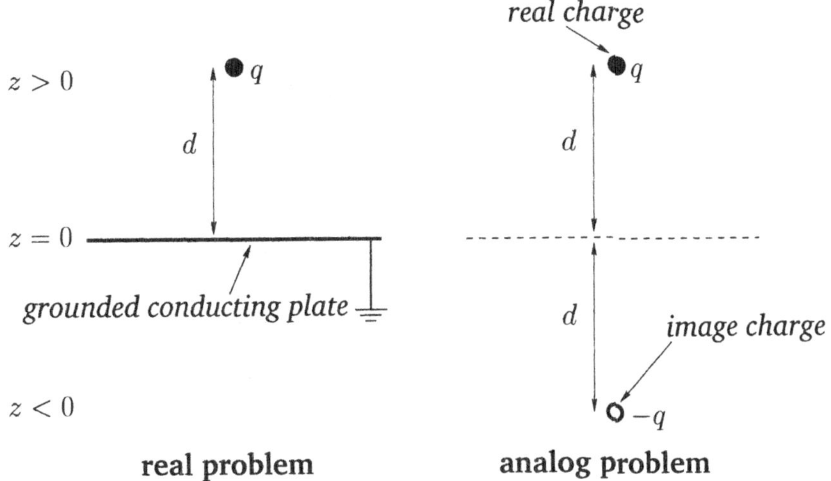

Figure 5.9: *The method of images for a charge and a grounded conducting plane.*

and

$$\phi_{\text{analog}}(x, y, z) \to 0 \qquad \text{as } x^2 + y^2 + z^2 \to \infty. \tag{5.135}$$

Moreover, in the region $z > 0$, ϕ_{analog} satisfies Poisson's equation for a point charge q located at $(0, 0, d)$. Thus, in this region, ϕ_{analog} is a solution to the problem posed earlier. Now, the uniqueness theorem tells us that there is only *one* solution to Poisson's equation which satisfies a given well-posed set of boundary conditions. So, ϕ_{analog} must be the correct potential in the region $z > 0$. Of course, ϕ_{analog} is completely wrong in the region $z < 0$. We know this because the grounded plate shields the region $z < 0$ from the point charge, so that $\phi = 0$ in this region. Note that we are leaning pretty heavily on the uniqueness theorem here! Without this theorem, it would be hard to convince a skeptical person that $\phi = \phi_{\text{analog}}$ is the correct solution in the region $z > 0$.

Now that we have found the potential in the region $z > 0$, we can easily work out the distribution of charges induced on the conducting plate. We already know that the relation between the electric field immediately above a conducting surface and the density of charge on the surface is

$$E_{\perp} = \frac{\sigma}{\epsilon_0}. \tag{5.136}$$

In this case,

$$E_\perp(x,y) = E_z(x,y,0_+) = -\frac{\partial\phi(x,y,0_+)}{\partial z} = -\frac{\partial\phi_{\rm analog}(x,y,0_+)}{\partial z}, \tag{5.137}$$

so

$$\sigma(x,y) = -\epsilon_0 \frac{\partial\phi_{\rm analog}(x,y,0_+)}{\partial z}. \tag{5.138}$$

Now, it follows from Equation (5.133) that

$$\frac{\partial\phi_{\rm analog}}{\partial z} = \frac{q}{4\pi\epsilon_0}\left\{\frac{-(z-d)}{[x^2+y^2+(z-d)^2]^{3/2}} + \frac{(z+d)}{[x^2+y^2+(z+d)^2]^{3/2}}\right\}, \tag{5.139}$$

so

$$\sigma(x,y) = -\frac{q\,d}{2\pi\,(x^2+y^2+d^2)^{3/2}}. \tag{5.140}$$

Clearly, the charge induced on the plate has the opposite sign to the point charge. The charge density on the plate is also symmetric about the z-axis, and is largest where the plate is closest to the point charge. The total charge induced on the plate is

$$Q = \int_{x-y\ \rm plane} \sigma\, dS, \tag{5.141}$$

which yields

$$Q = -\frac{q\,d}{2\pi}\int_0^\infty \frac{2\pi\, r\, dr}{(r^2+d^2)^{3/2}}, \tag{5.142}$$

where $r^2 = x^2 + y^2$. Thus,

$$Q = -\frac{q\,d}{2}\int_0^\infty \frac{dk}{(k+d^2)^{3/2}} = q\,d\left[\frac{1}{(k+d^2)^{1/2}}\right]_0^\infty = -q. \tag{5.143}$$

So, the total charge induced on the plate is equal and opposite to the point charge which induces it.

As we have just seen, our point charge induces charges of the opposite sign on the conducting plate. This, presumably, gives rise to a force of attraction between the charge and the plate. What is this force? Well, since the potentials, and, hence, the electric fields, in the vicinity of the point charge are the same in the real and analog problems, the forces on this charge must be the same as well. In the analog problem, there are two charges $\pm q$ a net distance $2\,d$ apart. The force on the charge at position (0, 0, d) (*i.e.*, the real charge) is

$$\mathbf{f} = -\frac{q^2}{16\pi\epsilon_0\, d^2}\, \mathbf{e}_z. \tag{5.144}$$

Hence, this is also the force on the charge in the real problem.

What, finally, is the potential energy of the system. For the analog problem this is simply

$$W_{\text{analog}} = -\frac{q^2}{8\pi\epsilon_0\, d}. \tag{5.145}$$

Note that in the analog problem the fields on opposite sides of the conducting plate are mirror images of one another. So are the charges (apart from the change in sign). This is why the technique of replacing conducting surfaces by imaginary charges is called the *method of images*. We know that the potential energy of a set of charges is equivalent to the energy stored in the electric field. Thus,

$$W = \frac{\epsilon_0}{2} \int_{\text{all space}} E^2\, dV. \tag{5.146}$$

Moreover, as we just mentioned, in the analog problem, the fields on either side of the x-y plane are mirror images of one another, so that $E^2(x, y, -z) = E^2(x, y, z)$. It follows that

$$W_{\text{analog}} = 2\,\frac{\epsilon_0}{2} \int_{z>0} E_{\text{analog}}^2\, dV. \tag{5.147}$$

Now, in the real problem

$$\mathbf{E} = \begin{cases} \mathbf{E}_{\text{analog}} & \text{for } z > 0 \\ \mathbf{0} & \text{for } z < 0 \end{cases}. \tag{5.148}$$

So,

$$W = \frac{\epsilon_0}{2} \int_{z>0} E^2\, dV = \frac{\epsilon_0}{2} \int_{z>0} E_{\text{analog}}^2\, dV = \frac{1}{2}\, W_{\text{analog}}, \tag{5.149}$$

giving

$$W = -\frac{q^2}{16\pi\epsilon_0\, d}. \tag{5.150}$$

There is another method by which we can obtain the above result. Suppose that the charge is gradually moved toward the plate along the z-axis, starting from infinity, until it reaches position (0, 0, d). How much work is required to achieve this? We know that the force of attraction acting on the charge is

$$f_z = -\frac{q^2}{16\pi\epsilon_0\, z^2}. \tag{5.151}$$

Thus, the work required to move this charge by dz is

$$dW = -f_z\, dz = \frac{q^2}{16\pi\epsilon_0\, z^2}\, dz. \tag{5.152}$$

So, the total work needed to move the charge from $z = \infty$ to $z = d$ is

$$W = \frac{1}{4\pi\epsilon_0}\int_\infty^d \frac{q^2}{4\,z^2}\, dz = \frac{1}{4\pi\epsilon_0}\left[-\frac{q^2}{4\,z}\right]_\infty^d = -\frac{q^2}{16\pi\epsilon_0\, d}. \tag{5.153}$$

Of course, this work is equivalent to the potential energy (5.150), and is, in turn, the same as the energy contained in the electric field.

As a second example of the method of images, consider a *grounded* conducting sphere of radius a centered on the origin. Suppose that a charge q is placed outside the sphere at $(b, 0, 0)$, where $b > a$—see Figure 5.10. What is the force of attraction between the sphere and

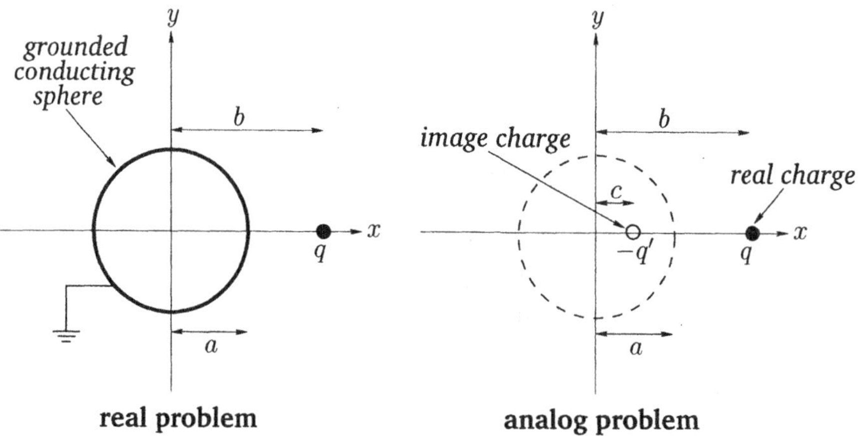

Figure 5.10: *The method of images for a charge and a grounded conducting sphere.*

the charge? In this case, we proceed by considering an analog problem in which the sphere is replaced by an image charge $-q'$ placed somewhere on the x-axis at $(c, 0, 0)$—see Figure 5.10. The electric potential throughout space in the analog problem is simply

$$\phi(x, y, z) = \frac{q}{4\pi\epsilon_0} \frac{1}{[(x-b)^2 + y^2 + z^2]^{1/2}} - \frac{q'}{4\pi\epsilon_0} \frac{1}{[(x-c)^2 + y^2 + z^2]^{1/2}}. \tag{5.154}$$

Now, the image charge must be chosen so as to make the surface $\phi = 0$ correspond to the surface of the sphere. Setting the above expression to zero, and performing a little algebra, we find that the $\phi = 0$ surface corresponds to

$$x^2 + \frac{2(c - \lambda b)}{\lambda - 1} x + y^2 + z^2 = \frac{c^2 - \lambda b^2}{\lambda - 1}, \tag{5.155}$$

where $\lambda = q'^2/q^2$. Of course, the surface of the sphere satisfies

$$x^2 + y^2 + z^2 = a^2. \tag{5.156}$$

The above two equations can be made identical by setting $\lambda = c/b$ and $a^2 = \lambda b^2$, or

$$q' = \frac{a}{b} q, \tag{5.157}$$

and

$$c = \frac{a^2}{b}. \tag{5.158}$$

According to the uniqueness theorem, the potential in the analog problem is now identical with that in the real problem in the region outside the sphere. (Of course, in the real problem, the potential inside the sphere is zero.) Hence, the force of attraction between the sphere and the original charge in the real problem is the same as the force of attraction between the image charge and the real charge in the analog problem. It follows that

$$f = \frac{q q'}{4\pi\epsilon_0 (b-c)^2} = \frac{q^2}{4\pi\epsilon_0} \frac{a b}{(b^2 - a^2)^2}. \tag{5.159}$$

What is the total charge induced on the grounded conducting sphere? Well, according to Gauss' law, the flux of the electric field out of a spherical Gaussian surface lying just outside the surface of the conducting sphere is equal to the enclosed charge divided by ϵ_0. In the real problem, the enclosed charge is the net charge induced on the surface of the sphere. In the analog problem, the enclosed charge is simply $-q'$. However, the electric fields outside the conducting sphere are identical in the real and analog problems. Hence, from Gauss' law, the charge enclosed by the Gaussian surface must also be the same in both problems. We thus conclude that the net charge induced on the surface of the conducting sphere is

$$-q' = -\frac{a}{b}\,q. \tag{5.160}$$

As another example of the method of images, consider an insulated *uncharged* conducting sphere of radius a, centered on the origin, in the presence of a charge q placed outside the sphere at $(b, 0, 0)$, where $b > a$—see Figure 5.11. What is the force of attraction between the sphere and the charge? Clearly, this new problem is very similar to the one which we just discussed. The only difference is that the surface of the sphere is now at some *unknown* fixed potential V, and also carries zero net charge. Note that if we add a second image charge q'', located at the origin, to the analog problem pictured in Figure 5.10 then the surface $r = a$ remains an equipotential surface. In fact, the potential of this surface becomes $V = q''/(4\pi\epsilon_0\, a)$. Moreover, the total charge enclosed by the surface is $-q' + q''$. This, of course, is the net charge induced on

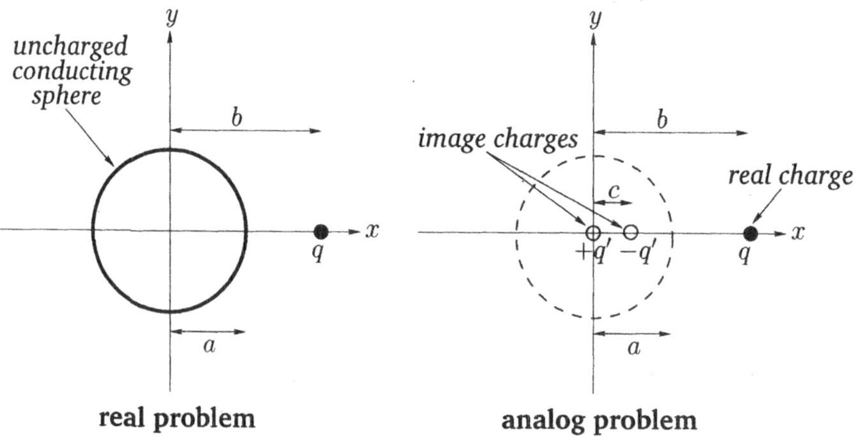

Figure 5.11: *The method of images for a charge and an uncharged conducting sphere.*

the surface of the sphere in the real problem. Hence, we can see that if $q'' = q' = (a/b)\,q$ then zero net charge is induced on the surface of the sphere. Thus, our modified analog problem is now a solution to the problem under discussion, in the region outside the sphere—see Figure 5.11. It follows that the surface of the sphere is at potential

$$V = \frac{q'}{4\pi\epsilon_0\, a} = \frac{q}{4\pi\epsilon_0\, b}. \tag{5.161}$$

Moreover, the force of attraction between the sphere and the original charge in the real problem is the same as the force of attraction between the image charges and the real charge in the analog problem. Hence, the force is given by

$$f = \frac{q\,q'}{4\pi\epsilon_0\,(b-c)^2} - \frac{q\,q'}{4\pi\epsilon_0\,b^2} = \frac{q^2}{4\pi\epsilon_0}\left(\frac{a}{b}\right)^3 \frac{(2\,b^2-a^2)}{(b^2-a^2)^2}. \tag{5.162}$$

As a final example of the method of images, consider two identical, infinitely long, conducting cylinders of radius a which run parallel to the z-axis, and lie a distance $2\,d$ apart. Suppose that one of the conductors is held at potential $+V$, whilst the other is held at potential $-V$—see Figure 5.12. What is the capacitance per unit length of the cylinders?

Consider an analog problem in which the conducting cylinders are replaced by two infinitely long charge lines, of charge per unit length $\pm\lambda$, which run parallel to the z-axis, and lie a distance $2\,p$ apart. Now, the potential in the x-y plane generated by a charge line λ running along the z-axis is

$$\phi(x, y) = -\frac{\lambda}{2\pi\epsilon_0}\,\ln r, \tag{5.163}$$

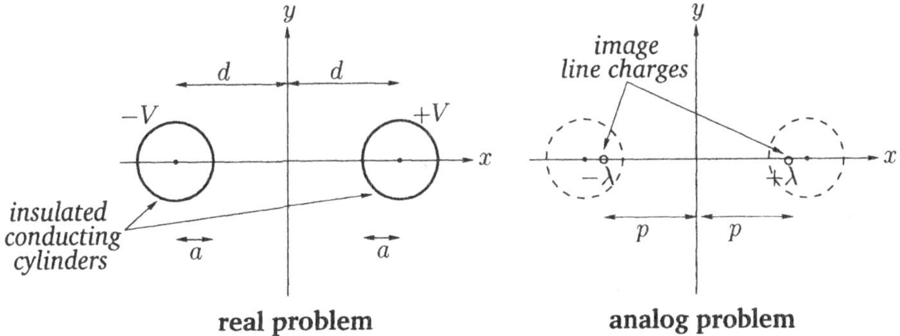

Figure 5.12: *The method of images for two parallel cylindrical conductors.*

where $r = \sqrt{x^2 + y^2}$ is the radial cylindrical polar coordinate. The corresponding electric field is radial, and satisfies

$$E_r(r) = -\frac{\partial \phi}{\partial r} = \frac{\lambda}{2\pi\epsilon_0 \, r}. \tag{5.164}$$

Incidentally, it is easily demonstrated from Gauss' law that this is the correct electric field. Hence, the potential generated by two charge lines $\pm\lambda$ located in the x-y plane at $(\pm p, 0)$, respectively, is

$$\phi(x, y) = \frac{\lambda}{4\pi\epsilon_0} \ln\left[\frac{(x+p)^2 + y^2}{(x-p)^2 + y^2}\right]. \tag{5.165}$$

Suppose that

$$\frac{(x+p)^2 + y^2}{(x-p)^2 + y^2} = \alpha, \tag{5.166}$$

where α is a constant. It follows that

$$x^2 - 2p\,\frac{(\alpha+1)}{(\alpha-1)}\,x + p^2 + y^2 = 0. \tag{5.167}$$

Completing the square, we obtain

$$(x - d)^2 + y^2 = a^2, \tag{5.168}$$

where

$$d = \frac{(\alpha+1)}{(\alpha-1)}\,p, \tag{5.169}$$

and

$$a^2 = d^2 - p^2. \tag{5.170}$$

Of course, Equation (5.168) is the equation of a cylindrical surface of radius a centered on $(d, 0)$. Moreover, it follows from Equations (5.165) and (5.166) that this surface lies at the constant potential

$$V = \frac{\lambda}{4\pi\epsilon_0} \ln \alpha. \tag{5.171}$$

Finally, it is easily demonstrated that the equipotential $\phi = -V$ corresponds to a cylindrical surface of radius a centered on $(-d, 0)$. Hence, we can make the analog problem match the real problem in the region outside the cylinders by choosing

$$\alpha = \frac{d+p}{d-p} = \frac{d+\sqrt{d^2-a^2}}{d-\sqrt{d^2-a^2}}. \tag{5.172}$$

Thus, we obtain

$$V = \frac{\lambda}{4\pi\epsilon_0}\ln\left(\frac{d+\sqrt{d^2-a^2}}{d-\sqrt{d^2-a^2}}\right). \tag{5.173}$$

Now, it follows from Gauss' law, and the fact that the electric fields in the real and analog problems are identical outside the cylinders, that the charge per unit length stored on the surfaces of the two cylinders is $\pm\lambda$. Moreover, the voltage difference between the cylinders is $2V$. Hence, the capacitance per unit length of the cylinders is $C = \lambda/(2V)$, yielding

$$C = 2\pi\epsilon_0 \bigg/ \ln\left(\frac{d+\sqrt{d^2-a^2}}{d-\sqrt{d^2-a^2}}\right). \tag{5.174}$$

This expression simplifies to give

$$C = \pi\epsilon_0 \bigg/ \ln\left(\frac{d}{a}+\sqrt{\frac{d^2}{a^2}-1}\right), \tag{5.175}$$

which can also be written

$$C = \frac{\pi\epsilon_0}{\cosh^{-1}(d/a)}, \tag{5.176}$$

since $\cosh^{-1} x \equiv \ln(x+\sqrt{x^2-1})$.

5.11 COMPLEX ANALYSIS

Let us now investigate another trick for solving Poisson's equation (actually it only solves Laplace's equation). Unfortunately, this method only works in *two dimensions*.

The complex variable is conventionally written

$$z = x + i\,y, \tag{5.177}$$

where x and y are both real, and are identified with the corresponding Cartesian coordinates. (Incidentally, z should not be confused with a z-coordinate: this is a strictly two-dimensional discussion.) We can write functions $F(z)$ of the complex variable just like we would write functions of a real variable. For instance,

$$F(z) = z^2, \tag{5.178}$$

$$F(z) = \frac{1}{z}. \tag{5.179}$$

For a given function, $F(z)$, we can substitute $z = x + i\,y$ and write

$$F(z) = U(x, y) + i\,V(x, y), \tag{5.180}$$

where U and V are two *real* two-dimensional functions. Thus, if

$$F(z) = z^2, \tag{5.181}$$

then

$$F(x + i\,y) = (x + i\,y)^2 = (x^2 - y^2) + 2\,i\,x\,y, \tag{5.182}$$

giving

$$U(x, y) = x^2 - y^2, \tag{5.183}$$

$$V(x, y) = 2\,x\,y. \tag{5.184}$$

We can define the derivative of a complex function in just the same manner as we would define the derivative of a real function: *i.e.*,

$$\frac{dF}{dz} = \lim_{|\delta z| \to \infty} \frac{F(z + \delta z) - F(z)}{\delta z}. \tag{5.185}$$

However, we now have a slight problem. If $F(z)$ is a "well-defined" function (we shall leave it to the mathematicians to specify exactly what being well-defined entails: suffice to say that most functions we can think of are well-defined) then it should not matter from which direction in the complex plane we approach z when taking the limit in Equation (5.185). There are, of course, many different directions we could approach z from, but if we look at a regular complex function, $F(z) = z^2$ (say), then

$$\frac{dF}{dz} = 2\,z \tag{5.186}$$

is perfectly well-defined, and is, therefore, completely independent of the details of how the limit is taken in Equation (5.185).

The fact that Equation (5.185) has to give the same result, no matter from which direction we approach z, means that there are some restrictions on the forms of the functions U and V in Equation (5.180). Suppose that we approach z along the real axis, so that $\delta z = \delta x$. We obtain

$$\begin{aligned} \frac{dF}{dz} &= \lim_{|\delta x|\to 0} \frac{U(x+\delta x, y) + i\,V(x+\delta x, y) - U(x,y) - i\,V(x,y)}{\delta x} \\ &= \frac{\partial U}{\partial x} + i\,\frac{\partial V}{\partial x}. \end{aligned} \tag{5.187}$$

Suppose that we now approach z along the imaginary axis, so that $\delta z = i\,\delta y$. We get

$$\begin{aligned} \frac{dF}{dz} &= \lim_{|\delta y|\to 0} \frac{U(x, y+\delta y) + i\,V(x, y+\delta y) - U(x,y) - i\,V(x,y)}{i\,\delta y} \\ &= -i\,\frac{\partial U}{\partial y} + \frac{\partial V}{\partial y}. \end{aligned} \tag{5.188}$$

But, if $F(z)$ is a well-defined function then its derivative must also be well-defined, which implies that the above two expressions are equivalent. This requires that

$$\frac{\partial U}{\partial x} = \frac{\partial V}{\partial y}, \tag{5.189}$$

$$\frac{\partial V}{\partial x} = -\frac{\partial U}{\partial y}. \tag{5.190}$$

These are called the *Cauchy-Riemann relations*, and are, in fact, sufficient to ensure that all possible ways of taking the limit (5.185) give the same answer.

So far, we have found that a general complex function $F(z)$ can be written

$$F(z) = U(x, y) + i\,V(x, y), \tag{5.191}$$

where $z = x + i y$. If $F(z)$ is well-defined then U and V *automatically* satisfy the Cauchy-Riemann relations. But, what has all of this got to do with electrostatics? Well, we can combine the two Cauchy-Riemann relations to give

$$\frac{\partial^2 U}{\partial x^2} = \frac{\partial}{\partial x}\frac{\partial V}{\partial y} = \frac{\partial}{\partial y}\frac{\partial V}{\partial x} = -\frac{\partial}{\partial y}\frac{\partial U}{\partial y}, \tag{5.192}$$

and

$$\frac{\partial^2 V}{\partial x^2} = -\frac{\partial}{\partial x}\frac{\partial U}{\partial y} = -\frac{\partial}{\partial y}\frac{\partial U}{\partial x} = -\frac{\partial}{\partial y}\frac{\partial V}{\partial y}, \tag{5.193}$$

which reduce to

$$\frac{\partial^2 U}{\partial x^2} + \frac{\partial^2 U}{\partial y^2} = 0, \tag{5.194}$$

$$\frac{\partial^2 V}{\partial x^2} + \frac{\partial^2 V}{\partial y^2} = 0. \tag{5.195}$$

Thus, both U and V *automatically* satisfy Laplace's equation in two dimensions: *i.e.*, both U and V are possible two-dimensional scalar potentials in free space.

Consider the two-dimensional gradients of U and V:

$$\nabla U = \left(\frac{\partial U}{\partial x}, \frac{\partial U}{\partial y}\right), \tag{5.196}$$

$$\nabla V = \left(\frac{\partial V}{\partial x}, \frac{\partial V}{\partial y}\right). \tag{5.197}$$

Now

$$\nabla U \cdot \nabla V = \frac{\partial U}{\partial x}\frac{\partial V}{\partial x} + \frac{\partial U}{\partial y}\frac{\partial V}{\partial y}. \tag{5.198}$$

However, it follows from the Cauchy-Riemann relations that

$$\nabla U \cdot \nabla V = \frac{\partial V}{\partial y}\frac{\partial V}{\partial x} - \frac{\partial V}{\partial x}\frac{\partial V}{\partial y} = 0. \tag{5.199}$$

Thus, the contours of U are everywhere *perpendicular* to the contours of V. It follows that if U maps out the contours of some free-space scalar potential then the contours of V indicate the directions of the associated electric field-lines, and *vice versa*.

For every well-defined complex function, we get two sets of free-space potentials, and the associated electric field-lines. For example, consider the function $F(z) = z^2$, for which

$$U(x, y) = x^2 - y^2, \tag{5.200}$$

$$V(x, y) = 2\,x\,y. \tag{5.201}$$

These are, in fact, the equations of two sets of orthogonal hyperboloids—see Figure 5.13. So, $U(x, y)$ (the solid lines in Figure 5.13) might

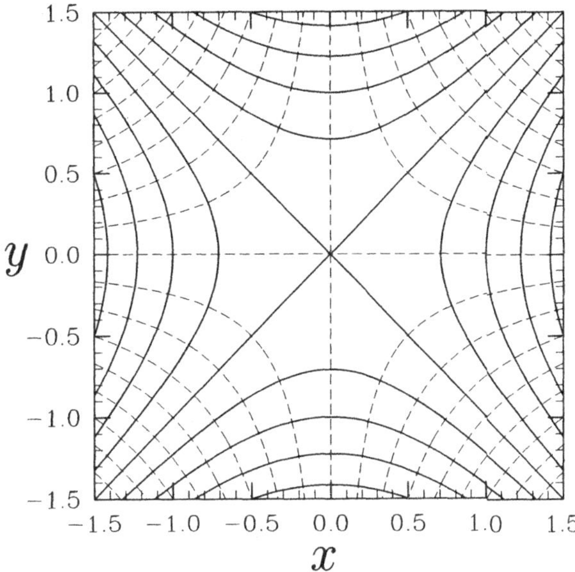

Figure 5.13: *Equally spaced contours of the real (solid lines) and imaginary (dashed lines) parts of* $F(z) = z^2$ *plotted in the complex plane.*

represent the contours of some scalar potential, and $V(x, y)$ (the dashed lines in Figure 5.13) the associated electric field-lines, or *vice versa.* But, how could we actually generate a hyperboloidal potential? This is easy. Consider the contours of U at level ± 1. These could represent the surfaces of four hyperboloid conductors maintained at potentials $\pm \mathcal{V}$, respectively. The scalar potential in the region between these conductors is given by $\mathcal{V}\, U(x, y)$, and the associated electric field-lines follow the contours of $V(x, y)$. Note that

$$E_x = -\frac{\partial \phi}{\partial x} = -\mathcal{V}\,\frac{\partial U}{\partial x} = -2\,\mathcal{V}\,x. \tag{5.202}$$

Thus, the x-component of the electric field is directly proportional to the distance from the x-axis. Likewise, the y-component of the field is directly proportional to the distance from the y-axis. This property can be exploited to make devices (called quadrupole electrostatic lenses) which are useful for focusing charged particle beams.

As a second example, consider the complex function

$$F(z) = z - \frac{c^2}{z}, \tag{5.203}$$

where c is real and positive. Writing $F(z) = U(x, y) + i\,V(x, y)$, we find that

$$U(x, y) = x - \frac{c^2\,x}{x^2 + y^2}, \tag{5.204}$$

$$V(x, y) = y + \frac{c^2\,y}{x^2 + y^2}. \tag{5.205}$$

Far from the origin, $U \to x$, which is the potential of a uniform electric field, of unit amplitude, pointing in the $-x$-direction. Moreover, the locus of $U = 0$ is $x = 0$, and

$$x^2 + y^2 = c^2, \tag{5.206}$$

which corresponds to a circle of radius c centered on the origin. Hence, we conclude that the potential

$$\phi(x, y) = -E_0\,U(x, y) = -E_0\,x + E_0\,c^2\,\frac{x}{x^2 + y^2} \tag{5.207}$$

corresponds to that outside a grounded, infinitely long, conducting cylinder of radius c, co-axial with the z-axis, which is placed in a uniform x-directed electric field of magnitude E_0. The corresponding electric field-lines run along contours of V—see Figure 5.14. Of course, the potential inside the cylinder (*i.e.*, $x^2 + y^2 < c^2$) is zero. Defining standard cylindrical polar coordinates, $r = \sqrt{x^2 + y^2}$ and $\theta = \tan^{-1}(y/x)$, the potential becomes

$$\phi(r, \theta) = -E_0\left(r\,\cos\theta - \frac{c^2\,\cos\theta}{r}\right). \tag{5.208}$$

Hence, the induced charge density on the surface of the cylinder is simply

$$\sigma(\theta) = \epsilon_0\,E_r(c, \theta) = -\epsilon_0\,\frac{\partial\phi(c, \theta)}{\partial r} = 2\,\epsilon_0\,E_0\,\cos\theta. \tag{5.209}$$

Note that zero net charge is induced on the surface. This implies that if the cylinder were insulated and uncharged, rather than being grounded, then the solution would not change.

As a final example, consider the complex function

$$F(z) = z^{1/2}. \tag{5.210}$$

Note that we need a branch-cut in the complex plane in order to make this function single-valued. Suppose that the cut is at $\arg(z) = \pi$, so that

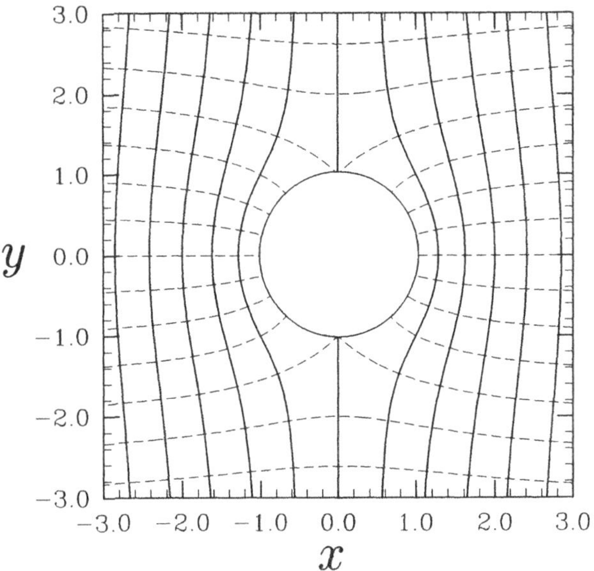

Figure 5.14: *Equally spaced contours of the real (solid lines) and imaginary (dashed lines) parts of* $F(z) = z - 1/z$ *plotted in the complex plane for* $|z| > 1$.

$-\pi \le \arg(z) \le \pi$. Adopting standard cylindrical polar coordinates, it is easily seen that

$$U(r, \theta) = r^{1/2} \cos(\theta/2), \tag{5.211}$$

$$V(r, \theta) = r^{1/2} \sin(\theta/2), \tag{5.212}$$

where $-\pi \le \theta \le \pi$. Now, the locus of $U = 0$ corresponds to $\theta = \pm\pi$. Hence, $U(r, \theta)$ represents the electric potential in the immediate vicinity of an earthed semi-infinite conducting plate occupying the negative x-axis. The corresponding electric field-lines run along contours of $V(r, \theta)$—see Figure 5.15. The surface charge density on the plate is easily obtained from

$$\begin{aligned} \sigma(r) &= \epsilon_0 \left[E_\theta(r, -\pi) - E_\theta(r, \pi)\right] \\ &= -\frac{\epsilon_0}{r}\left[\frac{\partial U(r, -\pi)}{\partial\theta} - \frac{\partial U(r, \pi)}{\partial\theta}\right] \\ &= -\epsilon_0\, r^{-1/2}. \end{aligned} \tag{5.213}$$

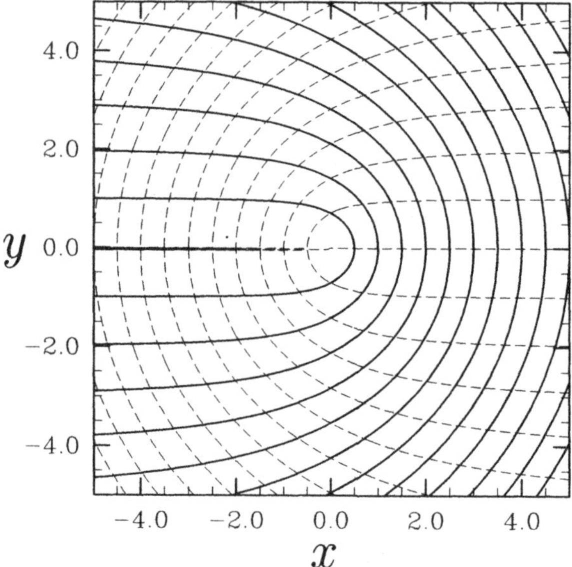

Figure 5.15: *Equally spaced contours of the real (solid lines) and imaginary (dashed lines) parts of* $F(z) = z^{1/2}$ *plotted in the complex plane for* $-\pi \leq \arg(z) \leq \pi$.

5.12 SEPARATION OF VARIABLES

The method of images and complex analysis are two rather elegant techniques for solving Poisson's equation. Unfortunately, they both have an extremely limited range of application. The next technique which we shall discuss—namely, the *separation of variables*—is somewhat messy, but possess a far wider range of application. Let us start by examining a well-known example.

Consider two semi-infinite, grounded, conducting plates lying parallel to the x-z plane, one at $y = 0$, and the other at $y = \pi$—see Figure 5.16. Suppose that the left boundary of the region between the plates, located at $x = 0$, is closed off by an infinite strip which is insulated from the two plates, and maintained at a specified potential $\phi_0(y)$. What is the potential in the region between the plates?

First of all, let us assume that the potential is z-independent, since everything else in the problem possesses this symmetry. This reduces the

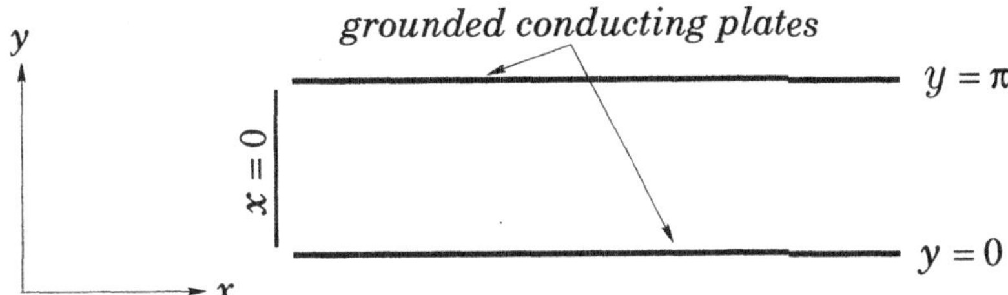

Figure 5.16: *Two semi-infinite grounded conducting plates.*

problem to two dimensions. Poisson's equation is written

$$\frac{\partial^2\phi}{\partial x^2} + \frac{\partial^2\phi}{\partial y^2} = 0 \tag{5.214}$$

in the vacuum region between the conductors. The boundary conditions are

$$\phi(x, 0) = 0, \tag{5.215}$$

$$\phi(x, \pi) = 0 \tag{5.216}$$

for $x > 0$, since the two plates are earthed, plus

$$\phi(0, y) = \phi_0(y) \tag{5.217}$$

for $0 \leq y \leq \pi$, and

$$\phi(x, y) \to 0 \qquad \text{as } x \to \infty. \tag{5.218}$$

The latter boundary condition is our usual one for the scalar potential at infinity.

The central assumption in the separation-of-variables method is that a multidimensional potential can be written as the product of one-dimensional potentials. Hence, in the present case, we would write

$$\phi(x, y) = X(x)\, Y(y). \tag{5.219}$$

The above solution is obviously a very special one, and is, therefore, only likely to satisfy a very small subset of possible boundary conditions. However, it turns out that by adding together lots of different solutions of this form we can match to general boundary conditions.

Substituting (5.219) into (5.214), we obtain

$$Y\,\frac{d^2X}{dx^2} + X\,\frac{d^2Y}{dy^2} = 0. \tag{5.220}$$

Let us now separate the variables: *i.e.*, let us collect all of the x-dependent terms on one side of the equation, and all of the y-dependent terms on the other side. Hence,

$$\frac{1}{X}\frac{d^2X}{dx^2} = -\frac{1}{Y}\frac{d^2Y}{dy^2}. \tag{5.221}$$

This equation has the form

$$f(x) = g(y), \tag{5.222}$$

where f and g are general functions. The only way in which the above equation can be satisfied, for general x and y, is if both sides are equal to the same constant. Thus,

$$\frac{1}{X}\frac{d^2X}{dx^2} = k^2 = -\frac{1}{Y}\frac{d^2Y}{dy^2}. \tag{5.223}$$

The reason why we write k^2, rather than $-k^2$, will become apparent later on. Equation (5.223) separates into two ordinary differential equations:

$$\frac{d^2X}{dx^2} = k^2\,X, \tag{5.224}$$

$$\frac{d^2Y}{dy^2} = -k^2\,Y. \tag{5.225}$$

We know the general solution to these equations:

$$X = A\exp(k\,x) + B\exp(-k\,x), \tag{5.226}$$

$$Y = C\sin(k\,y) + D\cos(k\,y), \tag{5.227}$$

giving

$$\phi(x, y) = [A\exp(k\,x) + B\exp(-k\,x)]\,[C\sin(k\,y) + D\cos(k\,y)]. \tag{5.228}$$

Here, A, B, C, and D are arbitrary constants. The boundary condition (5.218) is automatically satisfied if $A = 0$ and $k > 0$. Note that the choice k^2, instead of $-k^2$, in Equation (5.223) facilitates this by making ϕ either grow or decay monotonically in the x-direction instead of oscillating. The boundary condition (5.215) is automatically satisfied if $D = 0$. The boundary condition (5.216) is satisfied provided that

$$\sin(k\,\pi) = 0, \tag{5.229}$$

which implies that k is a positive integer, n (say). So, our solution reduces to

$$\phi(x, y) = C\,\exp(-n\,x)\,\sin(n\,y), \tag{5.230}$$

where B has been absorbed into C. Note that this solution is only able to satisfy the final boundary condition (5.217) provided that $\phi_0(y)$ is proportional to $\sin(n\,y)$. Thus, at first sight, it would appear that the method of separation of variables only works for a very special subset of boundary conditions. However, this is not the case.

Now comes the clever bit! Since Poisson's equation is *linear*, any linear combination of solutions is also a solution. We can therefore form a more general solution than (5.230) by adding together lots of solutions involving different values of n. Thus,

$$\phi(x, y) = \sum_{n=1}^{\infty} C_n\,\exp(-n\,x)\,\sin(n\,y), \tag{5.231}$$

where the C_n are constants. This solution automatically satisfies the boundary conditions (5.215), (5.216), and (5.218). The final boundary condition (5.217) reduces to

$$\phi(0, y) = \sum_{n=1}^{\infty} C_n\,\sin(n\,y) = \phi_0(y). \tag{5.232}$$

But, what choice of the C_n fits an arbitrary function $\phi_0(y)$? To answer this question, we can make use of two very useful properties of the functions $\sin(n\,y)$. Namely, that they are mutually *orthogonal*, and form a *complete set*. The orthogonality property of these functions manifests itself through the relation

$$\int_0^{\pi} \sin(n\,y)\,\sin(n'\,y)\,dy = \frac{\pi}{2}\,\delta_{nn'}, \tag{5.233}$$

where $\delta_{nn'}$—which is equal to 1 if $n = n'$, and 0, otherwise—is called a *Kroenecker delta function*. The completeness property of sine functions means that any general function $\phi_0(y)$ can always be adequately represented as a weighted sum of sine functions with various different n values. Multiplying both sides of Equation (5.232) by $\sin(n'y)$, and integrating over y, we obtain

$$\sum_{n=1}^{\infty} C_n \int_0^{\pi} \sin(n\,y)\,\sin(n'\,y)\,dy = \int_0^{\pi} \phi_0(y)\,\sin(n'\,y)\,dy. \tag{5.234}$$

The orthogonality relation yields

$$\frac{\pi}{2}\sum_{n=1}^{\infty} C_n\,\delta_{nn'} = \frac{\pi}{2}\,C_{n'} = \int_0^{\pi} \phi_0(y)\,\sin(n'\,y)\,dy, \tag{5.235}$$

so

$$C_n = \frac{2}{\pi}\int_0^{\pi} \phi_0(y)\,\sin(n\,y)\,dy. \tag{5.236}$$

Thus, we now have a general solution to the problem for any driving potential $\phi_0(y)$.

If the potential $\phi_0(y)$ is constant then

$$C_n = \frac{2\,\phi_0}{\pi}\int_0^{\pi} \sin(n\,y)\,dy = \frac{2\,\phi_0}{n\,\pi}\,[1-\cos(n\,\pi)], \tag{5.237}$$

giving

$$C_n = 0 \tag{5.238}$$

for even n, and

$$C_n = \frac{4\,\phi_0}{n\,\pi} \tag{5.239}$$

for odd n. Thus,

$$\phi(x,y) = \frac{4\,\phi_0}{\pi}\sum_{n=1,3,5,\cdots} \frac{\exp(-n\,x)\,\sin(n\,y)}{n}. \tag{5.240}$$

This potential is plotted in Figure 5.17.

In the above problem, we wrote the potential as the product of one-dimensional functions. Some of these functions grew and decayed monotonically (*i.e.*, the exponential functions), and the others oscillated (*i.e.*, the sinusoidal functions). The success of the separation-of-variables method depends crucially on the orthogonality and completeness of the

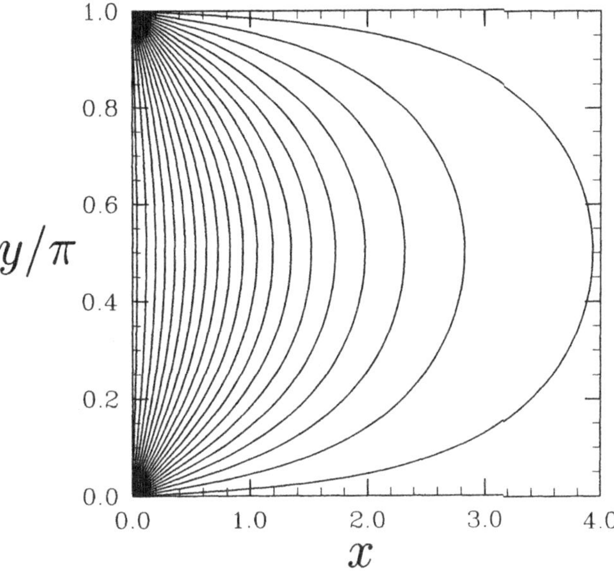

Figure 5.17: *Equally spaced contours of the potential specified in Equation (5.240). Only the first 50 terms in the series are retained.*

oscillatory functions. A set of functions $f_n(x)$ is *orthogonal* if the integral of the product of two different members of the set over some range is always zero: *i.e.*,

$$\int_a^b f_n(x)\, f_m(x)\, dx = 0, \tag{5.241}$$

for $n \neq m$. A set of functions is *complete* if any other function can be expanded as a weighted sum of them. It turns out that the scheme set out above can be generalized to more complicated geometries. For instance, in axisymmetric spherical geometry, the monotonic functions are power law functions of the radial variable, and the oscillatory functions are so-called Legendre polynomials involving the cosine of the polar angle θ. The latter functions are both mutually orthogonal and form a complete set. There are also cylindrical, ellipsoidal, hyperbolic, toroidal, *etc.*, coordinates. In all cases, the associated oscillating functions are mutually orthogonal and form a complete set. This implies that the separation-of-variables method is of quite general applicability.

Finally, as a very simple example of the solution of Poisson's equation in spherical geometry, let us consider the case of a grounded conducting sphere of radius a, centered on the origin, and placed in a uniform z-directed electric field of magnitude E_0. The scalar potential ϕ satisfies $\nabla^2\phi = 0$ for $r \geq a$, with the boundary conditions $\phi \to -E_0\, r\, \cos\theta$ (giving $\mathbf{E} \to E_0\, \mathbf{e}_z$) as $r \to \infty$, and $\phi = 0$ at $r = a$. Here, r and θ are spherical polar coordinates. Let us, first of all, assume that ϕ is independent of the azimuthal angle, since the boundary conditions possess this symmetry. Hence, $\phi = \phi(r, \theta)$. Next, let us try the simplified separable solution

$$\phi(r, \theta) = r^m \cos\theta. \tag{5.242}$$

It is easily demonstrated that the above solution only satisfies $\nabla^2\phi = 0$ provided $m = 1$ or $m = -2$. Thus, the most general solution of $\nabla^2\phi$ which satisfies the boundary condition at $r \to \infty$ is

$$\phi(r, \theta) = -E_0\, r\, \cos\theta + \alpha\, r^{-2}\, \cos\theta. \tag{5.243}$$

The boundary condition at $r = a$ is satisfied provided

$$\alpha = E_0\, a^3. \tag{5.244}$$

Hence, the potential takes the form

$$\phi(r, \theta) = -E_0 \left(r - \frac{a^3}{r^2} \right) \cos\theta. \tag{5.245}$$

Of course, $\phi = 0$ inside the sphere (*i.e.*, $r < a$). This potential is plotted in Figure 5.18 (for $a = 1$). The charge sheet density induced on the surface of the sphere is given by

$$\sigma(\theta) = \epsilon_0\, E_r(a, \theta) = -\epsilon_0\, \frac{\partial\phi(a, \theta)}{\partial r} = 3\, \epsilon_0\, E_0\, \cos\theta. \tag{5.246}$$

Note that zero net charge is induced on the surface of the sphere: This implies that the solution would be unchanged were the sphere insulated and uncharged, rather than grounded. Finally, it follows from Equations (5.243), (5.244), and Exercise 2.4 that the electric field outside the sphere consists of the original uniform field plus the field of an electric dipole of moment

$$\mathbf{p} = 4\pi\, a^3\, \epsilon_0\, E_0\, \mathbf{e}_z. \tag{5.247}$$

This is, of course, the dipole moment due to the charge separation induced on the surface of the sphere by the external field.

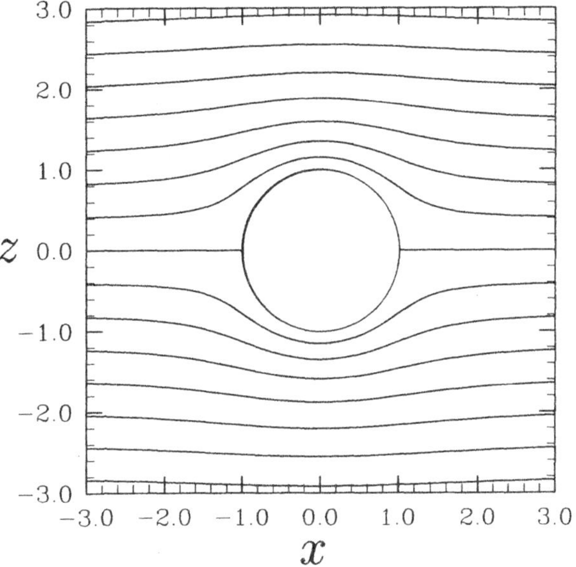

Figure 5.18: *Equally spaced contours of the axisymmetric potential* $\phi(r, \theta) = (r - 1/r^2)\cos\theta$ *plotted in the* x-z *plane for* $r > 1$.

5.13 EXERCISES

5.1. Eight identical point charges of magnitude q are placed at the vertices of a cube of dimension a. What is the electrostatic potential energy of this configuration of charges (excluding the self-energies of the charges)? Suppose that four of the charges are replaced by charges of magnitude $-q$ in such a manner that all of the nearest neighbors of a given charge are charges of the opposite sign. What now is the electrostatic potential energy of the configuration?

5.2. Find the electric field generated by a thin, uniform spherical shell of charge Q and radius a. Calculate the electrostatic potential energy of this charge distribution by integrating the energy density of the electric field over all space. Verify that the electrostatic energy is also given by

$$W = \frac{1}{2}\int_S \sigma\,\phi\,dS,$$

where ϕ is the scalar potential, σ the surface charge density, and the integral is taken over all surface charge distributions.

5.3. Suppose that a stationary charge distribution $\rho_1(\mathbf{r})$ generates the scalar potential field $\phi_1(\mathbf{r})$, and that an alternative charge distribution $\rho_2(\mathbf{r})$ generates the potential $\phi_2(\mathbf{r})$. Here, both charge distributions are assumed to be sufficiently localized that the potential fields they generate go to zero at large distances. Prove *Green's reciprocity theorem*:

$$\int \phi_1 \, \rho_2 \, dV = \int \phi_2 \, \rho_1 \, dV,$$

where the volume integral is over all space. Hint: Use Maxwell's equations and the divergence theorem.

5.4. Two grounded, infinite, parallel conducting plates are separated by a perpendicular distance d. A point charge q is placed between the plates. Demonstrate that the total charge induced on one of the plates is $(-q)$ times the fractional perpendicular distance of the point charge from the other plate. Hint: Use Green's reciprocity theorem.

5.5. Two grounded, concentric, thin spherical conducting shells have radii a and b, where $b > a$. A point charge q is placed between the shells at radius r (where $a < r < b$). Find the total charge induced on each shell. Hint: Use Green's reciprocity theorem.

5.6. Consider two insulated conductors, labeled 1 and 2. Let ϕ_1 be the potential of the first conductor when it is uncharged and the second conductor holds a charge Q. Likewise, let ϕ_2 be the potential of the second conductor when it is uncharged and the first conductor holds a charge Q. Use Green's reciprocity theorem to demonstrate that

$$\phi_1 = \phi_2.$$

5.7. Consider two insulated spherical conductors. Let the first have radius a. Let the second be sufficiently small that it can effectively be treated as a point charge, and let it also be located a distance $b > a$ from the center of the first. Suppose that the first conductor is uncharged, and that the second carries a charge q. What is the potential of the first conductor? Hint: Consider the result proved in the previous exercise.

5.8. Consider a set of N conductors distributed in a vacuum. Suppose that the ith conductor carries the charge Q_i and is at the scalar potential ϕ_i. It follows from the linearity of Maxwell's equations and Ohm's law that a linear relationship exists between the potentials and the charges: *i.e.*,

$$\phi_i = \sum_{j=1,N} p_{ij} \, Q_j.$$

Here, the p_{ij} are termed the *coefficients of potential*. Demonstrate that $p_{ij} = p_{ji}$ for all i, j. Hint: Consider the result proved in Exercise 5.6. Show that the total electrostatic potential energy of the charged conductors is

$$W = \frac{1}{2} \sum_{i,j=1,N} p_{ij}\, Q_i\, Q_j.$$

5.9. Find the coefficients of potential for two thin, concentric, spherical conducting shells of radius r_1 and r_2, where $r_2 > r_1$. Suppose that Q_1 and Q_2 are the charges on the inner and outer conductors, respectively. Calculate the electrostatic potential energy of the system.

5.10. Returning to the problem discussed in Exercise 5.7, suppose that the first conductor is now earthed, rather than being insulated and uncharged. What charge is induced on the first conductor by the charge on the second? Hint: Consider the concept of coefficient of potential.

5.11. Consider two separate conductors, the first of which is insulated and uncharged, and the second of which is earthed. Prove that the first conductor is also at zero (*i.e.*, earth) potential. Hint: Consider the previous hint.

5.12. Consider a flat annular plate (*e.g.*, a washer) of uniform thickness δ, inner radius a, and outer radius b. Let the plate be fabricated from metal of uniform resistivity η. Suppose that an electrical current I is fed into the plate symmetrically at its inner radius, and extracted symmetrically at its outer radius. What is the resistance of the plate? What is the rate of ohmic heating of the plate?

5.13. Consider an infinite uniform network of identical resistors of resistance R. Let four resistors come together at each junction of the network (*i.e.*, let the network have a square lattice). Suppose that current is fed into a given junction and extracted from a nearest neighbor junction. What is the effective resistance of the network?

5.14. According to the uniqueness theorem, Poisson's equation $\nabla^2\phi = -\rho/\epsilon_0$ can only have one solution if ρ is given in some volume V, and ϕ is specified on the bounding surface S. Demonstrate that two solutions can differ by, at most, a constant if the normal derivative of the potential, rather than the potential itself, is specified on the bounding surface.

5.15. Consider a point charge q which is placed inside a thin, grounded, spherical conducting shell of radius a a distance r from its center. Use the method of images to find the surface charge density induced on the inside of the shell. What is the net charge induced on the inside of the shell? What is the magnitude and direction of the force of attraction between the the charge and the shell? What electric field is induced outside the shell. How would these results be modified if the sphere were (a) uncharged and insulated, or (b) maintained at the constant potential V?

5.16. Using the method of images, show that the force of attraction, or repulsion, between a point charge q and an insulated conducting sphere of radius a carrying a charge Q is

$$f = \frac{q}{4\pi\epsilon_0}\left[\frac{Q + (a/d)\,q}{d^2} - \frac{a\,q}{d\,(d - a^2/d^2)}\right],$$

where $d > a$ is the distance of the charge from the center of the sphere. Demonstrate that when Q and q are of the same sign then the force is attractive provided that

$$\frac{Q}{d} < \frac{a\,d^2}{(d^2 - a^2)^2} - \frac{a}{d}.$$

5.17. An infinitely long conducting cylinder carries a charge per unit length λ and runs parallel to an infinite grounded conducting plane. Let the radius of the cylinder be a, and let the perpendicular distance between the cylinder's axis and the plane be d (where $d > a$). What is the force of attraction per unit length between the cylinder and the plane? What is the charge per unit length induced on the plane? Use the method of images.

5.18. A point charge q is located between two parallel, grounded, infinite conducting planes separated by a perpendicular distance d. Suppose that the perpendicular distance of the charge from one of the planes is x. Find the locations of the infinite number of image charges, and, hence, express the force exerted on the charge as an infinite series. Plot the magnitude of this force as a function of x/d. Find expressions for the net charges induced on the two planes. Plot these expressions as functions of x/d.

5.19. Two semi-infinite grounded conducting planes meet at right angles. A charge q is located a perpendicular distance a from one, and b from another. Use the method of images to find the magnitude and direction of the force of attraction between the planes and the charge. What is the net charge induced on the planes?

5.20. Two semi-infinite grounded conducting planes meet at sixty degrees. A charge q is located the same perpendicular distance a from both. Use the method of images to find the magnitude and direction of the force of attraction between the planes and the charge. What is the net charge induced on the planes?

5.21. Find the function $F(z)$, where z is the complex variable, whose real part can be interpreted as the scalar potential associated with (a) a uniform electric field of magnitude E_0 directed along the x-axis, (b) a uniform electric field of magnitude E_0 directed along the y-axis, and (c) a line charge of charge per unit length λ located at the origin.

5.22. Two semi-infinite conducting plates meet at 90°, and are both held at the constant potential V. Use complex analysis to find the variation of the surface charge density with perpendicular distance from the vertex on both sides of the plates.

5.23. Two semi-infinite conducting plates meet at 60°, and are both held at the constant potential V. Use complex analysis to find the variation of the surface charge density with perpendicular distance from the vertex on both sides of the plates.

5.24. Consider the complex function $F(z)$ defined implicitly by the equation

$$\mathrm{i}\,z = \mathrm{i}\,F(z) + e^{\mathrm{i}\,F(z)}.$$

Suppose that the real part of this function is interpreted as an electric potential. Plot the contours of this potential. What problem in electrostatics does this potential best describe?

5.25. Consider an empty cubic box of dimension a with conducting walls. Two opposite walls are held at the constant potential V, whilst the other walls are earthed. Find an expression for the electric potential inside the box. (Assume that the box is centered on the origin, that the walls are all normal to one of the Cartesian axes, and that the non-grounded walls are normal to the x-axis.) Suppose that the two walls normal to the x-axis are held at potentials $\pm V$. What now is the potential inside the box? Use separation of variables.

Chapter 6

DIELECTRIC AND MAGNETIC MEDIA

6.1 INTRODUCTION

In this chapter, we shall use Maxwell's equations to investigate the interaction of dielectric and magnetic media with quasi-static electric and magnetic fields.

6.2 POLARIZATION

The terrestrial environment is characterized by dielectric media (*e.g.*, air, water) which are, for the most part, electrically neutral, since they are made up of neutral atoms and molecules. However, if these atoms and molecules are placed in an external electric field then they tend to *polarize*: *i.e.*, their positively and negatively charged components displace with respect to one another. Suppose that if a given neutral molecule is placed in an external electric field $\mathbf{E}$ then the center of charge of its constituent electrons, whose total charge is (say) q, displaces by $\mathbf{d}$ with respect to the center of charge of its constituent atomic nuclii. The *dipole moment* of the molecule is defined as $\mathbf{p} = q\,\mathbf{d}$. If a dielectric medium is made up of N such molecules per unit volume then the *electric polarization*, $\mathbf{P}$, of the medium (*i.e.*, the dipole moment per unit volume) is given by $\mathbf{P} = N\,\mathbf{p}$. More generally,

$$\mathbf{P}(\mathbf{r}) = \sum_i N_i \,\langle \mathbf{p}_i \rangle, \qquad (6.1)$$

where $\langle \mathbf{p}_i \rangle$ is the average dipole moment of the ith type of molecule making up the medium, and N_i the average number of such molecules per unit volume, in the vicinity of point $\mathbf{r}$.

Now, we saw previously, in Exercise 3.4, that the scalar electric potential field generated by an electric dipole of moment $\mathbf{p}$ situated at the origin is

$$\phi(\mathbf{r}) = \frac{\mathbf{p}\cdot\mathbf{r}}{4\pi\epsilon_0\, r^3}. \qquad (6.2)$$

Hence, from the principle of superposition, the scalar potential field generated by a dielectric medium of dipole moment per unit volume $\mathbf{P}(\mathbf{r})$ is

$$\phi(\mathbf{r}) = \frac{1}{4\pi\epsilon_0}\int \frac{\mathbf{P}(\mathbf{r}')\cdot(\mathbf{r}-\mathbf{r}')}{|\mathbf{r}-\mathbf{r}'|^3}\,d^3\mathbf{r}', \tag{6.3}$$

where the volume integral is over all space. However, it follows from Equations (3.15) and (3.148) that

$$\phi(\mathbf{r}) = \frac{1}{4\pi\epsilon_0}\int \mathbf{P}(\mathbf{r}')\cdot\nabla'\left(\frac{1}{|\mathbf{r}-\mathbf{r}'|}\right)d^3\mathbf{r}'. \tag{6.4}$$

Finally, making use of Equation (3.152), we obtain

$$\phi(\mathbf{r}) = \frac{1}{4\pi\epsilon_0}\int \frac{\rho_b(\mathbf{r}')}{|\mathbf{r}-\mathbf{r}'|}\,d^3\mathbf{r}', \tag{6.5}$$

assuming that $|\mathbf{P}|/r \to 0$ as $r \to \infty$, where $\rho_b = -\nabla\cdot\mathbf{P}$. Thus, by comparison with Equation (3.17), we can see that minus the divergence of the polarization field is equivalent to a charge density.

As explained above, any divergence of the polarization field $\mathbf{P}(\mathbf{r})$ of a dielectric medium gives rise to an effective charge density $\rho_b(\mathbf{r})$, where

$$\rho_b = -\nabla\cdot\mathbf{P}. \tag{6.6}$$

This charge density is attributable to *bound charges* (*i.e.*, charges which arise from the polarization of neutral atoms), and is usually distinguished from the charge density $\rho_f(\mathbf{r})$ due to *free charges*, which typically represents a net surplus or deficit of electrons in the medium. Thus, the total charge density ρ in the medium is

$$\rho = \rho_f + \rho_b. \tag{6.7}$$

It must be emphasized that both terms in this equation represent real physical charge. Nevertheless, it is useful to make the distinction between bound and free charges, especially when it comes to working out the energy associated with electric fields in dielectric media.

Gauss' law takes the differential form

$$\nabla\cdot\mathbf{E} = \frac{\rho}{\epsilon_0} = \frac{\rho_f + \rho_b}{\epsilon_0}. \tag{6.8}$$

This expression can be rearranged to give

$$\nabla\cdot\mathbf{D} = \rho_f, \tag{6.9}$$

where

$$\mathbf{D} = \epsilon_0\,\mathbf{E} + \mathbf{P} \tag{6.10}$$

is termed the *electric displacement*, and has the same dimensions as **P** (*i.e.*, dipole moment per unit volume). Gauss' theorem tells us that

$$\oint_S \mathbf{D}\cdot d\mathbf{S} = \int_V \rho_f\, dV. \tag{6.11}$$

In other words, the flux of **D** out of some closed surface S is equal to the total *free charge* enclosed within that surface. Unlike the electric field **E** (which is the electric force acting on a unit charge), or the polarization **P** (which is the dipole moment per unit volume), the electric displacement **D** has no clear physical interpretation. In fact, the only reason for introducing this quantity is that it enables us to calculate electric fields in the presence of dielectric media without having to know the distribution of bound charges beforehand. However, this is only possible if we have a *constitutive relation* connecting **E** and **D**.

6.3 ELECTRIC SUSCEPTIBILITY AND PERMITTIVITY

In a large class of dielectric materials, there exits an approximately linear relationship between **P** and **E**. If the material is isotropic then

$$\mathbf{P} = \epsilon_0\,\chi_e\,\mathbf{E}, \tag{6.12}$$

where χ_e is termed the *electric susceptibility*. It follows that

$$\mathbf{D} = \epsilon_0\,\epsilon\,\mathbf{E}, \tag{6.13}$$

where

$$\epsilon = 1 + \chi_e \tag{6.14}$$

is termed the *relative dielectric constant* or *relative permittivity* of the medium. (Likewise, ϵ_0 is termed the *permittivity of free space*.) Note that ϵ is dimensionless. Values of ϵ for some common dielectric materials are given in Table 6.1. It can be seen that dielectric constants are generally greater than unity, and can be significantly greater than unity for liquids and solids.

It follows from Equations (6.9) and (6.13) that

$$\nabla\cdot\mathbf{E} = \frac{\rho_f}{\epsilon_0\,\epsilon}. \tag{6.15}$$

Material	Dielectric constant
Air (1 atm)	1.00059
Paper	3.5
Concrete	4.5
Glass	5–10
Silicon	11.68
Water	80.4

Table 6.1 *Low-frequency dielectric constants of some common materials.*

Thus, the electric fields produced by free charges in a *uniform* dielectric medium are analogous to those produced by the same charges in a vacuum, except that they are all reduced by a factor ϵ. This reduction can be understood in terms of a polarization of the atoms or molecules in the dielectric medium which produces electric fields which oppose those generated by the free charges. One immediate consequence of this effect is that the capacitance of a capacitor is increased by a factor ϵ if the empty space between its electrodes is filled with a dielectric medium of dielectric constant ϵ (assuming that fringing fields can be neglected).

It must be understood that Equations (6.12)–(6.15) merely represent an *approximation* which is generally found to hold under terrestrial conditions in *isotropic* media (provided that the electric field intensity is not too large). For anisotropic media (*e.g.*, crystals), Equation (6.13) generalizes to

$$\mathbf{D} = \epsilon_0 \, \boldsymbol{\epsilon} \cdot \mathbf{E}, \tag{6.16}$$

where $\boldsymbol{\epsilon}$ is a second-rank tensor known as the *dielectric tensor*. For strong electric fields, $\mathbf{D}$ ceases to vary linearly with $\mathbf{E}$.

6.4 BOUNDARY CONDITIONS FOR E AND D

When the space surrounding a set of charges contains dielectric material of *non-uniform* dielectric constant then the electric field no longer has the same functional form as in a vacuum. Suppose, for example, that space is occupied by two dielectric media, labeled 1 and 2, whose uniform

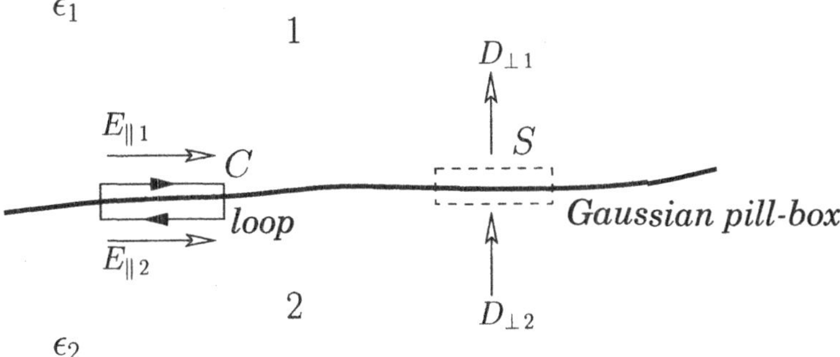

Figure 6.1: *The boundary between two different dielectric media.*

dielectric constants are ϵ_1 and ϵ_2, respectively. What are the boundary conditions on **E** and **D** at the interface between the two media?

Let us apply Equation 6.11 to a Gaussian pill-box S enclosing part of the interface—see Figure 6.1. The thickness of the pill-box is allowed to tend toward zero, so that the only contribution to the outward flux of **D** comes from the flat faces of the box, which are parallel to the interface. Assuming that there is no free charge inside the pill-box (which is reasonable in the limit in which the volume of the box tends to zero), then Equation (6.11) yields

$$D_{\perp 1} - D_{\perp 2} = 0, \tag{6.17}$$

where $D_{\perp 1}$ is the component of the electric displacement in medium 1 which is normal to the interface, *etc.* According to Equation (6.13), the boundary condition on the normal component of the electric field is

$$\epsilon_1 E_{\perp 1} - \epsilon_2 E_{\perp 2} = 0. \tag{6.18}$$

Integrating Faraday's law,

$$\nabla \times \mathbf{E} = -\frac{\partial \mathbf{B}}{\partial t}, \tag{6.19}$$

around a narrow rectangular loop C which straddles the interface—see Figure 6.1—yields

$$(E_{\parallel 1} - E_{\parallel 2})\, l = -A\, \frac{\partial B_\perp}{\partial t}, \tag{6.20}$$

where l is the length of the long side of the loop, A the area of the loop, and $B_\perp$ the magnetic field normal to the loop. In the limit in which the length of the short side of the loop tends to zero, A also goes to zero, and we obtain the familiar boundary condition

$$E_{\parallel 1} - E_{\parallel 2} = 0. \tag{6.21}$$

Generally speaking, there is a *bound charge sheet* on the interface between two dielectric media. The charge density of this sheet follows from Gauss' law:

$$\sigma_b = \epsilon_0 \, (E_{\perp 1} - E_{\perp 2}) = (1/\epsilon_1 - 1/\epsilon_2) \, D_\perp. \tag{6.22}$$

This can also be written

$$\sigma_b = (\mathbf{P}_1 - \mathbf{P}_2) \cdot \mathbf{n}, \tag{6.23}$$

where $\mathbf{n}$ is the unit normal at the interface (pointing from medium 1 to medium 2), and $\mathbf{P}_{1,2}$ are the electric polarizations in the two media.

6.5 BOUNDARY VALUE PROBLEMS WITH DIELECTRICS

Consider a point charge q embedded in a semi-infinite dielectric medium of uniform dielectric constant ϵ_1, and located a distance d away from a plane interface which separates this medium from another semi-infinite dielectric medium of dielectric constant ϵ_2. Let the interface coincide with the plane $z = 0$, and let the point charge lie on the positive z-axis. In order to solve this problem, we need to find solutions to the equations

$$\epsilon_1 \, \nabla \cdot \mathbf{E} = \frac{q \, \delta(\mathbf{r} - \mathbf{r}_0)}{\epsilon_0}, \tag{6.24}$$

where $\mathbf{r}_0 = (0, 0, d)$, for $z > 0$,

$$\epsilon_2 \, \nabla \cdot \mathbf{E} = 0 \tag{6.25}$$

for $z < 0$, and

$$\nabla \times \mathbf{E} = \mathbf{0} \tag{6.26}$$

everywhere, subject to the boundary conditions

$$\epsilon_1 \, E_z(x, y, 0_+) = \epsilon_2 \, E_z(x, y, 0_-), \tag{6.27}$$

$$E_x(x, y, 0_+) = E_x(x, y, 0_-), \tag{6.28}$$

$$E_y(z, y, 0_+) = E_y(x, y, 0_-). \tag{6.29}$$

We can solve this problem by employing a slightly modified form of the method of images—see Section 5.10. Since $\nabla \times \mathbf{E} = \mathbf{0}$ everywhere,

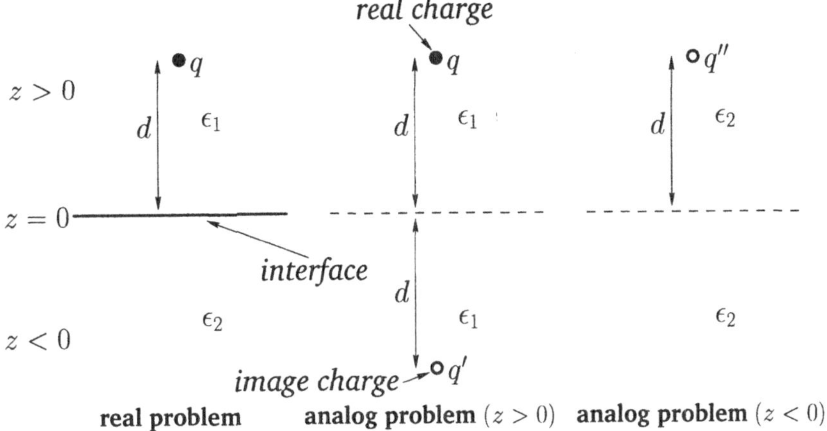

Figure 6.2: *The method of images for a charge near the plane interface between two dielectric media.*

the electric field can be written in terms of a scalar potential: *i.e.*, $\mathbf{E} = -\nabla\phi$. Consider the region $z > 0$. We shall assume that the scalar potential in this region is the same as that in an analog problem in which the whole of space is filled with a dielectric medium of dielectric constant ϵ_1, and, in addition to the real charge q at position $(0, 0, d)$, there is a second charge q' at the image position $(0, 0, -d)$—see Figure 6.2. If this is the case, then the potential at some general point in the region $z > 0$ is given by

$$\phi(r, z) = \frac{1}{4\pi\epsilon_0\,\epsilon_1}\left(\frac{q}{R_1} + \frac{q'}{R_2}\right), \tag{6.30}$$

where $R_1 = \sqrt{r^2 + (d - z)^2}$, $R_2 = \sqrt{r^2 + (d + z)^2}$, and $r = \sqrt{x^2 + y^2}$. Note that the potential (6.30) is clearly a solution of Equation (6.24) in the region $z > 0$: *i.e.*, it gives $\nabla \cdot \mathbf{E} = 0$, with the appropriate singularity at the position of the point charge q.

Consider the region $z < 0$. Let us assume that the scalar potential in this region is the same as that in an analog problem in which the whole of space is filled with a dielectric medium of dielectric constant ϵ_2, and a charge q'' is located at $(0, 0, d)$—see Figure 6.2. If this is the case, then the potential in the region $z < 0$ is given by

$$\phi(r, z) = \frac{1}{4\pi\epsilon_0\,\epsilon_2}\frac{q''}{R_1}. \tag{6.31}$$

The above potential is clearly a solution of Equation (6.25) in the region $z < 0$: *i.e.*, it gives $\nabla\cdot\mathbf{E} = 0$, with no singularities.

It now remains to choose q' and q'' in such a manner that the boundary conditions (6.27)–(6.29) are satisfied. The boundary conditions (6.28) and (6.29) are obviously satisfied if the scalar potential is continuous at the interface between the two dielectric media: *i.e.*, if

$$\phi(r, 0_+) = \phi(r, 0_-). \tag{6.32}$$

The boundary condition (6.27) implies a jump in the normal derivative of the scalar potential across the interface: *i.e.*,

$$\epsilon_1 \frac{\partial\phi(r, 0_+)}{\partial z} = \epsilon_2 \frac{\partial\phi(r, 0_-)}{\partial z}. \tag{6.33}$$

The first matching condition yields

$$\frac{q + q'}{\epsilon_1} = \frac{q''}{\epsilon_2}, \tag{6.34}$$

whereas the second gives

$$q - q' = q''. \tag{6.35}$$

Here, use has been made of

$$\frac{\partial}{\partial z}\left(\frac{1}{R_1}\right)_{z=0} = -\frac{\partial}{\partial z}\left(\frac{1}{R_2}\right)_{z=0} = \frac{d}{(r^2 + d^2)^{3/2}}. \tag{6.36}$$

Equations (6.34) and (6.35) imply that

$$q' = -\left(\frac{\epsilon_2 - \epsilon_1}{\epsilon_2 + \epsilon_1}\right) q, \tag{6.37}$$

$$q'' = \left(\frac{2\,\epsilon_2}{\epsilon_2 + \epsilon_1}\right) q. \tag{6.38}$$

Now, the bound charge density is given by $\rho_b = -\nabla\cdot\mathbf{P}$, however, we have $\mathbf{P} = \epsilon_0\,\chi_e\,\mathbf{E}$ inside both dielectric media. Hence, $\nabla\cdot\mathbf{P} = \epsilon_0\,\chi_e\,\nabla\cdot\mathbf{E} = 0$, except at the location of the original point charge. We conclude that there is zero bound charge density in either dielectric medium. However, there is a bound charge sheet on the interface between the two media. In fact, the density of this sheet is given by

$$\sigma_b(r) = \epsilon_0\,(E_{z\,1} - E_{z\,2})_{z=0}. \tag{6.39}$$

Hence,

$$\sigma_b(r) = \epsilon_0 \frac{\partial \phi(r,0_-)}{\partial z} - \epsilon_0 \frac{\partial \phi(r,0_+)}{\partial z} = -\frac{q}{2\pi\,\epsilon_1} \frac{(\epsilon_2 - \epsilon_1)}{(\epsilon_2 + \epsilon_1)} \frac{d}{(r^2 + d^2)^{3/2}}. \tag{6.40}$$

Incidentally, it is easily demonstrated that the net charge on the interface is q'/ϵ_1. In the limit $\epsilon_2 \gg \epsilon_1$, the dielectric with dielectric constant ϵ_2 behaves like a conductor (*i.e.*, $\mathbf{E} \to \mathbf{0}$ in the region $z < 0$), and the bound surface charge density on the interface approaches that obtained in the case where the plane $z = 0$ coincides with a conducting surface—see Section 5.10.

As a second example, consider a dielectric sphere of radius a, and uniform dielectric constant ϵ, placed in a uniform z-directed electric field of magnitude E_0. Suppose that the sphere is centered on the origin. Now, we can always write $\mathbf{E} = -\nabla\phi$ for an electrostatic problem. In the present case, $\nabla \cdot \mathbf{E} = 0$ both inside and outside the sphere, since there are no free charges, and the bound charge density is zero in a uniform dielectric medium (or a vacuum). Hence, the scalar potential satisfies Laplace's equation, $\nabla^2\phi = 0$, throughout space. Adopting spherical polar coordinates, (r, θ, φ), aligned along the z-axis, the boundary conditions are that $\phi \to -E_0\, r\, \cos\theta$ as $r \to \infty$, and that ϕ is well-behaved at $r = 0$. At the surface of the sphere, $r = a$, the continuity of $E_\parallel$ implies that ϕ is continuous. Furthermore, the continuity of $D_\perp = \epsilon_0\, \epsilon\, E_\perp$ leads to the matching condition

$$\left.\frac{\partial \phi}{\partial r}\right|_{r=a+} = \epsilon \left.\frac{\partial \phi}{\partial r}\right|_{r=a-}. \tag{6.41}$$

Let us try axisymmetric separable solutions of the form $r^m \cos\theta$. It is easily demonstrated that such solutions satisfy Laplace's equation provided that $m = 1$ or $m = -2$. Hence, the most general solution to Laplace's equation outside the sphere, which satisfies the boundary condition at $r \to \infty$, is

$$\phi(r,\theta) = -E_0\, r\, \cos\theta + E_0\, \alpha\, \frac{a^3 \cos\theta}{r^2}. \tag{6.42}$$

Likewise, the most general solution inside the sphere, which satisfies the boundary condition at $r = 0$, is

$$\phi(r,\theta) = -E_1\, r\, \cos\theta. \tag{6.43}$$

The continuity of ϕ at $r = a$ yields

$$E_0 - E_0\,\alpha = E_1. \tag{6.44}$$

Likewise, the matching condition (6.41) gives

$$E_0 + 2\,E_0\,\alpha = \epsilon\,E_1. \tag{6.45}$$

Hence, we obtain

$$\alpha = \frac{\epsilon - 1}{\epsilon + 2}, \tag{6.46}$$

$$E_1 = \frac{3\,E_0}{\epsilon + 2}. \tag{6.47}$$

Note that the electric field inside the sphere is *uniform*, parallel to the external electric field outside the sphere, and of magnitude E_1. Moreover, $E_1 < E_0$, provided that $\epsilon > 1$—see Figure 6.3. The density of the bound charge sheet on the surface of the sphere is

$$\sigma_b(\theta) = -\epsilon_0 \left(\left.\frac{\partial\phi}{\partial r}\right|_{r=a+} - \left.\frac{\partial\phi}{\partial r}\right|_{r=a-} \right) = 3\,\epsilon_0 \left(\frac{\epsilon - 1}{\epsilon + 2}\right) E_0\,\cos\theta. \tag{6.48}$$

Finally, the electric field outside the sphere consists of the original uniform field, plus the field of an electric dipole of moment

$$\mathbf{p} = 4\pi\,a^3\,\epsilon_0 \left(\frac{\epsilon - 1}{\epsilon + 2}\right) E_0\,\mathbf{e}_z. \tag{6.49}$$

This is simply the net induced dipole moment, $\mathbf{p} = (4/3)\,\pi\,a^3\,\mathbf{P}$, of the sphere, where $\mathbf{P} = \epsilon_0\,(\epsilon - 1)\,E_1\,\mathbf{e}_z$.

As a final example, consider a spherical cavity, of radius a, inside a uniform dielectric medium of dielectric constant ϵ, in the presence of a z-directed electric field of magnitude E_0. This problem is analogous to the previous one, except that the matching condition (6.41) becomes

$$\epsilon \left.\frac{\partial\phi}{\partial r}\right|_{r=a+} = \left.\frac{\partial\phi}{\partial r}\right|_{r=a-}. \tag{6.50}$$

Hence, $\epsilon \to 1/\epsilon$, and we obtain

$$\alpha = -\frac{\epsilon - 1}{1 + 2\,\epsilon}, \tag{6.51}$$

$$E_1 = \frac{3\,\epsilon\,E_0}{1 + 2\,\epsilon}. \tag{6.52}$$

Note that the field inside the cavity is *uniform*, parallel to the external electric field outside the sphere, and of magnitude E_1. Moreover, $E_1 > E_0$,

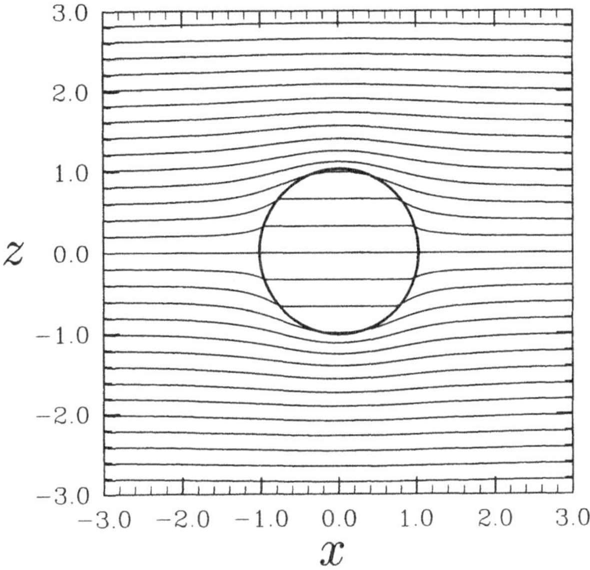

Figure 6.3: *Equally spaced coutours of the axisymmetric potential* $\phi(r, \theta)$ *generated by a dielectric sphere of unit radius and dielectric constant* $\epsilon = 3$ *placed in a uniform z-directed electric field.*

provided that $\epsilon > 1$—see Figure 6.4. The density of the bound charge sheet on the inside surface of the cavity is

$$\sigma_b(\theta) = -\epsilon_0 \left(\left.\frac{\partial \phi}{\partial r}\right|_{r=a+} - \left.\frac{\partial \phi}{\partial r}\right|_{r=a-} \right) = -3\,\epsilon_0 \left(\frac{\epsilon - 1}{1 + 2\,\epsilon} \right) E_0\, \cos\theta. \tag{6.53}$$

Hence, it follows from Equation (6.23) that the polarization immediately outside the cavity is

$$\mathbf{P} = 3\,\epsilon_0 \left(\frac{\epsilon - 1}{1 + 2\,\epsilon} \right) E_0\, \mathbf{e}_z. \tag{6.54}$$

This is less than the polarization field a long way from the cavity by a factor $3/(1 + 2\,\epsilon)$. In other words, the cavity induces a slight depolarization of the dielectric medium in its immediate vicinity. The electric field inside the cavity can be written

$$\mathbf{E}_1 = \mathbf{E}_0 + \frac{\mathbf{P}}{3\epsilon_0}, \tag{6.55}$$

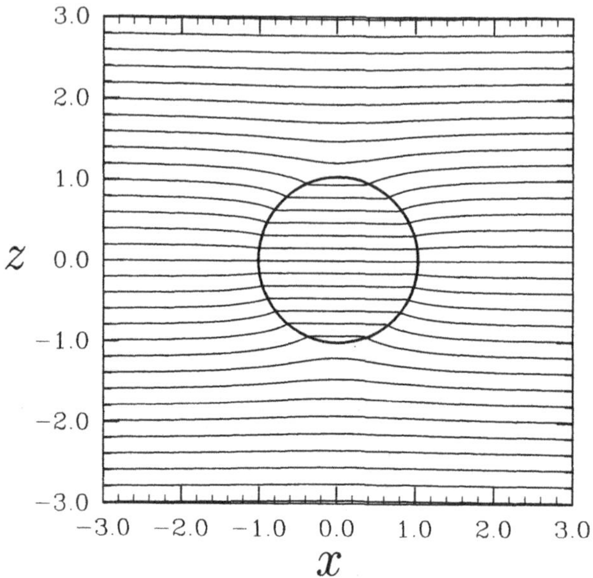

Figure 6.4: *Equally spaced coutours of the axisymmetric potential $\phi(r, \theta)$ generated by a cavity of unit radius inside a dielectric medium of dielectric constant $\epsilon = 3$ placed in a uniform z-directed electric field.*

where $\mathbf{E}_0$ is the external field, and $\mathbf{P}$ the polarization field immediately outside the cavity.

6.6 ENERGY DENSITY WITHIN A DIELECTRIC MEDIUM

Consider a system of free charges embedded in a dielectric medium. The increase in the total energy when a small amount of free charge $\delta\rho_f$ is added to the system is given by

$$\delta W = \int \phi\, \delta\rho_f\, d^3\mathbf{r}, \tag{6.56}$$

where the integral is taken over all space, and $\phi(\mathbf{r})$ is the electrostatic potential. Here, it is assumed that both the original charges and the dielectric medium are held fixed, so that no mechanical work is performed. It follows from Equation (6.9) that

$$\delta W = \int \phi\, \nabla\cdot\delta\mathbf{D}\, d^3\mathbf{r}, \tag{6.57}$$

where $\delta\mathbf{D}$ is the change in the electric displacement associated with the charge increment. Now the above equation can also be written

$$\delta W = \int \nabla\cdot(\phi\,\delta\mathbf{D})\,d^3\mathbf{r} - \int \nabla\phi\cdot\delta\mathbf{D}\,d^3\mathbf{r}, \tag{6.58}$$

giving

$$\delta W = \oint_S \phi\,\delta\mathbf{D}\cdot d\mathbf{S} - \int_V \nabla\phi\cdot\delta\mathbf{D}\,d^3\mathbf{r}, \tag{6.59}$$

where use has been made of Gauss' theorem. Here, V is some volume bounded by the closed surface S. If the dielectric medium is of finite spatial extent then the surface term is eliminated by integrating over all space. We thus obtain

$$\delta W = -\int \nabla\phi\cdot\delta\mathbf{D}\,d^3\mathbf{r} = \int \mathbf{E}\cdot\delta\mathbf{D}\,d^3\mathbf{r}. \tag{6.60}$$

Now, this energy increment cannot be integrated unless $\mathbf{E}$ is a known function of $\mathbf{D}$. Let us adopt the conventional approach, and assume that $\mathbf{D} = \epsilon_0\,\epsilon\,\mathbf{E}$, where the dielectric constant ϵ is *independent* of the electric field. The change in energy associated with taking the displacement field from zero to $\mathbf{D}(\mathbf{r})$ at all points in space is given by

$$W = \int_0^{\mathbf{D}} \delta W = \int_0^{\mathbf{D}} \int \mathbf{E}\cdot\delta\mathbf{D}\,d^3\mathbf{r}, \tag{6.61}$$

or

$$W = \int\int_0^E \frac{\epsilon_0\,\epsilon\,\delta(E^2)}{2}\,d^3\mathbf{r} = \frac{1}{2}\int \epsilon_0\,\epsilon\,E^2\,d^3\mathbf{r}, \tag{6.62}$$

which reduces to

$$W = \frac{1}{2}\int \mathbf{E}\cdot\mathbf{D}\,d^3\mathbf{r}. \tag{6.63}$$

Thus, the electrostatic energy density inside a dielectric medium is given by

$$U = \frac{1}{2}\,\mathbf{E}\cdot\mathbf{D}. \tag{6.64}$$

This is a standard result, and is often quoted in textbooks. Nevertheless, it is important to realize that the above formula is only valid for dielectric media in which the electric displacement $\mathbf{D}$ varies *linearly* with the electric field $\mathbf{E}$. Note, finally, that Equation (6.64) is consistent with the previously obtained expression (5.20).

6.7 FORCE DENSITY WITHIN A DIELECTRIC MEDIUM

Equation (6.63) was derived by considering a virtual process in which true charges are added to a system of charges and dielectrics which are held fixed, so that no mechanical work is done against physical displacements. Let us now consider a different virtual process in which the physical coordinates of the charges and dielectric are given a virtual displacement $\delta\mathbf{r}$ at each point in space, but no free charges are added to the system. Since we are dealing with a conservative system, the energy expression (6.63) can still be employed, despite the fact that it was derived in terms of another virtual process. The variation in the total electrostatic energy δW when the system undergoes a virtual displacement $\delta\mathbf{r}$ is related to the electrostatic force density $\mathbf{f}$ acting within the dielectric medium via

$$\delta W = -\int \mathbf{f}\cdot\delta\mathbf{r}\, d^3\mathbf{r}. \tag{6.65}$$

So, if the medium is moving, and has a velocity field $\mathbf{u}$, then the rate at which electrostatic energy is drained from the $\mathbf{E}$ and $\mathbf{D}$ fields is given by

$$\frac{dW}{dt} = -\int \mathbf{f}\cdot\mathbf{u}\, d^3\mathbf{r}. \tag{6.66}$$

Let us now consider the electrostatic energy increment due to a change $\delta\rho_f$ in the free charge distribution, and a change $\delta\epsilon$ in the dielectric constant, both of which are caused by the virtual displacement. From Equation (6.63),

$$\delta W = \frac{1}{2\epsilon_0}\int \left[D^2\,\delta(1/\epsilon) + 2\,\mathbf{D}\cdot\delta\mathbf{D}/\epsilon\right] d^3\mathbf{r}, \tag{6.67}$$

or

$$\delta W = -\frac{\epsilon_0}{2}\int E^2\,\delta\epsilon\, d^3\mathbf{r} + \int \mathbf{E}\cdot\delta\mathbf{D}\, d^3\mathbf{r}. \tag{6.68}$$

Here, the first term on the right-hand side represents the energy increment due to the change in dielectric constant associated with the virtual displacement, whereas the second term corresponds to the energy increment caused by the displacement of free charges. The second term can be written

$$\int \mathbf{E}\cdot\delta\mathbf{D}\, d^3\mathbf{r} = -\int \nabla\phi\cdot\delta\mathbf{D}\, d^3\mathbf{r} = \int \phi\,\nabla\cdot\delta\mathbf{D}\, d^3\mathbf{r} = \int \phi\,\delta\rho_f\, d^3\mathbf{r}, \tag{6.69}$$

where surface terms have been neglected. Thus, Equation (6.68) implies that

$$\frac{dW}{dt} = \int \left(\phi \, \frac{\partial \rho_f}{\partial t} - \frac{\epsilon_0}{2} \, E^2 \, \frac{\partial \epsilon}{\partial t} \right) d^3\mathbf{r}. \tag{6.70}$$

In order to arrive at an expression for the force density f, we need to express the time derivatives $\partial\rho/\partial t$ and $\partial\epsilon/\partial t$ in terms of the velocity field $\mathbf{u}$. This can be achieved by adopting a dielectric equation of state: *i.e.*, a relation which gives the dependence of the dielectric constant ϵ on the mass density ρ_m. Let us assume that $\epsilon(\rho_m)$ is a known function. It follows that

$$\frac{D\epsilon}{Dt} = \frac{d\epsilon}{d\rho_m} \frac{D\rho_m}{Dt}, \tag{6.71}$$

where

$$\frac{D}{Dt} \equiv \frac{\partial}{\partial t} + \mathbf{u}\cdot\nabla \tag{6.72}$$

is the total time derivative (*i.e.*, the time derivative in a frame of reference which is locally co-moving with the dielectric). The hydrodynamic equation of continuity of the dielectric is [see Equation (2.130)]

$$\frac{\partial \rho_m}{\partial t} + \nabla\cdot(\rho_m \, \mathbf{u}) = 0, \tag{6.73}$$

which implies that

$$\frac{D\rho_m}{Dt} = -\rho_m \nabla\cdot\mathbf{u}. \tag{6.74}$$

Hence, it follows that

$$\frac{\partial \epsilon}{\partial t} = -\frac{d\epsilon}{d\rho_m} \, \rho_m \nabla\cdot\mathbf{u} - \mathbf{u}\cdot\nabla\epsilon. \tag{6.75}$$

The conservation equation for the free charges is written

$$\frac{\partial \rho_f}{\partial t} + \nabla\cdot(\rho_f \, \mathbf{u}) = 0. \tag{6.76}$$

Thus, we can express Equation (6.70) in the form

$$\frac{dW}{dt} = \int \left[-\phi \, \nabla\cdot(\rho_f \, \mathbf{u}) + \frac{\epsilon_0}{2} \, E^2 \, \frac{d\epsilon}{d\rho_m} \, \rho_m \, \nabla\cdot\mathbf{u} + \left(\frac{\epsilon_0}{2} \, E^2 \, \nabla\epsilon \right)\cdot\mathbf{u} \right] d^3\mathbf{r}. \tag{6.77}$$

Integrating the first term on the right-hand side by parts, and neglecting any surface contributions, we obtain

$$-\int \phi\, \nabla\cdot(\rho_f\, \mathbf{u})\, d^3\mathbf{r} = \int \rho_f\, \nabla\phi\cdot\mathbf{u}\, d^3\mathbf{r}. \tag{6.78}$$

Likewise,

$$\int \frac{\epsilon_0}{2}\, E^2\, \frac{d\epsilon}{d\rho_m}\, \rho_m\, \nabla\cdot\mathbf{u}\, d^3\mathbf{r} = -\int \frac{\epsilon_0}{2}\, \nabla\left(E^2\, \frac{d\epsilon}{d\rho_m}\, \rho_m\right)\cdot\mathbf{u}\, d^3\mathbf{r}. \tag{6.79}$$

Hence, Equation (6.77) becomes

$$\frac{dW}{dt} = \int \left[-\rho_f\, \mathbf{E} + \frac{\epsilon_0}{2}\, E^2\, \nabla\epsilon - \frac{\epsilon_0}{2}\, \nabla\left(E^2\, \frac{d\epsilon}{d\rho_m}\, \rho_m\right)\right]\cdot\mathbf{u}\, d^3\mathbf{r}. \tag{6.80}$$

Comparing with Equation (6.66), we can see that the force density inside the dielectric medium is given by

$$\mathbf{f} = \rho_f\, \mathbf{E} - \frac{\epsilon_0}{2}\, E^2\, \nabla\epsilon + \frac{\epsilon_0}{2}\, \nabla\left(E^2\, \frac{d\epsilon}{d\rho_m}\, \rho_m\right). \tag{6.81}$$

The first term on the right-hand side of the above expression is the standard electrostatic force density. The second term represents a force which appears whenever an inhomogeneous dielectric is placed in an electric field. The last term, which is known as the *electrostriction* term, corresponds to a force acting on a dielectric placed in an inhomogeneous electric field. Note that the magnitude of the electrostriction force depends explicitly on the dielectric equation of state of the material, through $d\epsilon/d\rho_m$. The electrostriction term gives zero net force acting on any finite region of dielectric, if we can integrate over a large enough portion of the dielectric that its extremities lie in a field-free region. For this reason, the term is frequently neglected, since it usually does not contribute to the total force acting on a dielectric body. Note, however, that if the electrostriction term is omitted then we obtain an incorrect pressure variation within the dielectric, despite the fact that the total force is correct.

6.8 THE CLAUSIUS-MOSSOTTI RELATION

Let us now investigate what a dielectric equation of state actually looks like. Suppose that a dielectric medium is made up of identical molecules which develop a dipole moment

$$\mathbf{p} = \alpha\,\epsilon_0\,\mathbf{E} \tag{6.82}$$

when placed in an electric field **E**. The constant α (which has units of volume) is called the *molecular polarizability*. Note that α, which is solely a property of the molecule, is typically of order the molecular volume. If N is the number density of molecules then the polarization of the medium is

$$\mathbf{P} = N\,\mathbf{p} = N\,\alpha\,\epsilon_0\,\mathbf{E}, \tag{6.83}$$

or

$$\mathbf{P} = \frac{N_A\,\rho_m\,\alpha}{M}\,\epsilon_0\,\mathbf{E}, \tag{6.84}$$

where ρ_m is the mass density, N_A is Avogadro's number, and M is the molecular weight. But, how does the electric field experienced by an individual molecule relate to the average electric field in the medium? This is not a trivial question, since we expect the electric field to vary strongly (on atomic length-scales) inside the medium.

Suppose that the dielectric is polarized with a uniform mean electric field $\mathbf{E}_0 = E_0\,\mathbf{e}_z$. Consider one of the molecules which constitute the dielectric. Let us draw a sphere of radius a about this particular molecule. This is intended to represent the boundary between the microscopic and the macroscopic range of phenomena affecting the molecule. We shall treat the dielectric outside the sphere as a continuous medium, and the dielectric inside the sphere as a collection of polarized molecules. Note that, unlike the case of a spherical cavity in a dielectric medium, there is no depolarization of the dielectric immediately outside the sphere, since there is dielectric material inside the sphere. Thus, from Equation (6.55), the total field inside the sphere is

$$\mathbf{E} = \mathbf{E}_0 + \frac{\mathbf{P}}{3\epsilon_0}, \tag{6.85}$$

where **P** is given by Equation (6.84). The second term on the right-hand side of the above equation is the field at the molecule due to the surface charge on the inside of the sphere.

The field due to the individual molecules within the sphere is obtained by summing over the dipole fields of these molecules. The electric field at a distance **r** from a dipole of moment **p** is (see Exercise 3.4)

$$\mathbf{E} = -\frac{1}{4\pi\epsilon_0}\left[\frac{\mathbf{p}}{r^3} - \frac{3\,(\mathbf{p}\cdot\mathbf{r})\,\mathbf{r}}{r^5}\right]. \tag{6.86}$$

It is assumed that the dipole moment of each molecule within the sphere is the same, and also that the molecules are evenly distributed throughout the sphere. This being the case, the value of E_z at the molecule due to all of the other molecules within in the sphere,

$$E_z = -\frac{1}{4\pi\epsilon_0}\sum\left[\frac{p_z\,r^2 - 3\,(p_x\,x\,z + p_y\,y\,z + p_z\,z^2)}{r^5}\right], \tag{6.87}$$

is zero, since

$$\sum z^2 = \frac{1}{3}\sum r^2 \tag{6.88}$$

and

$$\sum x\,z = \sum y\,z = 0. \tag{6.89}$$

Here, the sum is over all of the molecules in the sphere. Furthermore, it is easily demonstrated that $E_\theta = E_\varphi = 0$ (where θ and φ are spherical polar coordinates). Hence, the electric field at the molecule due to the other molecules within the sphere vanishes.

It is clear that the net electric field seen by an individual molecule is

$$\mathbf{E} = \mathbf{E}_0 + \frac{\mathbf{P}}{3\epsilon_0}. \tag{6.90}$$

This is *larger* than the average electric field $\mathbf{E}_0$ in the dielectric. The above analysis indicates that this effect is ascribable to the long range (rather than the short range) interactions of the molecule with the other molecules in the medium. Making use of Equation (6.84), and the definition $\mathbf{P} = \epsilon_0\,(\epsilon - 1)\,\mathbf{E}_0$, we obtain

$$\frac{\epsilon - 1}{\epsilon + 2} = \frac{N_A\,\rho_m\,\alpha}{3\,M}. \tag{6.91}$$

This formula is called the *Clausius-Mossotti* relation, and is found to work fairly well for relatively dilute dielectric media whose dielectric constants are close to unity. Incidentally, the right-hand side of this expression is approximately the volume fraction of space occupied by the molecules making up the medium in question (and, should, therefore, be less than unity). Finally, the Clausius-Mossotti relation yields

$$\frac{d\epsilon}{d\rho_m} = \frac{(\epsilon - 1)\,(\epsilon + 2)}{3\,\rho_m}. \tag{6.92}$$

6.9 DIELECTRIC LIQUIDS IN ELECTROSTATIC FIELDS

Consider the behaviour of an uncharged dielectric liquid placed in an electrostatic field. If $p(\mathbf{r})$ is the pressure in the liquid when it is in equilibrium with the electrostatic force density $\mathbf{f}(\mathbf{r})$, then force balance requires that

$$\nabla p = \mathbf{f}. \tag{6.93}$$

It follows from Equation (6.81) that

$$\nabla p = -\frac{\epsilon_0}{2} E^2 \nabla \epsilon + \frac{\epsilon_0}{2} \nabla\left(E^2 \frac{d\epsilon}{d\rho_m} \rho_m\right) = \frac{\epsilon_0 \rho_m}{2} \nabla\left(E^2 \frac{d\epsilon}{d\rho_m}\right), \tag{6.94}$$

since $\nabla\epsilon = (d\epsilon/d\rho_m)\nabla\rho_m$. We can integrate this equation to give

$$\int_{p_1}^{p_2} \frac{dp}{\rho_m} = \frac{\epsilon_0}{2}\left(\left[E^2 \frac{d\epsilon}{d\rho_m}\right]_2 - \left[E^2 \frac{d\epsilon}{d\rho_m}\right]_1\right), \tag{6.95}$$

where 1 and 2 refer to two general points within the liquid. Here, it is assumed that the liquid possesses an equation of state, so that $p = p(\rho_m)$. If the liquid is essentially incompressible (*i.e.*, $\rho_m \simeq$ constant) then

$$p_2 - p_1 = \frac{\epsilon_0 \rho_m}{2}\left[E^2 \frac{d\epsilon}{d\rho_m}\right]_1^2. \tag{6.96}$$

Moreover, if the liquid obeys the Clausius-Mossotti relation then

$$p_2 - p_1 = \left[\frac{\epsilon_0 E^2}{2} \frac{(\epsilon - 1)(\epsilon + 2)}{3}\right]_1^2. \tag{6.97}$$

According to Equations (6.47) and (6.97), if a sphere of dielectric liquid is placed in a uniform electric field $\mathbf{E}_0$ then the pressure inside the liquid takes the constant value

$$p = \frac{3}{2}\epsilon_0\left(\frac{\epsilon - 1}{\epsilon + 2}\right) E_0^2. \tag{6.98}$$

Now, it is fairly clear that the electrostatic forces acting on the dielectric are all concentrated at the edge of the sphere, and are directed radially inward: *i.e.*, the dielectric is *compressed* by the external electric field. This is a somewhat surprising result, since the electrostatic forces acting on a rigid conducting sphere are also concentrated at the edge of the sphere, but are directed radially outward. We might expect these two cases to give the same result in the limit $\epsilon \to \infty$. The reason that this does not occur is because a dielectric liquid is slightly *compressible*, and is, therefore, subject to an electrostriction force. There is no electrostriction force for the case of a completely rigid body. In fact, the force density inside a rigid dielectric (for which $\nabla \cdot \mathbf{u} = 0$) is given by Equation (6.81) with the third term on the right-hand side (the electrostriction term) missing. It is easily seen that the force exerted by an electric field on a rigid dielectric is directed outward, and approaches that exerted on a rigid conductor in the limit $\epsilon \to 0$.

As is well-known, when a pair of charged (parallel plane) capacitor plates are dipped into a dielectric liquid, the liquid is drawn up between the plates to some extent. Let us examine this effect. We can, without loss of generality, assume that the transition from dielectric to vacuum takes place in a continuous manner. Consider the electrostatic pressure difference between a point A lying just above the surface of the liquid in the region between the plates, and a point B lying just above the surface of the liquid some region well away from the capacitor where $E \simeq 0$—see Figure 6.5. The pressure difference is given by

$$p_A - p_B = -\int_A^B \mathbf{f} \cdot d\mathbf{l}, \tag{6.99}$$

where $d\mathbf{l}$ is an element of some path linking points A and B. Note, however, that the Clausius-Mossotti relation yields $d\epsilon/d\rho_m = 0$ at both A and B, since $\epsilon = 1$ in a vacuum [see Equation (6.92)]. Thus, it is clear from Equation (6.81) that the electrostriction term makes no contribution to

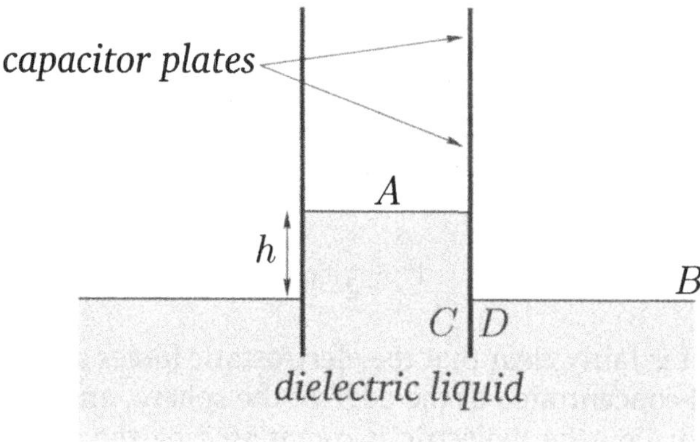

Figure 6.5: *Two capacitor plates dipped in a dielectric liquid.*

the line integral (6.99). It follows that

$$p_A - p_B = \frac{\epsilon_0}{2} \int_A^B E^2 \nabla\epsilon \cdot d\mathbf{l}. \tag{6.100}$$

The only contribution to this integral comes from the vacuum/dielectric interface in the vicinity of point A (since ϵ is constant inside the liquid, and $E \simeq 0$ in the vicinity of point B). Suppose that the electric field at point A has normal and tangential (to the surface) components $E_\perp$ and $E_\parallel$, respectively. Making use of the boundary conditions that $D_\perp$ and $E_\parallel$ are constant across a vacuum/dielectric interface, we obtain

$$p_A - p_B = \frac{\epsilon_0}{2} \left[E_\parallel^2 (\epsilon - 1) + \frac{D_\perp^2}{\epsilon_0^2} \int_1^\epsilon \frac{d\epsilon}{\epsilon^2} \right], \tag{6.101}$$

giving

$$p_A - p_B = \frac{\epsilon_0 (\epsilon - 1)}{2} \left[E_\parallel^2 + \frac{D_\perp^2}{\epsilon_0^2 \, \epsilon} \right]. \tag{6.102}$$

This electrostatic pressure difference can be equated to the hydrostatic pressure difference $\rho_m \, g \, h$ to determine the height h that the liquid rises between the plates. At first sight, the above analysis appears to suggest that the dielectric liquid is drawn upward by a surface force acting on the vacuum/dielectric interface in the region between the plates. In fact, this is far from being the case. A brief examination of Equations (6.93) and (6.97) shows that this surface force is actually directed *downward*. Indeed, according to Equation (6.81), the force which causes the liquid to rise between the plates is a volume force which develops in the region of non-uniform electric field at the base of the capacitor, where the field splays out between the plates. Thus, although we can determine the height to which the fluid rises between the plates without reference to the electrostriction force, it is, somewhat paradoxically, this force which is actually responsible for supporting the liquid against gravity.

Let us consider another paradox concerning the electrostatic forces exerted in a dielectric medium. Suppose that we have two charges embedded in a uniform dielectric medium of dielectric constant ϵ. The electric field generated by each charge is the same as that in vacuum, except that it is reduced by a factor ϵ. Therefore, we would expect that the force exerted by one charge on another is the same as that in vacuum, except that it is also reduced by a factor ϵ. Let us examine how this reduction in force comes about. Consider a simple example. Suppose that we take a parallel plate capacitor, and insert a block of solid dielectric between the plates. Suppose, further, that there is a small vacuum gap between the faces of the block and each of the capacitor plates. Let $\pm\sigma$ be the surface charge densities on each of the capacitor plates, and let $\pm\sigma_b$ be the bound surface charge densities which develop on the outer faces of the intervening dielectric block. The two layers of polarization charge produce equal and opposite electric fields on each plate, and their effects therefore cancel each other. Thus, from the point of view of electrical interaction alone, there would appear to be no change in the force exerted by one capacitor plate on the other when a dielectric slab is placed between them (assuming that σ remains constant during this process). That is, the force per unit area (which is attractive) remains

$$f = \frac{\sigma^2}{2\epsilon_0} \tag{6.103}$$

—see Equation (5.71). However, in experiments in which a capacitor is submerged in a dielectric liquid, the force per unit area exerted by one plate on another is observed to decrease to

$$f = \frac{\sigma^2}{2\epsilon_0\,\epsilon}. \tag{6.104}$$

This apparent paradox can be explained by taking into account the difference in liquid pressure in the field-filled space between the plates, and the field-free region outside the capacitor. This pressure difference is balanced by internal elastic forces in the case of a solid dielectric, but is transmitted to the plates in the case of the liquid. We can compute the pressure difference between a point C on the inside surface of one of the capacitor plates, and a point D on the outside surface of the same plate using Equation (6.100)—see Figure 6.5. If we neglect end effects, then the electric field is normal to the plates in the region between the plates, and is zero everywhere else. Thus, the only contribution to the line integral (6.100) comes from the plate/dielectric interface in the vicinity of point C. Adopting Equation (6.102), we find that

$$p_C - p_D = \frac{\epsilon_0}{2}\left(1 - \frac{1}{\epsilon}\right) E_\perp^2 = \frac{\sigma^2}{2\epsilon_0}\left(1 - \frac{1}{\epsilon}\right), \tag{6.105}$$

where $E_\perp = \sigma/\epsilon_0$ is the normal field strength between the plates in the absence of the dielectric. The sum of this pressure force (which is repulsive) and the attractive electrostatic force per unit area (6.103) yields a net attractive force per unit area of

$$f = \frac{\sigma^2}{2\epsilon_0\,\epsilon} \tag{6.106}$$

acting between the plates. Thus, any decrease in the forces exerted by charges on one another when they are immersed, or embedded, in a dielectric medium can only be understood in terms of mechanical forces transmitted between the charges by the medium itself.

6.10 POLARIZATION CURRENT

We have seen that the bound charge density is related to the polarization field via

$$\rho_b = -\nabla \cdot \mathbf{P}. \tag{6.107}$$

Now, it is clear, from this equation, that if the polarization field inside some dielectric material changes in time then the distribution of bound charges will also change. Hence, in order to conserve charge, a net current must flow. This current is known as the *polarization current*. Charge conservation implies that

$$\nabla \cdot \mathbf{j}_p + \frac{\partial \rho_b}{\partial t} = 0, \tag{6.108}$$

where $\mathbf{j}_p$ is the polarization current density. It follows from the previous two equations that

$$\mathbf{j}_p = \frac{\partial \mathbf{P}}{\partial t}. \tag{6.109}$$

Note that the polarization current is a real current, despite the fact that it is generated by the rearrangement of bound charges. There is, however, no drifting of real charges over length-scales longer than atomic or molecular length-scales associated with this current.

6.11 MAGNETIZATION

All matter is built up out of molecules, and each molecule consists of electrons in motion around stationary nuclii. The currents associated with this type of electron motion are termed *molecular currents*. Each molecular current is a tiny closed circuit of atomic dimensions, and may therefore be appropriately described as a magnetic dipole. Suppose that a given molecule has a magnetic dipole moment $\mathbf{m}$. If there are N such molecules per unit volume then the *magnetization* $\mathbf{M}$ (*i.e.*, the magnetic dipole moment per unit volume) is given by $\mathbf{M} = N\,\mathbf{m}$. More generally,

$$\mathbf{M}(\mathbf{r}) = \sum_i N_i \, \langle \mathbf{m}_i \rangle, \tag{6.110}$$

where $\langle \mathbf{m}_i \rangle$ is the average magnetic dipole moment of the ith type of molecule, and N_i is the average number of such molecules per unit volume, in the vicinity of point $\mathbf{r}$.

Now, we saw earlier, in Exercise 3.20, that the vector potential field generated by a magnetic dipole of moment $\mathbf{m}$ situated at the origin is

$$\mathbf{A}(\mathbf{r}) = \frac{\mu_0}{4\pi} \frac{\mathbf{m} \times \mathbf{r}}{r^3}. \tag{6.111}$$

Hence, from the principle of superposition, the vector potential field generated by a magnetic medium of magnetic moment per unit volume $\mathbf{M}(\mathbf{r})$ is

$$\mathbf{A}(\mathbf{r}) = \frac{\mu_0}{4\pi} \int \frac{\mathbf{M}(\mathbf{r}') \times (\mathbf{r} - \mathbf{r}')}{|\mathbf{r} - \mathbf{r}'|^3} \, d^3\mathbf{r}', \tag{6.112}$$

where the volume integral is taken over all space. However, it follows from Equations (3.15) and (3.148) that

$$\mathbf{A}(\mathbf{r}) = \frac{\mu_0}{4\pi} \int \mathbf{M}(\mathbf{r}') \times \nabla' \left(\frac{1}{|\mathbf{r} - \mathbf{r}'|} \right) d^3\mathbf{r}'. \tag{6.113}$$

Now, it is easily demonstrated that

$$\int \mathbf{f} \times \nabla g \, d^3\mathbf{r} = \int g \, \nabla \times \mathbf{f} \, d^3\mathbf{r}, \tag{6.114}$$

provided that the integral is over all space, and $g\,|\mathbf{f}| \rightarrow 0$ as $|\mathbf{r}| \rightarrow \infty$. Hence, for a magnetization field of finite extent, we can write

$$\mathbf{A}(\mathbf{r}) = \frac{\mu_0}{2\pi} \int \frac{\mathbf{j}_m(\mathbf{r}')}{|\mathbf{r} - \mathbf{r}'|} \, d^3\mathbf{r}', \tag{6.115}$$

where

$$\mathbf{j}_m = \nabla \times \mathbf{M}. \tag{6.116}$$

It follows, by comparison with Equation (3.215), that the curl of the magnetization field constitutes a current density. The associated current is known as the *magnetization current*.

The total current density, $\mathbf{j}$, in a general medium takes the form

$$\mathbf{j} = \mathbf{j}_t + \nabla \times \mathbf{M} + \frac{\partial \mathbf{P}}{\partial t}, \tag{6.117}$$

where the three terms on the right-hand side represent the *true current density* (*i.e.*, that part of the current density which is due to the movement of free charges), the magnetization current density, and the polarization current density, respectively. It must be emphasized that all three terms represent real physical currents, although only the first term is due to the motion of real charges (over more than molecular dimensions).

Now, the differential form of Ampère's law is

$$\nabla \times \mathbf{B} = \mu_0 \mathbf{j} + \mu_0 \epsilon_0 \frac{\partial \mathbf{E}}{\partial t}, \tag{6.118}$$

which can also be written

$$\nabla \times \mathbf{B} = \mu_0 \mathbf{j}_t + \mu_0 \nabla \times \mathbf{M} + \mu_0 \frac{\partial \mathbf{D}}{\partial t}, \tag{6.119}$$

where use has been made of Equation (6.117) and the definition $\mathbf{D} = \epsilon_0 \mathbf{E} + \mathbf{P}$. The above expression can be rearranged to give

$$\nabla \times \mathbf{H} = \mathbf{j}_t + \frac{\partial \mathbf{D}}{\partial t}, \tag{6.120}$$

where

$$\mathbf{H} = \frac{\mathbf{B}}{\mu_0} - \mathbf{M} \tag{6.121}$$

is termed the *magnetic intensity*, and has the same dimensions as $\mathbf{M}$ (*i.e.*, magnetic dipole moment per unit volume). In a steady-state situation, Stokes' theorem tell us that

$$\oint_C \mathbf{H} \cdot d\mathbf{l} = \int_S \mathbf{j}_t \cdot d\mathbf{S}. \tag{6.122}$$

In other words, the line integral of $\mathbf{H}$ around some closed loop is equal to the flux of the true current through any surface attached to that loop. Unlike the magnetic field $\mathbf{B}$ (which specifies the magnetic force $q\,\mathbf{v} \times \mathbf{B}$ acting on a charge q moving with velocity $\mathbf{v}$), or the magnetization $\mathbf{M}$ (which is the magnetic dipole moment per unit volume), the magnetic intensity **H** has no clear physical interpretation. The only reason for introducing this quantity is that it enables us to calculate magnetic fields in the presence of magnetic materials without having to know the distribution of magnetization currents beforehand. However, this is only possible if we possess a constitutive relation connecting **B** and **H**.

6.12 MAGNETIC SUSCEPTIBILITY AND PERMEABILITY

In a large class of magnetic materials, there exists an approximately linear relationship between **M** and **H**. If the material is isotropic then

$$\mathbf{M} = \chi_m \, \mathbf{H}, \tag{6.123}$$

where χ_m is called the *magnetic susceptibility*. If χ_m is positive then the material is called *paramagnetic*, and the magnetic field is strengthened by the presence of the material. On the other hand, if χ_m is negative then the material is *diamagnetic*, and the magnetic field is weakened in the presence of the material. The magnetic susceptibilities of paramagnetic and diamagnetic materials are generally extremely small. A few example values of χ_m are given in Table 6.2.

A linear relationship between **M** and **H** also implies a linear relationship between **B** and **H**. In fact, we can write

$$\mathbf{B} = \mu_0 \, \mu \, \mathbf{H}, \tag{6.124}$$

Material	**Magnetic susceptibility**
Aluminum	$+2.3 \times 10^{-5}$
Copper	-9.8×10^{-6}
Diamond	-2.2×10^{-5}
Tungsten	$+6.8 \times 10^{-5}$
Hydrogen (1 atm)	-2.1×10^{-9}
Oxygen (1 atm)	$+2.1 \times 10^{-6}$
Nitrogen (1 atm)	-5.0×10^{-9}

Table 6.2 *Low-frequency magnetic susceptibilities of some common materials.*

where

$$\mu = 1 + \chi_m \tag{6.125}$$

is termed the *relative magnetic permeability* of the material in question. (Likewise, μ_0 is termed the *permeability of free space*.) Note that μ is dimensionless. It is clear from Table 6.2 that the relative permeabilities of common diamagnetic and paramagnetic materials do not differ substantially from unity. In fact, to all intents and purposes, the magnetic properties of such materials can be safely neglected.

6.13 FERROMAGNETISM

There exists, however, a third class of magnetic materials called *ferromagnetic* materials. Such materials are characterized by a possible permanent magnetization, and generally have a profound effect on magnetic fields (*i.e.*, $\mu \gg 1$). Unfortunately, ferromagnetic materials *do not* generally exhibit a linear dependence between **M** and **H**, or between **B** and **H**, so that we cannot employ Equations (6.123) and (6.124) with constant values of χ_m and μ. It is still expedient to use Equation (6.124) as the definition of μ, with $\mu = \mu(\mathbf{H})$. However, this practice can lead to difficulties under certain circumstances. The permeability of a ferromagnetic material, as defined by Equation (6.124), can vary through the entire range of possible values from zero to infinity, and may be either positive or negative. The most sensible approach is to consider each problem involving ferromagnetic materials separately, try to determine which region of the **B**-**H** diagram is important for the particular case in hand, and then make approximations appropriate to this region.

First, let us consider an unmagnetized sample of ferromagnetic material. If the magnetic intensity, which is initially zero, is increased *monotonically*, then the **B**-**H** relationship traces out a curve such as that shown schematially in Figure 6.6. This is called a *magnetization curve*. It is evident that the permeabilities μ derived from the curve (according to the rule $\mu = \mu_0^{-1}\, B/H$) are always positive, and show a wide range of values. The maximum permeability occurs at the "knee" of the curve. In some materials, this maximum permeability is as large as 10^5. The reason for the knee is that the magnetization **M** reaches a maximum value in the material, so that

$$\mathbf{B} = \mu_0\,(\mathbf{H} + \mathbf{M}) \tag{6.126}$$

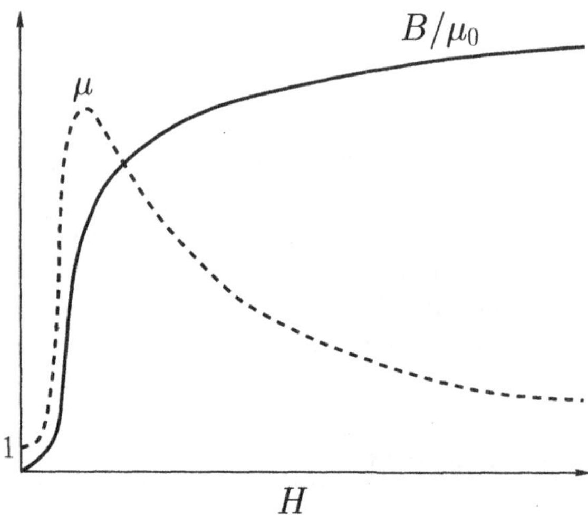

Figure 6.6: *A typical magnetization curve for a ferromagnet.*

continues to increase at large **H** only because of the μ_0 **H** term. The maximum value of **M** is called the *saturation magnetization* of the material. Incidentally, it is clear from the above equation that $\mu = 1$ for a fully saturated magnetic material in which $|\mathbf{H}| \gg |\mathbf{M}|$.

Next, consider a ferromagnetic sample magnetized by the above procedure. If the magnetic intensity **H** is decreased then the **B**-**H** relation does not return along the curve shown in Figure 6.6, but instead moves along a new curve, which is sketched in Figure 6.7, to the point R. Thus, the magnetization, once established, does not disappear with the removal of **H**. In fact, it takes a reversed magnetic intensity to reduce the magnetization to zero. If **H** continues to build up in the reversed direction, then **M** (and, hence, **B**) becomes increasingly negative. Finally, when **H** increases again the operating point follows the lower curve in Figure 6.7. Thus, the **B**-**H** curve for increasing **H** is quite different to that for decreasing **H**. This phenomenon is known as *hysteresis*.

The curve sketched in Figure 6.7 called the *hysteresis loop* of the ferromagnetic material in question. The value of **B** at the point R is called the *retentivity* or *remanence*. The magnitude of **H** at the point C is called the *coercivity*. It is evident that μ is negative in the second and fourth quadrants of the loop, and positive in the first and third quadrants. The shape of the hysteresis loop depends not only on the nature of the ferromagnetic material, but also on the maximum value of $|\mathbf{H}|$ to which the material has been subjected. However, once this maximum value,

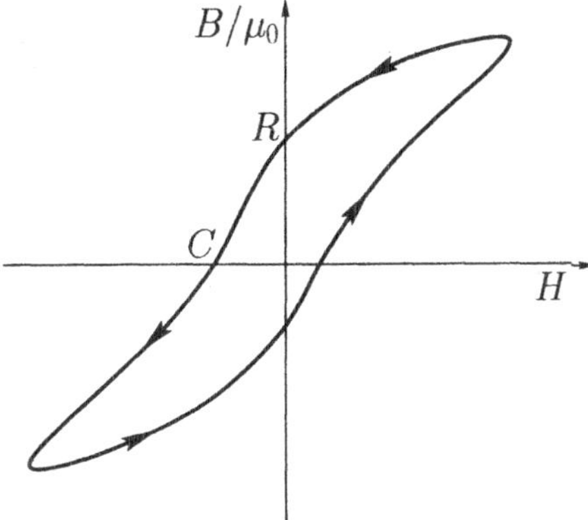

Figure 6.7: *A typical hysteresis loop for a ferromagnet.*

$|\mathbf{H}|_{max}$, becomes sufficiently large to produce saturation in the material, the hysteresis loop does not change shape with any further increase in $|\mathbf{H}|_{max}$.

Ferromagnetic materials are used either to channel magnetic flux (*e.g.*, around transformer circuits) or as sources of magnetic field (*e.g.*, permanent magnets). For use as a permanent magnet, the material is first magnetized by placing it in a strong magnetic field. However, once the magnet is removed from the external field it is subject to a demagnetizing **H**. Thus, it is vitally important that a permanent magnet should possess both a large remanence and a large coercivity. As will become clear later on, it is generally a good idea for the ferromagnetic materials used to channel magnetic flux around transformer circuits to possess small remanences and small coercivities.

6.14 BOUNDARY CONDITIONS FOR B AND H

What are the boundary conditions for **B** and **H** at the interface between two magnetic media? Well, the governing equations for a steady-state situation are

$$\nabla\cdot\mathbf{B} = 0, \tag{6.127}$$

and

$$\nabla\times\mathbf{H} = \mathbf{j}_t. \tag{6.128}$$

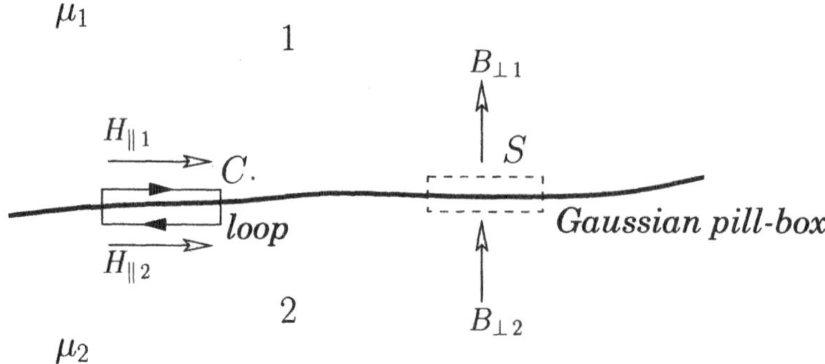

Figure 6.8: *The boundary between two different magnetic media.*

Integrating Equation (6.127) over a thin Gaussian pill-box S enclosing part of the interface between the two media gives

$$B_{\perp 1} - B_{\perp 2} = 0, \tag{6.129}$$

where $B_{\perp 1}$ denotes the component of **B** perpendicular to the interface in medium 1, *etc.*—see Figure 6.8. Integrating Equation (6.128) around a narrow loop C which straddles the interface yields

$$H_{\parallel 1} - H_{\parallel 2} = 0, \tag{6.130}$$

assuming that there is no true current sheet flowing at the interface—see Figure 6.8. Here, $H_{\parallel 1}$ denotes the component of **H** parallel to the interface in medium 1, *etc*. In general, there is a magnetization current sheet flowing at the interface between two magnetic materials whose density is

$$\mathbf{J}_m = (\mathbf{M}_1 - \mathbf{M}_2) \times \mathbf{n}, \tag{6.131}$$

where **n** is the unit normal to the interface (pointing from medium 1 to medium 2), and $\mathbf{M}_{1,2}$ are the magnetizations in the two media.

6.15 BOUNDARY VALUE PROBLEMS WITH FERROMAGNETS

Consider a ferromagnetic sphere of permanent magnetization $\mathbf{M} = M\,\mathbf{e}_z$, where M is a constant. What is the magnetic field generated by such a sphere? Suppose that the sphere is of radius a, and is centered on the origin. From Equation (6.120), we have

$$\nabla \times \mathbf{H} = \mathbf{0}, \tag{6.132}$$

since this is a time-independent problem with no true currents. It follows that

$$\mathbf{H} = -\nabla\phi_m, \tag{6.133}$$

where ϕ_m is termed the *magnetic scalar potential*. Now,

$$\mathbf{H} = \frac{\mathbf{B}}{\mu_0} - \mathbf{M}, \tag{6.134}$$

which implies that

$$\nabla \cdot \mathbf{H} = 0 \tag{6.135}$$

everywhere (apart from on the surface of the sphere), since $\nabla \cdot \mathbf{B} = 0$, and $\mathbf{M}$ is constant inside the sphere, and zero outside. It follows that the magnetic scalar potential satisfies Laplace's equation,

$$\nabla^2\phi_m = 0, \tag{6.136}$$

both inside and outside the sphere.

Adopting spherical polar coordinates, (r, θ, φ), aligned along the z-axis, the boundary conditions are that ϕ_m is well-behaved at $r = 0$, and $\phi_m \to 0$ as $r \to \infty$. Moreover, Equation (6.130) implies that ϕ_m must be continuous at $r = a$, whereas Equations (6.126) and (6.129) yield

$$-\left[\frac{\partial\phi_m}{\partial r}\right]_{a-}^{a+} = \mathbf{M} \cdot \mathbf{e}_r, \tag{6.137}$$

or

$$\left.\frac{\partial\phi_m}{\partial r}\right|_{r=a+} - \left.\frac{\partial\phi_m}{\partial r}\right|_{r=a-} = -M\cos\theta. \tag{6.138}$$

Let us try separable solutions of the form $r^m \cos\theta$. It is easily demonstrated that such solutions satisfy Laplace's equation provided that $m = 1$ or $m = -2$. Hence, the most general solution to Laplace's equation outside the sphere, which satisfies the boundary condition at $r \to \infty$, is

$$\phi_m(r, \theta) = C\,\frac{a^3\cos\theta}{r^2}. \tag{6.139}$$

Likewise, the most general solution inside the sphere, which satisfies the boundary condition at $r = 0$, is

$$\phi_m(r, \theta) = D\,r\cos\theta. \tag{6.140}$$

The continuity of ϕ_m at $r = a$ gives $C = D$, whereas the boundary condition (6.138) yields $C = M/3$. Hence, the magnetic scalar potential takes the form

$$\phi_m = \frac{M}{3}\frac{a^3}{r^2}\cos\theta \tag{6.141}$$

outside the sphere, and

$$\phi_m = \frac{M}{3}\,r\cos\theta \tag{6.142}$$

inside the sphere. It follows that

$$\mathbf{H} = -\mathbf{M}/3, \tag{6.143}$$

$$\mathbf{B}/\mu_0 = 2\,\mathbf{M}/3 \tag{6.144}$$

inside the sphere. Hence, both the magnetic field and the magnetic intensity are uniform and parallel or anti-parallel to the permanent magnetization within the sphere—see Figure 6.9. Note, however, that the sphere is subject to a *demagnetizing* magnetic intensity (*i.e.*, $\mathbf{H} \propto -\mathbf{M}$).

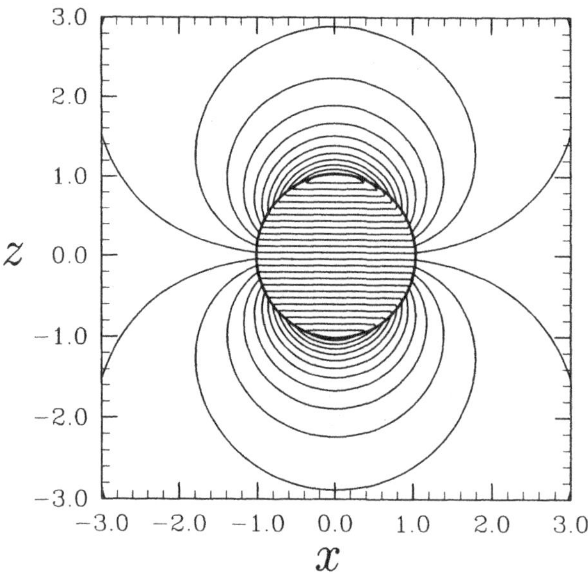

Figure 6.9: *Equally spaced coutours of the axisymmetric potential* $\phi_m(r,\theta)$ *generated by a ferromagnetic sphere of unit radius uniformly magnetized in the z-direction.*

It is easily demonstrated that the scalar magnetic potential due to a magnetic dipole of moment $\mathbf{m}$ at the origin is

$$\phi_m(\mathbf{r}) = \frac{1}{4\pi}\,\frac{\mathbf{m}\cdot\mathbf{r}}{r^3}. \tag{6.145}$$

Thus, it is clear that the magnetic field outside the sphere is the same as that of a magnetic dipole of moment

$$\mathbf{m} = \frac{4}{3}\,\pi\,a^3\,\mathbf{M} \tag{6.146}$$

at the origin. This, of course, is the permanent magnetic dipole moment of the sphere. Finally, the magnetization sheet current density at the surface of the sphere is given by

$$\mathbf{J}_m = \mathbf{M}\times\mathbf{e}_r = M\,\sin\theta\,\mathbf{e}_\varphi. \tag{6.147}$$

Consider a ferromagnetic sphere, of uniform permeability μ, placed in a uniform z-directed magnetic field of magnitude B_0. Suppose that the sphere is centered on the origin. In the absence of any true currents, we have $\nabla\times\mathbf{H} = \mathbf{0}$. Hence, we can again write $\mathbf{H} = -\nabla\phi_m$. Given that $\nabla\cdot\mathbf{B} = 0$, and $\mathbf{B} = \mu_0\,\mu\,\mathbf{H}$, it follows that $\nabla^2\phi_m = 0$ in any uniform magnetic medium (or a vacuum). Thus, $\nabla^2\phi_m = 0$ throughout space. Adopting spherical polar coordinates, (r, θ, φ), aligned along the z-axis, the boundary conditions are that $\phi_m \to -(B_0/\mu_0)\,r\,\cos\theta$ as $r\to\infty$, and that ϕ_m is well-behaved at $r = 0$. At the surface of the sphere, $r = a$, the continuity of $H_\parallel$ implies that ϕ_m is continuous. Furthermore, the continuity of $B_\perp = \mu_0\,\mu\,H_\perp$ leads to the matching condition

$$\left.\frac{\partial\phi_m}{\partial r}\right|_{r=a+} = \mu\left.\frac{\partial\phi_m}{\partial r}\right|_{r=a-}. \tag{6.148}$$

Let us again try separable solutions of the form $r^m\cos\theta$. The most general solution to Laplace's equation outside the sphere, which satisfies the boundary condition at $r\to\infty$, is

$$\phi_m(r,\theta) = -(B_0/\mu_0)\,r\,\cos\theta + (B_0/\mu_0)\,\alpha\,\frac{a^3\,\cos\theta}{r^2}. \tag{6.149}$$

Likewise, the most general solution inside the sphere, which satisfies the boundary condition at $r = 0$, is

$$\phi_m(r,\theta) = -(B_1/\mu_0\,\mu)\,r\,\cos\theta. \tag{6.150}$$

The continuity of ϕ_m at $r = a$ yields

$$B_0 - B_0\,\alpha = B_1/\mu. \tag{6.151}$$

Likewise, the matching condition (6.148) gives

$$B_0 + 2\,B_0\,\alpha = B_1. \tag{6.152}$$

Hence,

$$\alpha = \frac{\mu - 1}{\mu + 2}, \tag{6.153}$$

$$B_1 = \frac{3\,\mu\,B_0}{\mu + 2}. \tag{6.154}$$

Note that the magnetic field inside the sphere is *uniform*, parallel to the external magnetic field outside the sphere, and of magnitude B_1. Moreover, $B_1 > B_0$, provided that $\mu > 1$. The magnetization inside the sphere is also uniform and parallel to the external magnetic field. In fact,

$$\mathbf{M} = \frac{3\,(\mu - 1)}{\mu + 2}\,\frac{B_0}{\mu_0}\,\mathbf{e}_z. \tag{6.155}$$

The magnetic field outside the sphere is that due to the external field plus the field of a magnetic dipole of moment $\mathbf{m} = (4/3)\,\pi\,a^3\,\mathbf{M}$. This is, of course, the induced magnetic dipole moment of the sphere. Finally, the magnetization sheet current density at the surface of the sphere is $\mathbf{J}_m = \mathbf{M}\times\mathbf{e}_r = M\,\sin\theta\,\mathbf{e}_\varphi$.

As a final example, consider an electromagnet of the form sketched in Figure 6.10. A wire, carrying a current I, is wrapped N times around a thin toroidal iron core of radius a and permeability $\mu \gg 1$. The core contains a thin gap of width d. What is the magnetic field induced in the gap? Let us neglect any leakage of magnetic flux from the core, which is reasonable if $\mu \gg 1$. We expect the magnetic field, B_c, and the magnetic intensity, H_c, in the core to both be toroidal and essentially uniform. It is also reasonable to suppose that the magnetic field, B_g, and the magnetic intensity, H_g, in the gap are toroidal and uniform, since $d \ll a$. We have $B_c = \mu_0\,\mu\,H_c$ and $B_g = \mu_0\,H_g$. Moreover, since the magnetic field is normal to the interface between the core and the gap, the continuity of $B_\perp$ implies that

$$B_c = B_g. \tag{6.156}$$

Thus, the magnetic field-strength in the core is the same as that in the gap. However, the magnetic intensities in the core and the gap are quite

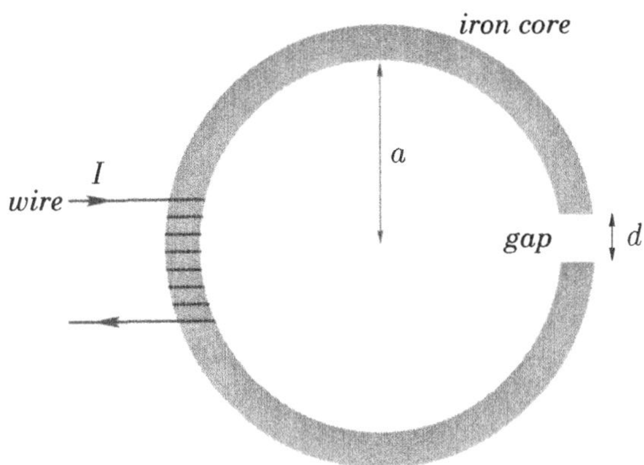

Figure 6.10: *An electromagnet.*

different: $H_c = B_c/\mu_0\,\mu = B_g/\mu_0\,\mu = H_g/\mu$. Integration of Equation (6.128) around the torus yields

$$\oint \mathbf{H}\cdot d\mathbf{l} = \int \mathbf{j}_t \cdot d\mathbf{S} = N\,I. \tag{6.157}$$

Hence,

$$(2\pi\,a - d)\,H_c + d\,H_g = N\,I. \tag{6.158}$$

It follows that

$$B_g = \frac{N\,\mu_0\,I}{(2\pi\,a - d)/\mu + d}. \tag{6.159}$$

Note that if $\mu \gg 1$ and $d \ll 2\pi\,a$ then the magnetic field in the gap is considerably larger than that which would be obtained if the core of the electromagnet were not ferromagnetic.

6.16 MAGNETIC ENERGY

Consider an electrical conductor. Suppose that a battery with an electromotive field $\mathbf{E}'$ is feeding energy into this conductor. The energy is either dissipated as heat, or is used to generate a magnetic field. Ohm's law inside the conductor gives

$$\mathbf{j}_t = \sigma\,(\mathbf{E} + \mathbf{E}'), \tag{6.160}$$

where $\mathbf{j}_t$ is the true current density, σ is the conductivity, and $\mathbf{E}$ is the inductive electric field. Taking the scalar product with $\mathbf{j}_t$, we obtain

$$\mathbf{E}'\cdot\mathbf{j}_t = \frac{j_t^2}{\sigma} - \mathbf{E}\cdot\mathbf{j}_t. \tag{6.161}$$

The left-hand side of this equation represents the rate at which the battery does work on the conductor. The first term on the right-hand side is the rate of ohmic heating inside the conductor. Thus, the remaining term must represent the rate at which energy is fed into the magnetic field. If all fields are quasi-static (*i.e.*, slowly varying) then the displacement current can be neglected, and the differential form of Ampère's law reduces to $\nabla\times\mathbf{H} = \mathbf{j}_t$. Substituting this expression into Equation (6.161), and integrating over all space, we get

$$\int \mathbf{E}'\cdot(\nabla\times\mathbf{H})\,d^3\mathbf{r} = \int \frac{(\nabla\times\mathbf{H})^2}{\sigma}\,d^3\mathbf{r} - \int \mathbf{E}\cdot(\nabla\times\mathbf{H})\,d^3\mathbf{r}. \tag{6.162}$$

The last term can be integrated by parts using the vector identity

$$\nabla\cdot(\mathbf{E}\times\mathbf{H}) \equiv \mathbf{H}\cdot(\nabla\times\mathbf{E}) - \mathbf{E}\cdot(\nabla\times\mathbf{H}). \tag{6.163}$$

Gauss' theorem plus the differential form of Faraday's law yield

$$\int \mathbf{E}\cdot(\nabla\times\mathbf{H})\,d^3\mathbf{r} = -\int \mathbf{H}\cdot\frac{\partial\mathbf{B}}{\partial t}\,d^3\mathbf{r} - \int (\mathbf{E}\times\mathbf{H})\cdot d\mathbf{S}. \tag{6.164}$$

Since $\mathbf{E}\times\mathbf{H}$ falls off at least as fast as $1/r^5$ in quasi-static electric and magnetic fields ($1/r^2$ comes from electric monopole fields, and $1/r^3$ from magnetic dipole fields), the surface integral in the above expression can be neglected. Of course, this is not the case for radiation fields, for which $\mathbf{E}$ and $\mathbf{H}$ both fall off like $1/r$. Thus, the "quasi-static" constraint effectively means that the fields vary sufficiently slowly that any radiation fields can be neglected.

The total power expended by the battery can now be written

$$\int \mathbf{E}'\cdot(\nabla\times\mathbf{H})\,d^3\mathbf{r} = \int \frac{(\nabla\times\mathbf{H})^2}{\sigma}\,d^3\mathbf{r} + \int \mathbf{H}\cdot\frac{\partial\mathbf{B}}{\partial t}\,d^3\mathbf{r}. \tag{6.165}$$

The first term on the right-hand side has already been identified as the energy loss rate due to ohmic heating, and the second as the rate at which energy is fed into the magnetic field. The variation δW in the magnetic field energy can therefore be written

$$\delta W = \int \mathbf{H}\cdot\delta\mathbf{B}\, d^3\mathbf{r}. \tag{6.166}$$

This result is analogous to the result (6.60) for the variation in the energy of an electrostatic field.

In order to make Equation (6.166) integrable, we must assume a functional relationship between **H** and **B**. For a medium which magnetizes linearly, the integration can be carried out in an analogous manner to that used to derive Equation (6.63), to give

$$W = \frac{1}{2}\int \mathbf{H}\cdot\mathbf{B}\, d^3\mathbf{r}. \tag{6.167}$$

Thus, the magnetostatic energy density inside a linear magnetic material is given by

$$U = \frac{1}{2}\,\mathbf{H}\cdot\mathbf{B}. \tag{6.168}$$

Unfortunately, most interesting magnetic materials, such as ferromagnets, exhibit a nonlinear relationship between **H** and **B**. For such materials, Equation (6.166) can only be integrated between definite states, and the result, in general, depends on the past history of the sample. For ferromagnets, the integral of Equation (6.166) has a finite, non-zero value when **B** is integrated around a complete magnetization cycle. This cyclic energy loss is given by

$$\Delta W = \int \oint \mathbf{H}\cdot d\mathbf{B}\, d^3\mathbf{r}. \tag{6.169}$$

In other words, the energy expended per unit volume when a magnetic material is carried around a magnetization cycle is equal to the *area* of its hysteresis loop, as plotted in a graph of B against H. Thus, it is particularly important to ensure that the magnetic materials used to form transformer cores possess hysteresis loops with comparatively small areas, otherwise the transformers are likely to be extremely lossy.

6.17 EXERCISES

6.1. An infinite slab of dielectric of uniform dielectric constant ϵ lies between the planes $z = -a$ and $z = a$. Suppose that the slab contains free charge of uniform charge density ρ_f. Find **E**, **D**, and **P** as functions of z. What is the bound charge sheet density on the two faces of the slab?

6.2. An infinite slab of uncharged dielectric of uniform dielectric constant ϵ lies between the planes $z = -a$ and $z = a$, and is placed in a uniform electric field **E** whose field-lines make an angle θ with the z-axis. What is the bound charge sheet density on the two faces of the slab?

6.3. An uncharged dielectric sphere of radius a, centered on the origin, possesses a polarization field $\mathbf{P} = p\,\mathbf{r}$, where p is a constant. Find the bound charge density inside the sphere, and the bound charge sheet density on the surface of the sphere. Find **E** and **D** both inside and outside the sphere.

6.4. An infinite dielectric of dielectric constant ϵ contains a uniform electric field $\mathbf{E}_0$. Find the electric field inside a needle-shaped cavity running parallel to $\mathbf{E}_0$. Find the field inside a wafer-shaped cavity aligned perpendicular to $\mathbf{E}_0$. Neglect end effects.

6.5. Consider a plane interface between two uniform dielectrics of dielectric constants ϵ_1 and ϵ_2. A straight electric field-line which passes across the interface is bent at an angle. Demonstrate that

$$\epsilon_1\,\tan\theta_2 = \epsilon_2\,\tan\theta_1,$$

where θ_1 is the angle the field-line makes with the normal to the interface in medium 1, *etc.*

6.6. A charge q lies at the center of an otherwise uncharged dielectric sphere of radius a and uniform dielectric constant ϵ. Find **D** and **E** throughout space. Find the bound charge sheet density on the surface of the sphere.

6.7. A cylindrical coaxial cable consists of an inner conductor of radius a, surrounded by a dielectric sheath of dielectric constant ϵ_1 and outer radius b, surrounded by a second dielectric sheath of dielectric constant ϵ_2 and outer radius c, surrounded by an outer conductor. All components of the cable are touching. What is the capacitance per unit length of the cable?

6.8. A long dielectric cylinder of radius a and uniform dielectric constant ϵ is placed in a uniform electric field $\mathbf{E}_0$ which runs perpendicular to the axis of the cylinder. Find the electric field both inside and outside the cylinder. Find the bound charge sheet density on the surface of the cylinder. Hint: Use separation of variables.

6.9. An electric dipole of moment $\mathbf{p} = p\,\mathbf{e}_z$ lies at the center of an uncharged dielectric sphere of radius a and uniform dielectric constant ϵ. Find **D** and **E** throughout space. Find the bound charge sheet density on the surface of the sphere. Hint: Use the separation of variables.

6.10. A parallel plate capacitor has plates of area A and spacing d. Half of the region between the plates is filled with a dielectric of uniform dielectric constant ϵ_1, and the other half is filled with a dielectric of uniform dielectric constant ϵ_2. If the interface between the two dielectric media is a plane parallel to the two plates, lying half-way between them, what is the capacitance of the capacitor? If the interface is perpendicular to the two plates, and bisects them, what is the capacitance of the capacitor?

6.11. Consider a parallel plate capacitor whose plates are of area A and spacing d. Find the force of attraction per unit area between the plates when:

(a) The region between the plates is empty and the capacitor is connected to a battery of voltage V.

(b) The capacitor is disconnected from the battery (but remains charged), and then fully immersed in a dielectric liquid of uniform dielectric constant ϵ.

(c) The dielectric liquid is replaced by a slab of solid dielectric of uniform dielectric constant ϵ which fills the region between the plates, but does not touch the plates.

(d) The uncharged capacitor is fully immersed in a dielectric liquid of uniform dielectric constant ϵ, and then charged to a voltage V.

(e) The region between the plates of the uncharged capacitor is filled by a solid dielectric of uniform dielectric constant ϵ, which does not touch the plates, and the capacitor is then charged to a voltage V.

6.12. A parallel plate capacitor has the region between its electrodes completely filled with a dielectric slab of uniform dielectric constant ϵ. The plates are of length l, width w, and spacing d. The capacitor is charged until its plates are at a potential difference V, and then disconnected. The dielectric slab is then partially withdrawn in the l dimension until only a length x remains between the plates. What is the potential difference between the plates? What is the force acting to pull the slab back toward its initial position? Neglect end effects. Hint: Use an energy argument to calculate the force.

6.13. A parallel plate capacitor with electrodes of area A, spacing d, which carry the fixed charges $\pm Q$, is dipped vertically into a large vat of dielectric liquid of uniform dielectric constant ϵ and mass density ρ_m. What height h does the liquid rise between the plates (relative to the liquid level outside the plates)? Neglect end effects.

6.14. A solenoid consists of a wire wrapped uniformly around a long, solid, cylindrical ferromagnetic core of radius a and uniform permeability μ. Suppose that there are N turns of the wire per unit length. What is the magnetic field-strength inside the core when a current I flows through the wire? Suppose that the core is replaced by a cylindrical annulus of the same material. What is the magnetic field-strength in the cylindrical cavity inside the annulus when a current I flows through the wire?

6.15. A long straight wire carries a current I and is surrounded by a ferromagnetic co-axial cylindrical annulus of uniform permeability μ, inner radius a, and outer radius b. Find the magnetic field everywhere. Find the magnetization current density both within the annulus and on the surfaces of the annulus.

6.16. An infinitely long cylinder of radius a, which is coaxial with the z-axis, has a uniform magnetization $\mathbf{M} = M\,\mathbf{e}_z$. Find the induced magnetic field both inside and outside the cylinder. Find the magnetization current density on the surface of the cylinder.

6.17. A very large piece of magnetic material of constant permeability μ contains a uniform magnetic field $\mathbf{B}_0$. Find the magnetic field inside a needle-shaped cavity running parallel to $\mathbf{B}_0$. Find the field inside a wafer-shaped cavity aligned perpendicular to $\mathbf{B}_0$. Neglect end effects.

6.18. A spherical annulus of magnetic material of inner radius a, outer radius b, and uniform permeability μ is placed in the uniform magnetic field $\mathbf{B} = B_0\,\mathbf{e}_z$. Find the magnetic scalar potential everywhere. What is the magnetic field inside the shell? Hint: Use the separation of variables.

6.19. A long ferromagnetic cylinder of radius a and uniform permeability μ is placed in a uniform magnetic field $\mathbf{B}_0$ which runs perpendicular to the axis of the cylinder. Find the magnetic field both inside and outside the cylinder. Find the magnetization current on the surface of the cylinder. Hint: Use separation of variables.

6.20. A magnetic dipole of moment $\mathbf{m} = m\,\mathbf{e}_z$ lies at the center of a ferromagnetic sphere of radius a and uniform permeability μ. Find $\mathbf{H}$ and $\mathbf{B}$ throughout space. Find the magnetization current density on the surface of the sphere. Hint: Use the separation of variables.

6.21. A magnet consists of a thin ring of magnetic material of radius a containing a narrow gap of width d, where $d \ll a$ (*i.e.*, the magnet has the same shape as the core of the electromagnet shown in Figure 6.10). The magnetic material possesses a uniform permanent magnetization $\mathbf{M} = M\,\mathbf{e}_\varphi$, where $\mathbf{e}_\varphi$ is a unit vector which runs toroidally around the ring. What is the strength of the magnetic field in the gap? If A is the cross-sectional area of the ring, how much magnetostatic energy does the magnet possess? Neglect field leakage.

6.22. Suppose that half of the permanently magnetized material making up the ring in the previous question is replaced by material of uniform permeability μ. What is the magnetic field-strength in the gap? How much magnetostatic energy does the magnet possess? Neglect field leakage.

Chapter 7

MAGNETIC INDUCTION

7.1 INTRODUCTION

In this chapter, we shall use Maxwell's equations to investigate magnetic induction and related phenomena.

7.2 INDUCTANCE

We have already learned about the concepts of voltage, resistance, and capacitance. Let us now investigate the concept of *inductance*. Electrical engineers like to reduce all pieces of electrical circuitary to an *equivalent circuit* consisting of pure voltage sources, pure inductors, pure capacitors, and pure resistors. Hence, once we understand inductors, we shall be ready to apply the laws of electromagnetism to general electrical circuits.

Consider two stationary loops of wire, labeled 1 and 2—see Figure 7.1. Let us run a steady current I_1 around the first loop to produce a magnetic field $\mathbf{B}_1$. Some of the field-lines of $\mathbf{B}_1$ will pass through the second loop. Let Φ_2 be the flux of $\mathbf{B}_1$ through loop 2,

$$\Phi_2 = \int_{\text{loop 2}} \mathbf{B}_1 \cdot d\mathbf{S}_2, \tag{7.1}$$

where $d\mathbf{S}_2$ is a surface element of loop 2. This flux is generally quite difficult to calculate exactly (unless the two loops have a particularly simple geometry). However, we can infer from the Biot-Savart law,

$$\mathbf{B}_1(\mathbf{r}) = \frac{\mu_0 I_1}{4\pi} \oint_{\text{loop 1}} \frac{d\mathbf{l}_1 \times (\mathbf{r} - \mathbf{r}_1)}{|\mathbf{r} - \mathbf{r}_1|^3}, \tag{7.2}$$

that the magnitude of $\mathbf{B}_1$ is proportional to the current I_1. This is ultimately a consequence of the linearity of Maxwell's equations. Here, $d\mathbf{l}_1$ is a line element of loop 1 located at position vector $\mathbf{r}_1$. It follows that

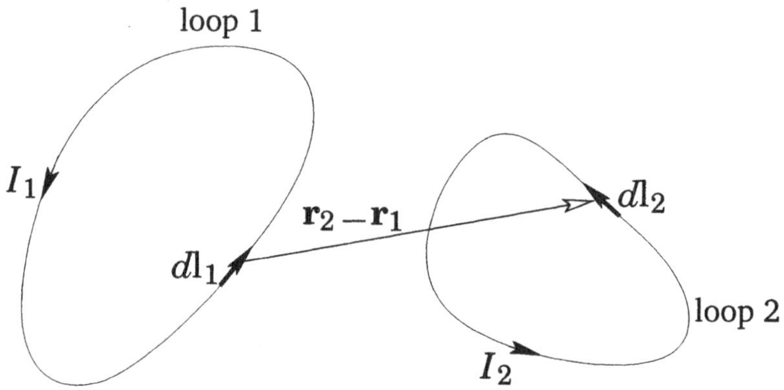

Figure 7.1: *Two current-carrying loops.*

the flux Φ_2 must also be proportional to I_1. Thus, we can write

$$\Phi_2 = M_{21}\, I_1, \tag{7.3}$$

where M_{21} is a constant of proportionality. This constant is called the *mutual inductance* of the two loops.

Let us write the magnetic field $\mathbf{B}_1$ in terms of a vector potential $\mathbf{A}_1$, so that

$$\mathbf{B}_1 = \nabla \times \mathbf{A}_1. \tag{7.4}$$

It follows from Stokes' theorem that

$$\Phi_2 = \int_{\text{loop 2}} \mathbf{B}_1 \cdot d\mathbf{S}_2 = \int_{\text{loop 2}} \nabla \times \mathbf{A}_1 \cdot d\mathbf{S}_2 = \oint_{\text{loop 2}} \mathbf{A}_1 \cdot d\mathbf{l}_2, \tag{7.5}$$

where $d\mathbf{l}_2$ is a line element of loop 2. However, we know that

$$\mathbf{A}_1(\mathbf{r}) = \frac{\mu_0\, I_1}{4\pi} \oint_{\text{loop 1}} \frac{d\mathbf{l}_1}{|\mathbf{r} - \mathbf{r}_1|}. \tag{7.6}$$

The above equation is just a special case of the more general law,

$$\mathbf{A}_1(\mathbf{r}) = \frac{\mu_0}{4\pi} \int_{\text{all space}} \frac{\mathbf{j}(\mathbf{r}')}{|\mathbf{r} - \mathbf{r}'|}\, d^3\mathbf{r}', \tag{7.7}$$

for $\mathbf{j}(\mathbf{r}_1) = d\mathbf{l}_1\, I_1/dl_1\, dA$ and $d^3\mathbf{r}' = dl_1\, dA$, where dA is the cross-sectional area of loop 1. Thus,

$$\Phi_2 = \frac{\mu_0\, I_1}{4\pi} \oint_{\text{loop 1}} \oint_{\text{loop 2}} \frac{d\mathbf{l}_1 \cdot d\mathbf{l}_2}{|\mathbf{r}_2 - \mathbf{r}_1|}, \tag{7.8}$$

where $\mathbf{r}_2$ is the position vector of the line element $d\mathbf{l}_2$ of loop 2, which implies that

$$M_{21} = \frac{\mu_0}{4\pi} \oint_{\text{loop 1}} \oint_{\text{loop 2}} \frac{d\mathbf{l}_1 \cdot d\mathbf{l}_2}{|\mathbf{r}_2 - \mathbf{r}_1|}. \tag{7.9}$$

In fact, mutual inductances are rarely worked out using the above formula, because it is usually much too difficult. However, this expression—which is known as the *Neumann formula*—tells us two important things. Firstly, the mutual inductance of two current loops is a purely *geometric* quantity, having to do with the sizes, shapes, and relative orientations of the loops. Secondly, the integral is unchanged if we switch the roles of loops 1 and 2. In other words,

$$M_{21} = M_{12}. \tag{7.10}$$

Hence, we can drop the subscripts, and just call both these quantities M. This is a rather surprising result. It implies that no matter what the shapes and relative positions of the two loops, the magnetic flux through loop 2 when we run a current I around loop 1 is *exactly* the same as the flux through loop 1 when we run the same current around loop 2.

We have seen that a current I flowing around some wire loop, 1, generates a magnetic flux linking some other loop, 2. However, flux is also generated through the first loop. As before, the magnetic field, and, therefore, the flux, Φ, is proportional to the current, so we can write

$$\Phi = L\, I. \tag{7.11}$$

The constant of proportionality L is called the *self-inductance*. Like M it only depends on the geometry of the loop.

Inductance is measured in SI units called henries (H): 1 henry is 1 volt-second per ampere. The henry, like the farad, is a rather unwieldy unit, since inductors in electrical circuits typically have inductances of order one micro-henry.

7.3 SELF-INDUCTANCE

Consider a long, uniformly wound, cylindrical solenoid of length l, and radius r, which has N turns per unit length, and carries a current I. The longitudinal (*i.e.*, directed along the axis of the solenoid) magnetic field within the solenoid is approximately uniform, and is given by

$$B = \mu_0 N I. \tag{7.12}$$

(This result is easily obtained by integrating Ampère's law over a rectangular loop whose long sides run parallel to the axis of the solenoid, one inside the solenoid, and the other outside, and whose short sides run perpendicular to the axis.) The magnetic flux though each turn of the solenoid wire is $B \pi r^2 = \mu_0 N I \pi r^2$. The total flux through the solenoid wire, which has $N l$ turns, is

$$\Phi = N l \mu_0 N I \pi r^2. \tag{7.13}$$

Thus, the self-inductance of the solenoid is

$$L = \frac{\Phi}{I} = \mu_0 N^2 \pi r^2 l. \tag{7.14}$$

Note that the self-inductance only depends on geometric quantities, such as the number of turns per unit length of the solenoid, and the cross-sectional area of the turns.

Suppose that the current I flowing through the solenoid changes. Let us assume that the change is sufficiently slow that we can neglect the displacement current, and retardation effects, in our calculations. This implies that the typical time-scale of the change must be much longer than the time for a light-ray to traverse the circuit. If this is the case then the above formulae remain valid.

A change in the current implies a change in the magnetic flux linking the solenoid wire, since $\Phi = L I$. According to Faraday's law, this change generates an emf in the wire. By Lenz's law, the emf is such as to oppose the change in the current—*i.e.*, it is a *back-emf*. Thus, we can write

$$V = -\frac{d\Phi}{dt} = -L \frac{dI}{dt}, \tag{7.15}$$

where V is the generated back-emf.

Suppose that our solenoid has an electrical resistance R. Let us connect the ends of the solenoid across the terminals of a battery of constant voltage V. What is going to happen? The equivalent circuit is shown in Figure 7.2. The inductance and resistance of the solenoid are represented by a perfect inductor, L, and a perfect resistor, R, connected in series. The voltage drop across the inductor and resistor is equal to the voltage of the battery, V. The voltage drop across the resistor is simply $I R$, whereas

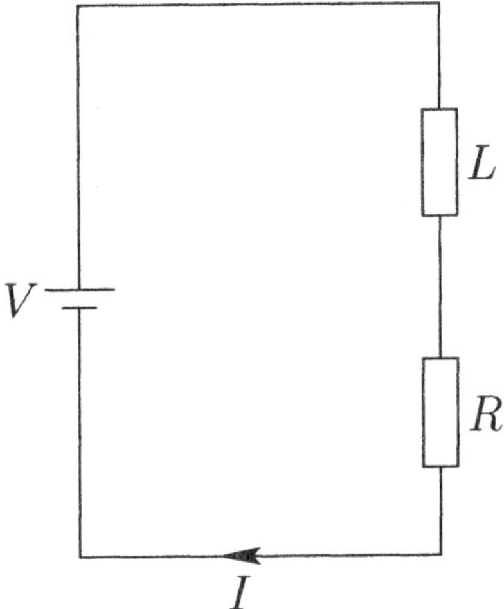

Figure 7.2: *The equivalent circuit of a solenoid connected to a battery.*

the voltage drop across the inductor (*i.e.*, the back-emf) is $L\,dI/dt$. Here, I is the current flowing through the solenoid. It follows that

$$V = I\,R + L\,\frac{dI}{dt}. \tag{7.16}$$

This is a differential equation for the current I. We can rearrange it to give

$$\frac{dI}{dt} + \frac{R}{L}\,I = \frac{V}{L}. \tag{7.17}$$

The general solution is

$$I(t) = \frac{V}{R} + k\exp(-R\,t/L). \tag{7.18}$$

The constant k is fixed by the boundary conditions. Suppose that the battery is connected at time $t = 0$, when $I = 0$. It follows that $k = -V/R$, so that

$$I(t) = \frac{V}{R}\,[1 - \exp(-R\,t/L)\,]. \tag{7.19}$$

This curve is shown in Figure 7.3. It can be seen that, after the battery is connected, the current ramps up, and attains its steady-state value V/R

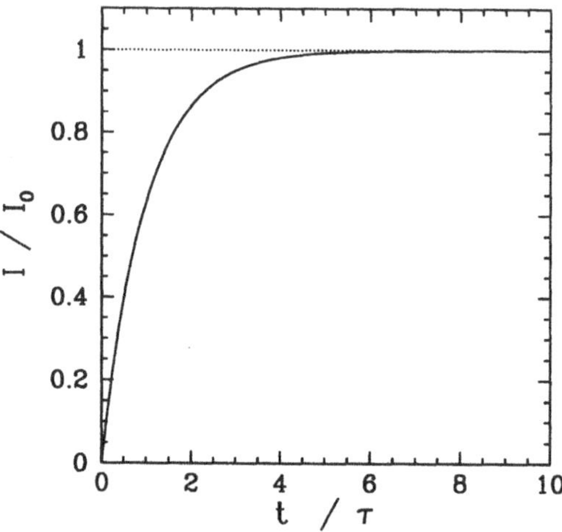

Figure 7.3: *Typical current rise profile in a circuit of the type shown in Figure 7.2. Here,* $I_0 = V/R$ *and* $\tau = L/R$.

(which comes from Ohm's law), on the characteristic time-scale

$$\tau = \frac{L}{R}. \tag{7.20}$$

This time-scale is sometimes called the *time constant* of the circuit, or (somewhat unimaginatively) the *L over R time* of the circuit.

We can now appreciate the significance of self-inductance. The back-emf generated in an inductor, as the current flowing through it tries to change, prevents the current from rising (or falling) much faster than the L/R time. This effect is sometimes advantageous, but is often a great nuisance. All circuit elements possess some self-inductance, as well as some resistance, and thus have a finite L/R time. This means that when we power up a DC circuit, the current does not jump up instantaneously to its steady-state value. Instead, the rise is spread out over the L/R time of the circuit. This is a good thing. If the current were to rise instantaneously then extremely large electric fields would be generated by the sudden jump in the induced magnetic field, leading, inevitably, to breakdown and electric arcing. So, if there were no such thing as self-inductance then every time we switched an electric circuit on or off there would be a blue flash due to arcing between conductors. Self-inductance can also be a bad thing. Suppose that we possess an expensive power supply which can generate a wide variety of complicated voltage waveforms,

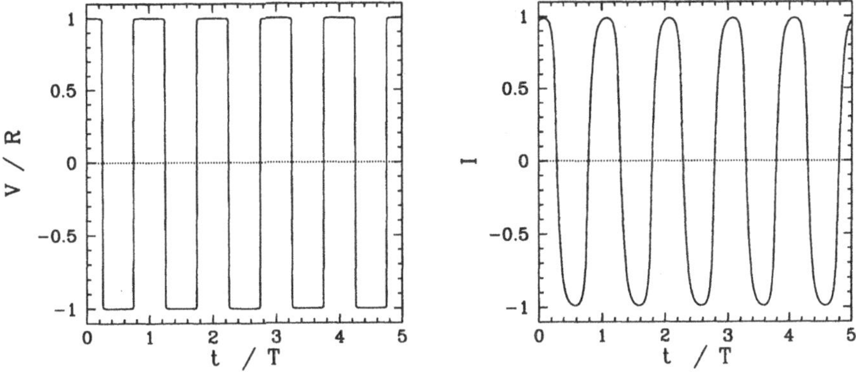

Figure 7.4: *Difference between the input waveform (left panel) and the output waveform (right panel) when a square-wave is sent down a wire whose* L/R *time is 1/20th of the square-wave period,* T.

and we wish to use it to send such a waveform down a wire (or transmission line). Of course, the wire or transmission line will possess both resistance and inductance, and will, therefore, have some characteristic L/R time. Suppose that we try to send a square-wave signal down the wire. Since the current in the wire cannot rise or fall faster than the L/R time, the leading and trailing edges of the signal will get smoothed out over an L/R time. The typical difference between the signal fed into the wire, and that which comes out of the other end, is illustrated in Figure 7.4. Clearly, there is little point in having an expensive power supply unless we also possess a low-inductance wire, or transmission line, so that the signal from the power supply can be transmitted to some load device without serious distortion.

7.4 MUTUAL INDUCTANCE

Consider, now, two long thin cylindrical solenoids, one wound on top of the other. The length of each solenoid is l, and the common radius is r. Suppose that the bottom coil has N_1 turns per unit length, and carries a current I_1. The magnetic flux passing through each turn of the top coil is $\mu_0 N_1 I_1 \pi r^2$, and the total flux linking the top coil is therefore $\Phi_2 = N_2 l \mu_0 N_1 I_1 \pi r^2$, where N_2 is the number of turns per unit length in the top coil. It follows that the mutual inductance of the two coils, defined $\Phi_2 = M\,I_1$, is given by

$$M = \mu_0\,N_1\,N_2\,\pi\,r^2\,l. \tag{7.21}$$

Recall that the self-inductance of the bottom coil is

$$L_1 = \mu_0\,N_1^2\,\pi\,r^2\,l, \tag{7.22}$$

and that of the top coil is

$$L_2 = \mu_0\,N_2^2\,\pi\,r^2\,l. \tag{7.23}$$

Hence, the mutual inductance can be written

$$M = \sqrt{L_1\,L_2}. \tag{7.24}$$

Note that this result depends on the assumption that *all* of the magnetic flux produced by one coil passes through the other coil. In reality, some of the flux leaks out, so that the mutual inductance is somewhat less than that given in the above formula. We can write

$$M = k\,\sqrt{L_1\,L_2}, \tag{7.25}$$

where the dimensionless constant k is called the *coefficient of coupling*, and lies in the range $0 \leq k \leq 1$.

Suppose that the two coils have resistances R_1 and R_2. If the bottom coil has an instantaneous current I_1 flowing through it, and a total voltage drop V_1, then the voltage drop due to its resistance is $I_1\,R_1$. The voltage drop due to the back-emf generated by the self-inductance of the coil is $L_1\,dI_1/dt$. There is also a back-emf due to inductive coupling with the top coil. We know that the flux through the bottom coil due to the instantaneous current I_2 flowing in the top coil is

$$\Phi_1 = M\,I_2. \tag{7.26}$$

Thus, by Faraday's law and Lenz's law, the back-emf induced in the bottom coil is

$$V = -M\,\frac{dI_2}{dt}. \tag{7.27}$$

The voltage drop across the bottom coil due to its mutual inductance with the top coil is minus this expression. Thus, the circuit equation for the bottom coil is

$$V_1 = R_1\,I_1 + L_1\frac{dI_1}{dt} + M\,\frac{dI_2}{dt}. \tag{7.28}$$

Likewise, the circuit equation for the top coil is

$$V_2 = R_2\, I_2 + L_2 \frac{dI_2}{dt} + M\, \frac{dI_1}{dt}. \tag{7.29}$$

Here, V_2 is the total voltage drop across the top coil.

Suppose that we suddenly connect a battery of constant voltage V_1 to the bottom coil, at time $t = 0$. The top coil is assumed to be open-circuited, or connected to a voltmeter of very high internal resistance, so that $I_2 = 0$. What is the voltage generated in the top coil? Since $I_2 = 0$, the circuit equation for the bottom coil is

$$V_1 = R_1\, I_1 + L_1 \frac{dI_1}{dt}, \tag{7.30}$$

where V_1 is constant, and $I_1(t = 0) = 0$. We have already seen the solution to this equation:

$$I_1 = \frac{V_1}{R_1}\,[1 - \exp(-R_1\, t/L_1)]\,. \tag{7.31}$$

The circuit equation for the top coil is

$$V_2 = M\, \frac{dI_1}{dt}, \tag{7.32}$$

giving

$$V_2 = V_1\, \frac{M}{L_1}\, \exp(-R_1\, t/L_1). \tag{7.33}$$

It follows from Equation (7.25) that

$$V_2 = V_1\, k\, \sqrt{\frac{L_2}{L_1}}\, \exp(-R_1\, t/L_1). \tag{7.34}$$

Since $L_1/L_2 = N_1^2/N_2^2$, we obtain

$$V_2 = V_1\, k\, \frac{N_2}{N_1}\, \exp(-R_1\, t/L_1). \tag{7.35}$$

Note that $V_2(t)$ is discontinuous at $t = 0$. This is not a problem, since the resistance of the top circuit is infinite, so there is no discontinuity in the current (and, hence, in the magnetic field). But, what about the displacement current, which is proportional to $\partial \mathbf{E}/\partial t$? Surely, this is discontinuous at $t = 0$ (which is clearly unphysical)? The crucial point, here, is that we have specifically neglected the displacement current in all of our previous analysis, so it does not make much sense to start worrying about it now. If we had retained the displacement current in our calculations then we would have found that the voltage in the top circuit jumps up, at $t = 0$, on a time-scale similar to the light traverse time across the circuit (*i.e.*, the jump is instantaneous, to all intents and purposes, but the displacement current remains finite).

Now,

$$\frac{V_2(t=0)}{V_1} = k\,\frac{N_2}{N_1}, \tag{7.36}$$

so if $N_2 \gg N_1$ then the voltage in the bottom circuit is considerably amplified in the top circuit. This effect is the basis for old-fashioned car ignition systems. A large voltage spike is induced in a secondary circuit (connected to a coil with very many turns) whenever the current in a primary circuit (connected to a coil with not so many turns) is either switched on or off. The primary circuit is connected to the car battery (whose voltage is typically 12 volts). The switching is done by a set of points, which are mechanically opened and closed as the engine turns. The large voltage spike induced in the secondary circuit, as the points are either opened or closed, causes a spark to jump across a gap in this circuit. This spark ignites a petrol/air mixture in one of the cylinders. We might think that the optimum configuration is to have only one turn in the primary circuit, and lots of turns in the secondary circuit, so that the ratio N_2/N_1 is made as large as possible. However, this is not the case. Most of the magnetic flux generated by a single turn primary coil is likely to miss the secondary coil altogether. This means that the coefficient of coupling k is small, which reduces the voltage induced in the secondary circuit. Thus, we need a reasonable number of turns in the primary coil in order to localize the induced magnetic flux, so that it links effectively with the secondary coil.

7.5 MAGNETIC ENERGY

Suppose that, at $t = 0$, a coil of inductance, L, and resistance R, is connected across the terminals of a battery of voltage V. The circuit equation is

$$V = L\,\frac{dI}{dt} + R\,I. \tag{7.37}$$

Now, the power output of the battery is $V I$. [Every charge q that goes around the circuit falls through a potential difference $q V$. In order to raise it back to the starting potential, so that it can perform another circuit, the battery must do work $q V$. The work done per unit time (*i.e.*, the power) is $n q V$, where n is the number of charges per unit time passing a given point on the circuit. But, $I = n q$, so the power output is $V I$.] Thus, the net work done by the battery in raising the current in the circuit from zero at time $t = 0$ to I_T at time $t = T$ is

$$W = \int_0^T V I \, dt. \tag{7.38}$$

Using the circuit equation (7.37), we obtain

$$W = L \int_0^T I \frac{dI}{dt} \, dt + R \int_0^T I^2 \, dt, \tag{7.39}$$

giving

$$W = \frac{1}{2} L I_T^2 + R \int_0^T I^2 \, dt. \tag{7.40}$$

The second term on the right-hand side of the above equation represents the irreversible conversion of electrical energy into heat energy by the resistor. The first term is the amount of energy stored in the inductor at time T. This energy can be recovered after the inductor is disconnected from the battery. Suppose that the battery is disconnected at time T. The circuit equation is now

$$0 = L \frac{dI}{dt} + R I, \tag{7.41}$$

giving

$$I = I_T \exp\left[-\frac{R}{L}(t - T)\right], \tag{7.42}$$

where we have made use of the boundary condition $I(T) = I_T$. Thus, the current decays away exponentially. The energy stored in the inductor is dissipated as heat in the resistor. The total heat energy appearing in the resistor after the battery is disconnected is

$$\int_T^\infty I^2 R \, dt = \frac{1}{2} L I_T^2, \tag{7.43}$$

where use has been made of Equation (7.42). Thus, the heat energy appearing in the resistor is equal to the energy stored in the inductor.

This energy is actually stored in the magnetic field generated around the inductor.

Consider, again, our circuit with two coils wound on top of one another. Suppose that each coil is connected to its own battery. The circuit equations are thus

$$V_1 = R_1 I_1 + L_1 \frac{dI_1}{dt} + M \frac{dI_2}{dt},$$

$$V_2 = R_2 I_2 + L_2 \frac{dI_2}{dt} + M \frac{dI_1}{dt}, \tag{7.44}$$

where V_1 is the voltage of the battery in the first circuit, *etc.* The net work done by the two batteries in increasing the currents in the two circuits, from zero at time 0, to I_1 and I_2 at time T, respectively, is

$$\begin{aligned} W &= \int_0^T (V_1 I_1 + V_2 I_2)\, dt \\ &= \int_0^T (R_1 I_1^2 + R_2 I_2^2)\, dt + \frac{1}{2} L_1 I_1^2 + \frac{1}{2} L_2 I_2^2 \\ &\quad + M \int_0^T \left(I_1 \frac{dI_2}{dt} + I_2 \frac{dI_1}{dt} \right) dt. \end{aligned} \tag{7.45}$$

Thus,

$$\begin{aligned} W &= \int_0^T (R_1 I_1^2 + R_2 I_2^2)\, dt \\ &\quad + \frac{1}{2} L_1 I_1^2 + \frac{1}{2} L_2 I_2^2 + M I_1 I_2. \end{aligned} \tag{7.46}$$

Clearly, the total magnetic energy stored in the two coils is

$$W_B = \frac{1}{2} L_1 I_1^2 + \frac{1}{2} L_2 I_2^2 + M I_1 I_2. \tag{7.47}$$

Note that the mutual inductance term increases the stored magnetic energy if I_1 and I_2 are of the same sign—*i.e.*, if the currents in the two coils flow in the same direction, so that they generate magnetic fields which reinforce one another. Conversely, the mutual inductance term decreases the stored magnetic energy if I_1 and I_2 are of the opposite sign. However, the total stored energy can never be negative, otherwise the coils would constitute a power source (a negative stored energy is equivalent to a positive generated energy). Thus,

$$\frac{1}{2} L_1 I_1^2 + \frac{1}{2} L_2 I_2^2 + M I_1 I_2 \geq 0, \tag{7.48}$$

which can be written

$$\frac{1}{2}\left(\sqrt{L_1}\, I_1 + \sqrt{L_2}\, I_2\right)^2 - I_1\, I_2(\sqrt{L_1\, L_2} - M) \geq 0, \tag{7.49}$$

assuming that $I_1\, I_2 < 0$. It follows that

$$M \leq \sqrt{L_1\, L_2}. \tag{7.50}$$

The equality sign corresponds to the situation in which all of the magnetic flux generated by one coil passes through the other. If some of the flux misses then the inequality sign is appropriate. In fact, the above formula is valid for any two inductively coupled circuits, and effectively sets an upper limit on their mutual inductance.

We intimated previously that the energy stored in an inductor is actually stored in the surrounding magnetic field. Let us now obtain an explicit formula for the energy stored in a magnetic field. Consider an ideal cylindrical solenoid. The energy stored in the solenoid when a current I flows through it is

$$W = \frac{1}{2}\, L\, I^2, \tag{7.51}$$

where L is the self-inductance. We know that

$$L = \mu_0\, N^2\, \pi\, r^2\, l, \tag{7.52}$$

where N is the number of turns per unit length of the solenoid, r the radius, and l the length. The magnetic field inside the solenoid is approximately uniform, with magnitude

$$B = \mu_0\, N\, I, \tag{7.53}$$

and is approximately zero outside the solenoid. Equation (7.51) can be rewritten

$$W = \frac{B^2}{2\mu_0}\, V, \tag{7.54}$$

where $V = \pi\, r^2\, l$ is the volume of the solenoid. The above formula strongly suggests that a magnetic field possesses an energy density

$$U = \frac{B^2}{2\mu_0}. \tag{7.55}$$

Let us now examine a more general proof of the above formula. Consider a system of N circuits (labeled $i = 1$ to N), each carrying a current I_i. The magnetic flux through the ith circuit is written [*cf.*, Equation (7.5)]

$$\Phi_i = \int \mathbf{B} \cdot d\mathbf{S}_i = \oint \mathbf{A} \cdot d\mathbf{l}_i, \tag{7.56}$$

where $\mathbf{B} = \nabla \times \mathbf{A}$, and $d\mathbf{S}_i$ and $d\mathbf{l}_i$ denote a surface element and a line element of this circuit, respectively. The back-emf induced in the ith circuit follows from Faraday's law:

$$V_i = -\frac{d\Phi_i}{dt}. \tag{7.57}$$

The rate of work of the battery which maintains the current I_i in the ith circuit against this back-emf is

$$P_i = I_i \frac{d\Phi_i}{dt}. \tag{7.58}$$

Thus, the total work required to raise the currents in the N circuits from zero at time 0, to I_{0i} at time T, is

$$W = \sum_{i=1}^{N} \int_0^T I_i \frac{d\Phi_i}{dt} \, dt. \tag{7.59}$$

The above expression for the work done is, of course, equivalent to the total energy stored in the magnetic field surrounding the various circuits. This energy is independent of the manner in which the currents are set up. Suppose, for the sake of simplicity, that the currents are ramped up linearly, so that

$$I_i = I_{0i} \frac{t}{T}. \tag{7.60}$$

The fluxes are proportional to the currents, so they must also ramp up linearly: *i.e.*,

$$\Phi_i = \Phi_{0i} \frac{t}{T}. \tag{7.61}$$

It follows that

$$W = \sum_{i=1}^{N} \int_0^T I_{0i} \, \Phi_{0i} \frac{t}{T^2} \, dt, \tag{7.62}$$

giving

$$W = \frac{1}{2}\sum_{i=1}^{N} I_{0i}\,\Phi_{0i}. \tag{7.63}$$

So, if instantaneous currents I_i flow in the the N circuits, which link instantaneous fluxes Φ_i, then the instantaneous stored energy is

$$W = \frac{1}{2}\sum_{i=1}^{N} I_i\,\Phi_i. \tag{7.64}$$

Equations (7.56) and (7.64) imply that

$$W = \frac{1}{2}\sum_{i=1}^{N} I_i \oint \mathbf{A}\cdot d\mathbf{l}_i. \tag{7.65}$$

It is convenient, at this stage, to replace our N line currents by N current distributions of small, but finite, cross-sectional area. Equation (7.65) transforms to

$$W = \frac{1}{2}\int_V \mathbf{A}\cdot\mathbf{j}\,dV, \tag{7.66}$$

where V is a volume which contains all of the circuits. Note that for an element of the ith circuit, $\mathbf{j} = I_i\,d\mathbf{l}_i/dl_i\,A_i$ and $dV = dl_i\,A_i$, where A_i is the cross-sectional area of the circuit. Now, $\mu_0\,\mathbf{j} = \nabla\times\mathbf{B}$ (we are neglecting the displacement current in this calculation), so

$$W = \frac{1}{2\mu_0}\int_V \mathbf{A}\cdot\nabla\times\mathbf{B}\,dV. \tag{7.67}$$

According to vector field theory,

$$\nabla\cdot(\mathbf{A}\times\mathbf{B}) \equiv \mathbf{B}\cdot\nabla\times\mathbf{A} - \mathbf{A}\cdot\nabla\times\mathbf{B}, \tag{7.68}$$

which implies that

$$W = \frac{1}{2\mu_0}\int_V [-\nabla\cdot(\mathbf{A}\times\mathbf{B}) + \mathbf{B}\cdot\nabla\times\mathbf{A}]\,dV. \tag{7.69}$$

Using Gauss' theorem, and $\mathbf{B} = \nabla\times\mathbf{A}$, we obtain

$$W = -\frac{1}{2\mu_0}\oint_S \mathbf{A}\times\mathbf{B}\cdot d\mathbf{S} + \frac{1}{2\mu_0}\int_V B^2\,dV, \tag{7.70}$$

where S is the bounding surface of some volume V. Let us take this surface to infinity. It is easily demonstrated that the magnetic field generated by a current loop falls of like r^{-3} at large distances. The vector potential falls off like r^{-2}. However, the area of surface S only increases like r^2. It follows that the surface integral is negligible in the limit $r \to \infty$. Thus, the above expression reduces to

$$W = \int \frac{B^2}{2\mu_0}\, dV, \tag{7.71}$$

where the integral is over all space. Since this expression is valid for any magnetic field whatsoever, we can safely conclude that the energy density of a general magnetic field generated by a system of electrical circuits is given by

$$U = \frac{B^2}{2\mu_0}. \tag{7.72}$$

Note, that the above expression is consistent with Equation (6.168) which we previously obtained during our investigation of magnetic media.

7.6 ALTERNATING CURRENT CIRCUITS

Alternating current (AC) circuits are made up of voltage sources and *three* different types of passive elements: *i.e.*, resistors, inductors, and capacitors. Resistors satisfy Ohm's law,

$$V = I\,R, \tag{7.73}$$

where R is the resistance, I the current flowing through the resistor, and V the voltage drop across the resistor (in the direction in which the current flows). Inductors satisfy

$$V = L\,\frac{dI}{dt}, \tag{7.74}$$

where L is the inductance. Finally, capacitors obey

$$V = \frac{q}{C} = \int_0^t I\,dt \bigg/ C, \tag{7.75}$$

where C is the capacitance, q is the charge stored on the plate with the most positive potential, and $I = 0$ for $t < 0$. Note that any passive component of a real electrical circuit can always be represented as a combination of ideal resistors, inductors, and capacitors.

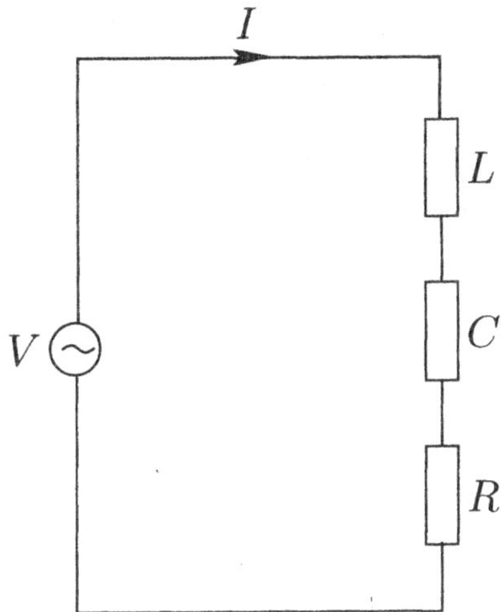

Figure 7.5: *An LCR circuit.*

Let us consider the classic LCR circuit, which consists of an inductor, L, a capacitor, C, and a resistor, R, all connected in series with a voltage source, V—see Figure 7.5. The circuit equation is obtained by setting the input voltage V equal to the sum of the voltage drops across the three passive elements in the circuit. Thus,

$$V = I\,R + L\,\frac{dI}{dt} + \int_0^t I\,dt \Big/ C. \tag{7.76}$$

This is an integro-differential equation which, in general, is quite difficult to solve. Suppose, however, that both the voltage and the current oscillate at some fixed angular frequency ω, so that

$$V(t) = V_0\,\exp(i\,\omega\,t), \tag{7.77}$$

$$I(t) = I_0\,\exp(i\,\omega\,t), \tag{7.78}$$

where the physical solution is understood to be the *real part* of the above expressions. The assumed behaviour of the voltage and current is clearly relevant to electrical circuits powered by the mains voltage (which oscillates at 60 hertz).

Equations (7.76)–(7.78) yield

$$V_0\,\exp(i\,\omega\,t) = I_0\,\exp(i\,\omega\,t)\,R + L\,i\,\omega\,I_0\,\exp(i\,\omega\,t) + \frac{I_0\,\exp(i\,\omega\,t)}{i\,\omega\,C}, \tag{7.79}$$

giving

$$V_0 = I_0 \left(i\,\omega\,L + \frac{1}{i\,\omega\,C} + R \right). \tag{7.80}$$

It is helpful to define the *impedance* of the circuit:

$$Z = \frac{V}{I} = i\,\omega\,L + \frac{1}{i\,\omega\,C} + R. \tag{7.81}$$

Impedance is a generalization of the concept of resistance. In general, the impedance of an AC circuit is a *complex* quantity.

The average power output of the voltage source is

$$P = \langle V(t)\,I(t)\rangle, \tag{7.82}$$

where the average is taken over one period of the oscillation. Let us, first of all, calculate the power using real (rather than complex) voltages and currents. We can write

$$V(t) = |V_0|\,\cos(\omega\,t), \tag{7.83}$$

$$I(t) = |I_0|\,\cos(\omega\,t - \theta), \tag{7.84}$$

where θ is the phase-lag of the current with respect to the voltage. It follows that

$$P = |V_0|\,|I_0| \int_{\omega t=0}^{\omega t=2\pi} \cos(\omega\,t)\,\cos(\omega\,t - \theta)\,\frac{d(\omega\,t)}{2\pi} \tag{7.85}$$

$$= |V_0|\,|I_0| \int_{\omega t=0}^{\omega t=2\pi} \cos(\omega\,t)\,[\cos(\omega\,t)\,\cos\theta + \sin(\omega\,t)\,\sin\theta]\,\frac{d(\omega\,t)}{2\pi},$$

giving

$$P = \frac{1}{2}|V_0|\,|I_0|\cos\theta, \tag{7.86}$$

since $\langle\cos(\omega\,t)\,\sin(\omega\,t)\rangle = 0$ and $\langle\cos(\omega\,t)\,\cos(\omega\,t)\rangle = 1/2$. In complex representation, the voltage and the current are written

$$V(t) = |V_0|\,\exp(i\,\omega\,t), \tag{7.87}$$

$$I(t) = |I_0|\,\exp[i\,(\omega\,t - \theta)]. \tag{7.88}$$

Now,

$$\frac{1}{2}(V\,I^* + V^*\,I) = |V_0|\,|I_0|\,\cos\theta. \tag{7.89}$$

It follows that

$$P = \frac{1}{4}(V\,I^* + V^*\,I) = \frac{1}{2}\,\mathrm{Re}(V\,I^*). \tag{7.90}$$

Making use of Equation (7.81), we find that

$$P = \frac{1}{2}\,\mathrm{Re}(Z)\,|I|^2 = \frac{1}{2}\frac{\mathrm{Re}(Z)\,|V|^2}{|Z|^2}. \tag{7.91}$$

Note that power dissipation is associated with the *real part* of the impedance. For the special case of an LCR circuit,

$$P = \frac{1}{2}\,R\,|I_0|^2. \tag{7.92}$$

We conclude that only the resistor dissipates energy in this circuit. The inductor and the capacitor both store energy, but they eventually return it to the circuit without dissipation.

According to Equation (7.81), the amplitude of the current which flows in an LCR circuit for a given amplitude of the input voltage is given by

$$|I_0| = \frac{|V_0|}{|Z|} = \frac{|V_0|}{\sqrt{(\omega\,L - 1/\omega\,C)^2 + R^2}}. \tag{7.93}$$

As can be seen from Figure 7.6, the response of the circuit is *resonant*, peaking at $\omega = 1/\sqrt{L\,C}$, and reaching $1/\sqrt{2}$ of the peak value at $\omega = 1/\sqrt{L\,C} \pm R/(2\,L)$ (assuming that $R \ll \sqrt{L/C}$). For this reason, LCR circuits are used in analog radio tuners to filter out signals whose frequencies fall outside a given band.

The phase-lag of the current with respect to the voltage is given by

$$\theta = \arg(Z) = \tan^{-1}\left(\frac{\omega\,L - 1/\omega\,C}{R}\right). \tag{7.94}$$

As can be seen from Figure 7.6, the phase-lag varies from $-\pi/2$ for frequencies significantly below the resonant frequency, to zero at the resonant frequency ($\omega = 1/\sqrt{L\,C}$), to $\pi/2$ for frequencies significantly above the resonant frequency.

It is clear that in conventional AC circuits the circuit equation reduces to a simple algebraic equation, and that the behavior of the circuit is summed up by the complex impedance, Z. The real part of Z tells us the power dissipated in the circuit, the magnitude of Z gives the ratio of the peak current to the peak voltage, and the argument of Z gives the phase-lag of the current with respect to the voltage.

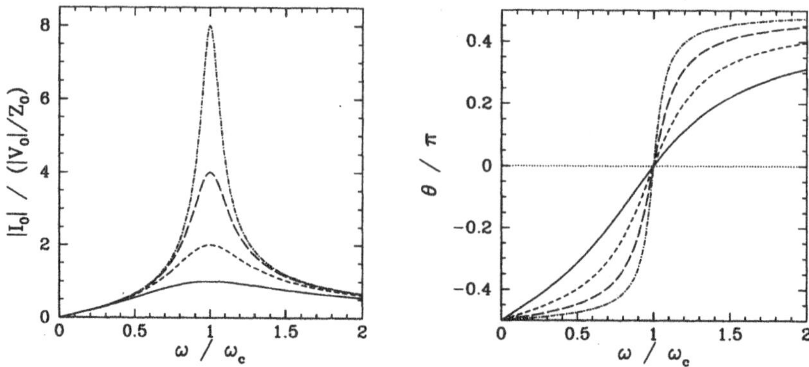

Figure 7.6: *The characteristics of an LCR circuit. The left-hand and right-hand panes show the amplitude and phase-lag of the current versus frequency, respectively. Here,* $\omega_c = 1/\sqrt{LC}$ *and* $Z_0 = \sqrt{L/C}$. *The solid, short-dashed, long-dashed, and dot-dashed curves correspond to* $R/Z_0 = 1$, *1/2, 1/4, and 1/8, respectively.*

7.7 TRANSMISSION LINES

The central assumption made in the analysis of conventional AC circuits is that the voltage (and, hence, the current) has the *same phase* throughout the circuit. Unfortunately, if the circuit is sufficiently large, or the frequency of oscillation, ω, is sufficiently high, then this assumption becomes invalid. The assumption of a constant phase throughout the circuit is reasonable if the wavelength of the oscillation, $\lambda = 2\pi c/\omega$, is much larger than the dimensions of the circuit. (Here, we assume that signals propagate around electrical circuits at about the velocity of light. This assumption will be justified later on.) This is generally not the case in electrical circuits which are associated with *communication*. The frequencies in such circuits tend to be very high, and the dimensions are, almost by definition, large. For instance, leased telephone lines (the type to which computers are connected) run at 56 kHz. The corresponding wavelength is about 5 km, so the constant-phase approximation clearly breaks down for long-distance calls. Computer networks generally run at about 100 MHz, corresponding to $\lambda \sim 3$ m. Thus, the constant-phase approximation also breaks down for most computer networks, since such networks are generally significantly larger than 3 m. It turns out that we need a special sort of wire, called a *transmission line*, to propagate signals around circuits whose dimensions greatly exceed the wavelength, λ. Let us investigate transmission lines.

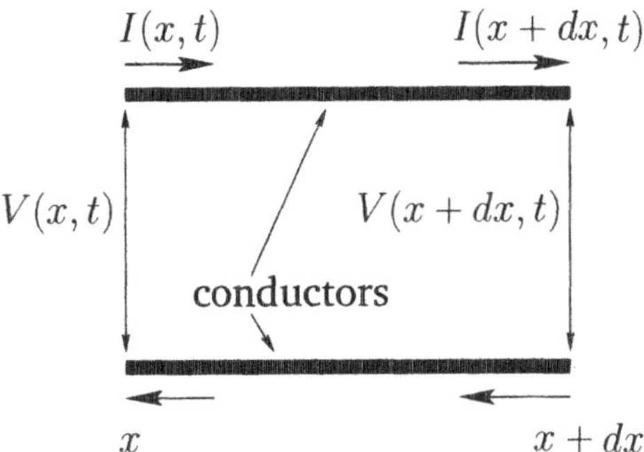

Figure 7.7: *A segment of a transmission line.*

An idealized transmission line consists of two parallel conductors of uniform cross-sectional area. The conductors possess a capacitance per unit length, C, and an inductance per unit length, L. Suppose that x measures the position along the line.

Consider the voltage difference between two neighboring points on the line, located at positions x and $x + \delta x$, respectively—see Figure 7.7. The self-inductance of the portion of the line lying between these two points is $L\,\delta x$. This small section of the line can be thought of as a conventional inductor, and, therefore, obeys the well-known equation

$$V(x,t) - V(x+\delta x,t) = L\,\delta x\,\frac{\partial I(x,t)}{\partial t}, \tag{7.95}$$

where $V(x,t)$ is the voltage difference between the two conductors at position x and time t, and $I(x,t)$ is the current flowing in one of the conductors at position x and time t [the current flowing in the other conductor is $-I(x,t)$]. In the limit $\delta x \to 0$, the above equation reduces to

$$\frac{\partial V}{\partial x} = -L\,\frac{\partial I}{\partial t}. \tag{7.96}$$

Consider the difference in current between two neighboring points on the line, located at positions x and $x + \delta x$, respectively—see Figure 7.7. The capacitance of the portion of the line lying between these two points is $C\,\delta x$. This small section of the line can be thought of as a conventional capacitor, and, therefore, obeys the well-known equation

$$\int_0^t I(x,t)\,dt - \int_0^t I(x+\delta x,t)\,dt = C\,\delta x\,V(x,t), \tag{7.97}$$

where $t = 0$ denotes a time at which the charge stored in either of the conductors in the region x to $x + \delta x$ is zero. In the limit $\delta x \to 0$, the above equation yields

$$\frac{\partial I}{\partial x} = -C\,\frac{\partial V}{\partial t}. \tag{7.98}$$

Equations (7.96) and (7.98) are generally known as the *Telegrapher's equations,* since an old-fashioned telegraph line can be thought of as a primitive transmission line (telegraph lines consist of a single wire—the other conductor is the Earth.)

Differentiating Equation (7.96) with respect to x, we obtain

$$\frac{\partial^2 V}{\partial x^2} = -L\,\frac{\partial^2 I}{\partial x\,\partial t}. \tag{7.99}$$

Differentiating Equation (7.98) with respect to t yields

$$\frac{\partial^2 I}{\partial x\,\partial t} = -C\,\frac{\partial^2 V}{\partial t^2}. \tag{7.100}$$

The above two equations can be combined to give

$$L\,C\,\frac{\partial^2 V}{\partial t^2} = \frac{\partial^2 V}{\partial x^2}. \tag{7.101}$$

This is clearly a wave equation, with wave velocity $v = 1/\sqrt{L\,C}$. An analogous equation can be written for the current, I.

Consider a transmission line which is connected to a generator at one end ($x = 0$), and a resistor, R, at the other ($x = l$). Suppose that the generator outputs a voltage $V_0\,\cos(\omega\,t)$. If follows that

$$V(0,t) = V_0\,\cos(\omega\,t). \tag{7.102}$$

The solution to the wave equation (7.101), subject to the above boundary condition, is

$$V(x,t) = V_0\cos(\omega\,t - k\,x), \tag{7.103}$$

where $k = \omega/v$. This clearly corresponds to a wave which propagates from the generator toward the resistor. Equations (7.96) and (7.103) yield

$$I(x,t) = \frac{V_0}{\sqrt{L/C}}\cos(\omega\,t - k\,x). \tag{7.104}$$

For self-consistency, the resistor at the end of the line must have a particular value:

$$R = \frac{V(l,t)}{I(l,t)} = \sqrt{\frac{L}{C}}. \tag{7.105}$$

The so-called *input impedance* of the line is defined

$$Z_{in} = \frac{V(0,t)}{I(0,t)} = \sqrt{\frac{L}{C}}. \tag{7.106}$$

Thus, a transmission line terminated by a resistor $R = \sqrt{L/C}$ acts very much like a conventional resistor $R = Z_{in}$ in the circuit containing the generator. In fact, the transmission line could be replaced by an effective resistor $R = Z_{in}$ in the circuit diagram for the generator circuit. The power loss due to this effective resistor corresponds to power which is extracted from the circuit, transmitted down the line, and absorbed by the terminating resistor.

The most commonly occurring type of transmission line is a coaxial cable, which consists of two coaxial cylindrical conductors of radii a and b (with $b > a$). We have already shown that the capacitance per unit length of such a cable is (see Section 5.6)

$$C = \frac{2\pi\,\epsilon_0}{\ln(b/a)}. \tag{7.107}$$

Let us now calculate the inductance per unit length. Suppose that the inner conductor carries a current I. According to Ampère's law, the magnetic field in the region between the conductors is given by

$$B_\theta = \frac{\mu_0\, I}{2\pi\, r}. \tag{7.108}$$

The flux linking unit length of the cable is

$$\Phi = \int_a^b B_\theta\, dr = \frac{\mu_0\, I}{2\pi} \ln(b/a). \tag{7.109}$$

Thus, the self-inductance per unit length is

$$L = \frac{\Phi}{I} = \frac{\mu_0}{2\pi} \ln(b/a). \tag{7.110}$$

So, the speed of propagation of a wave down a coaxial cable is

$$v = \frac{1}{\sqrt{L\,C}} = \frac{1}{\sqrt{\epsilon_0\,\mu_0}} = c. \tag{7.111}$$

Not surprisingly, the wave (which is a type of electromagnetic wave) propagates at the speed of light. The impedance of the cable is given by

$$Z_0 = \sqrt{\frac{L}{C}} = \left(\frac{\mu_0}{4\pi^2\,\epsilon_0}\right)^{1/2} \ln(b/a) = 60\,\ln(b/a)\ \text{ohms}. \tag{7.112}$$

If we fill the region between the two cylindrical conductors with a dielectric of dielectric constant ϵ, then, according to the discussion in Section 6.2, the capacitance per unit length of the transmission line goes up by a factor ϵ. However, the dielectric has no effect on magnetic fields, so the inductance per unit length of the line remains unchanged. It follows that the propagation speed of signals down a dielectric-filled coaxial cable is

$$v = \frac{1}{\sqrt{L\,C}} = \frac{c}{\sqrt{\epsilon}}. \tag{7.113}$$

As we shall discover later, this is simply the propagation velocity of electromagnetic waves through a dielectric medium of dielectric constant ϵ. The impedance of the cable becomes

$$Z_0 = 60\,\frac{\ln(b/a)}{\sqrt{\epsilon}}\ \text{ohms}. \tag{7.114}$$

We have seen that if a transmission line is terminated by a resistor whose resistance R matches the impedance Z_0 of the line then all of the power sent down the line is absorbed by the resistor. What happens if $R \neq Z_0$? The answer is that some of the power is reflected back down the line. Suppose that the beginning of the line lies at $x = -l$, and the end of the line is at $x = 0$. Let us consider a solution

$$V(x,t) = V_0\,\exp[i\,(\omega\,t - k\,x)] + K\,V_0\,\exp[i\,(\omega\,t + k\,x)]. \tag{7.115}$$

This corresponds to a voltage wave of amplitude V_0 which travels down the line, and is reflected at the end of the line, with reflection coefficient K. It is easily demonstrated from the Telegrapher's equations that the corresponding current waveform is

$$I(x,t) = \frac{V_0}{Z_0}\,\exp[i\,(\omega\,t - k\,x)] - \frac{K\,V_0}{Z_0}\,\exp[i\,(\omega\,t + k\,x)]. \tag{7.116}$$

Since the line is terminated by a resistance R at $x = 0$, we have, from Ohm's law,

$$\frac{V(0,t)}{I(0,t)} = R. \tag{7.117}$$

This yields an expression for the coefficient of reflection,

$$K = \frac{R - Z_0}{R + Z_0}. \tag{7.118}$$

The input impedance of the line is given by

$$Z_{in} = \frac{V(-l, t)}{I(-l, t)} = Z_0 \frac{R \cos(k\,l) + i\,Z_0 \sin(k\,l)}{Z_0 \cos(k\,l) + i\,R \sin(k\,l)}. \tag{7.119}$$

Clearly, if the resistor at the end of the line is properly matched, so that $R = Z_0$, then there is no reflection (*i.e.*, $K = 0$), and the input impedance of the line is Z_0. If the line is short-circuited, so that $R = 0$, then there is total reflection at the end of the line (*i.e.*, $K = -1$), and the input impedance becomes

$$Z_{in} = i\,Z_0 \tan(k\,l). \tag{7.120}$$

This impedance is purely imaginary, implying that the transmission line absorbs no net power from the generator circuit. In fact, the line acts rather like a pure inductor or capacitor in the generator circuit (*i.e.*, it can store, but cannot absorb, energy). If the line is open-circuited, so that $R \to \infty$, then there is again total reflection at the end of the line (*i.e.*, $K = 1$), and the input impedance becomes

$$Z_{in} = i\,Z_0 \tan(k\,l - \pi/2). \tag{7.121}$$

Thus, the open-circuited line acts like a closed-circuited line which is shorter by one quarter of a wavelength. For the special case where the length of the line is exactly one quarter of a wavelength (*i.e.*, $k\,l = \pi/2$), we find that

$$Z_{in} = \frac{Z_0^2}{R}. \tag{7.122}$$

Thus, a quarter-wave line looks like a pure resistor in the generator circuit. Finally, if the length of the line is much less than the wavelength (*i.e.*, $k\,l \ll 1$) then we enter the constant-phase regime, and $Z_{in} \simeq R$ (*i.e.*, we can forget about the transmission line connecting the terminating resistor to the generator circuit).

Suppose that we wish to build a radio transmitter. We can use a standard half-wave antenna (*i.e.*, an antenna whose length is half the wavelength of the transmitted radio waves) to emit the radiation. In electrical circuits, such an antenna acts like a resistor of resistance 73 ohms (it is more usual to say that the antenna has an impedance of

73 ohms—see Section 9.2). Suppose that we buy a 500 kW generator to supply the power to the antenna. How do we transmit the power from the generator to the antenna? We use a transmission line, of course. (It is clear that if the distance between the generator and the antenna is of order the dimensions of the antenna (*i.e.*, $\lambda/2$) then the constant-phase approximation breaks down, and so we have to use a transmission line.) Since the impedance of the antenna is fixed at 73 ohms, we need to use a 73 ohm transmission line (*i.e.*, $Z_0 = 73$ ohms) to connect the generator to the antenna, otherwise some of the power we send down the line is reflected (*i.e.*, not all of the power output of the generator is converted into radio waves). If we wish to use a coaxial cable to connect the generator to the antenna then it is clear from Equation (7.114) that the radii of the inner and outer conductors need to be such that $b/a = 3.38 \exp(\sqrt{\epsilon})$.

Suppose, finally, that we upgrade our transmitter to use a full-wave antenna (*i.e.*, an antenna whose length equals the wavelength of the emitted radiation). A full-wave antenna has a different impedance than a half-wave antenna. Does this mean that we have to rip out our original coaxial cable, and replace it by one whose impedance matches that of the new antenna? Not necessarily. Let Z_0 be the impedance of the coaxial cable, and Z_1 the impedance of the antenna. Suppose that we place a quarter-wave transmission line (*i.e.*, one whose length is one quarter of a wavelength) of characteristic impedance $Z_{1/4} = \sqrt{Z_0 Z_1}$ between the end of the cable and the antenna. According to Equation (7.122) (with $Z_0 \to \sqrt{Z_0 Z_1}$ and $R \to Z_1$), the input impedance of the quarter-wave line is $Z_{in} = Z_0$, which matches that of the cable. The output impedance matches that of the antenna. Consequently, there is no reflection of the power sent down the cable to the antenna. A quarter-wave line of the appropriate impedance can easily be fabricated from a short length of coaxial cable of the appropriate b/a.

7.8 EXERCISES

7.1. A planar wire loop of resistance R and cross-sectional area A is placed in a uniform magnetic field of strength B. Let the normal to the loop subtend an angle θ with the direction of the magnetic field. Suppose that the loop is made to rotate steadily, such that $\theta = \omega t$. Use Faraday's law to find the emf induced around the loop. What is the current circulating around the loop. Find the torque exerted on the loop by the magnetic field. Demonstrate that the mean rate of work required to maintain the rotation of the loop against this torque is equal to the mean ohmic power loss in the loop. Hint: It may be helpful to treat the loop as a magnetic dipole. Neglect the self-inductance of the loop.

7.2. Consider a long, uniformly wound, cylindrical solenoid of length l, radius r, and turns per unit length N. Suppose that the solenoid is wound around a ferromagnetic core of permeability μ. What is the self-inductance of the solenoid?

7.3. A cable consists of a long cylindrical conductor of radius a which carries current uniformly distributed over its cross-section. The current returns in a thin insulated sheath on the surface of the cable. Find the self-inductance per unit length of the cable.

7.4. Consider two coplanar and concentric circular wire loops of radii a and b, where $a \ll b$. What is the mutual inductance of the loops? Suppose that the smaller loop is shifted a distance z out of the plane of the larger loop (whilst remaining coaxial with the larger loop). What now is the mutual inductance of the two loops?

7.5. Two small current loops are sufficiently far apart that they interact like two magnetic dipoles. Suppose that the loops have position vectors $\mathbf{r}_1$ and $\mathbf{r}_2$, cross-sectional areas A_1 and A_2, and unit normals $\mathbf{n}_1$ and $\mathbf{n}_2$, respectively. What is the mutual inductance of the loops?

7.6. A circular loop of wire of radius a lies in the plane of a long straight wire, with its center a perpendicular distance $b > a$ from the wire. Find the mutual inductance of the two wires.

7.7. An electric circuit consists of a resistor, R, a capacitor, C, and an inductor, L, connected in series with a switch, and a battery of constant voltage V. Suppose that the switch is turned on at $t = 0$. What current subsequently flows in the circuit? Consider the three cases $\omega_0 > \nu$, $\omega_0 = \nu$, and $\omega_0 < \nu$ separately, where $\omega_0 = 1/\sqrt{LC}$, and $\nu = R/(2L)$.

7.8. A coil of self-inductance L and resistance R is connected in series with a switch, and a battery of constant voltage V. The switch is closed, and the steady current $I = V/R$ is established in the circuit. The switch is then opened at $t = 0$. Find the current as a function of time, for $t > 0$.

7.9. A steady voltage is suddenly applied to a coil of self-inductance L_1 in the presence of a nearby closed second coil of self-inductance L_2. Suppose that the mutual inductance of the two coils is M. Demonstrate that the presence of the second coil effectively decreases the initial self-inducatance of the first coil from L_1 to $L_1 - M^2/L_2$.

7.10. An alternating circuit consists of a resistor, R_1, and an inductor, L_1, in series with an alternating voltage source of peak voltage V_1 and angular frequency ω.

This circuit is inductively coupled to a closed wire loop of self-inductance L_2 and resistance R_2. Let M be the mutual inductance of the two circuits. Find the impedance of the first circuit. Demonstrate that the presence of the second circuit causes the effective resistance of the first circuit to increase to

$$R_1 + \frac{\omega^2 M^2 R_2}{(R_2^2 + \omega^2 L^2)},$$

and its effective inductance to decrease to

$$L_1 - \frac{\omega^2 M^2 L_2}{(R_2^2 + \omega^2 L^2)}.$$

7.11. An alternating circuit consists of a coil and a capacitor connected in parallel across an alternating voltage source of angular frequency ω. Suppose that the coil has self-inductance L, and resistance R. Find the impedance of the circuit. Demonstrate that if $L \gg C R^2$ then the amplitude of the current drawn from the voltage source goes through a minimum at $\omega = 1/\sqrt{L C}$.

7.12. Repeat the calculation of Exercise 7.1, taking into account the self-inductance, L, of the loop.

7.13. Find the characteristic impedance of a transmission line consisting of two identical parallel cylindrical wires of radius a and spacing d.

7.14. Suppose that a transmission line has an inductance per unit length, L, a capacitance per unit length, C, and a resistance per unit length, R. Demonstrate that a signal sent down the line decays exponentially on the characteristic length-scale $l = 2 L/(R v)$, where v is the propagation velocity. You may assume that l is much longer than the wavelength of the signal.

7.15. Three coaxial cables of impedance Z_0 have their central conductors connected via three identical resistors of resistance R, as shown in the diagram. The outer conductors are all earthed. What must the value R be in order to ensure that there is no reflection of signals coming into the junction from any cable?

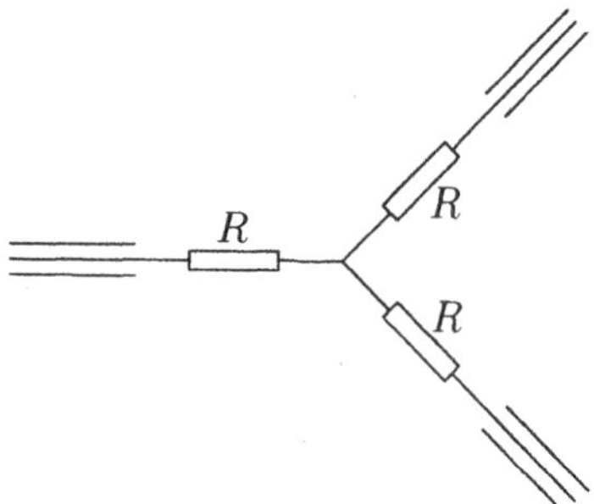

Chapter 8

ELECTROMAGNETIC ENERGY AND MOMENTUM

8.1 INTRODUCTION

In this chapter, we shall demonstrate that Maxwell's equations conserve both energy and momentum.

8.2 ENERGY CONSERVATION

We have seen that the energy density of an electric field is given by [see Equation (5.20)]

$$U_E = \frac{\epsilon_0 E^2}{2}, \tag{8.1}$$

whereas the energy density of a magnetic field satisfies [see Equation (7.55)]

$$U_B = \frac{B^2}{2\mu_0}. \tag{8.2}$$

This suggests that the energy density of a general electromagnetic field is

$$U = \frac{\epsilon_0 E^2}{2} + \frac{B^2}{2\mu_0}. \tag{8.3}$$

We are now in a position to demonstrate that the classical theory of electromagnetism conserves energy. We have already come across one conservation law in electromagnetism: *i.e.*,

$$\frac{\partial \rho}{\partial t} + \nabla \cdot \mathbf{j} = 0. \tag{8.4}$$

This is the equation of charge conservation. Integrating over some volume V, bounded by a surface S, and making use of Gauss' theorem, we obtain

$$-\frac{\partial}{\partial t}\int_V \rho\, dV = \oint_S \mathbf{j}\cdot d\mathbf{S}. \tag{8.5}$$

In other words, the rate of decrease of the charge contained in volume V equals the net flux of charge across surface S. This suggests that an energy conservation law for electromagnetism should have the form

$$-\frac{\partial}{\partial t}\int_V U\, dV = \oint_S \mathbf{u}\cdot d\mathbf{S}. \tag{8.6}$$

Here, U is the energy density of the electromagnetic field, and $\mathbf{u}$ is the flux of electromagnetic energy (*i.e.*, energy $|\mathbf{u}|$ per unit time, per unit cross-sectional area, passes a given point in the direction of $\mathbf{u}$). According to the above equation, the rate of decrease of the electromagnetic energy in volume V equals the net flux of electromagnetic energy across surface S.

However, Equation (8.6) is incomplete, because electromagnetic fields can gain or lose energy by interacting with matter. We need to factor this into our analysis. We saw earlier (see Section 5.3) that the rate of heat dissipation per unit volume in a conductor (the so-called ohmic heating rate) is $\mathbf{E}\cdot\mathbf{j}$. This energy is extracted from electromagnetic fields, so the rate of energy loss of the fields in volume V due to interaction with matter is $\int_V \mathbf{E}\cdot\mathbf{j}\, dV$. Thus, Equation (8.6) generalizes to

$$-\frac{\partial}{\partial t}\int_V U\, dV = \oint_S \mathbf{u}\cdot d\mathbf{S} + \int_V \mathbf{E}\cdot\mathbf{j}\, dV. \tag{8.7}$$

From Gauss' theorem, the above equation is equivalent to

$$\frac{\partial U}{\partial t} + \nabla\cdot\mathbf{u} = -\mathbf{E}\cdot\mathbf{j}. \tag{8.8}$$

Let us now see if we can derive an expression of this form from Maxwell's equations.

We start from the differential form of Ampère's law (including the displacement current):

$$\nabla\times\mathbf{B} = \mu_0\,\mathbf{j} + \epsilon_0\mu_0\frac{\partial\mathbf{E}}{\partial t}. \tag{8.9}$$

Dotting this equation with the electric field yields

$$-\mathbf{E}\cdot\mathbf{j} = -\frac{\mathbf{E}\cdot\nabla\times\mathbf{B}}{\mu_0} + \epsilon_0\,\mathbf{E}\cdot\frac{\partial\mathbf{E}}{\partial t}. \tag{8.10}$$

This can be rewritten

$$-\mathbf{E}\cdot\mathbf{j} = -\frac{\mathbf{E}\cdot\nabla\times\mathbf{B}}{\mu_0} + \frac{\partial}{\partial t}\left(\frac{\epsilon_0\,E^2}{2}\right). \tag{8.11}$$

Now, from vector field theory,

$$\nabla\cdot(\mathbf{E}\times\mathbf{B}) \equiv \mathbf{B}\cdot\nabla\times\mathbf{E} - \mathbf{E}\cdot\nabla\times\mathbf{B}, \tag{8.12}$$

so

$$-\mathbf{E}\cdot\mathbf{j} = \nabla\cdot\left(\frac{\mathbf{E}\times\mathbf{B}}{\mu_0}\right) - \frac{\mathbf{B}\cdot\nabla\times\mathbf{E}}{\mu_0} + \frac{\partial}{\partial t}\left(\frac{\epsilon_0\,E^2}{2}\right). \tag{8.13}$$

The differential form of Faraday's law yields

$$\nabla\times\mathbf{E} = -\frac{\partial\mathbf{B}}{\partial t}, \tag{8.14}$$

so

$$-\mathbf{E}\cdot\mathbf{j} = \nabla\cdot\left(\frac{\mathbf{E}\times\mathbf{B}}{\mu_0}\right) + \mu_0^{-1}\,\mathbf{B}\cdot\frac{\partial\mathbf{B}}{\partial t} + \frac{\partial}{\partial t}\left(\frac{\epsilon_0\,E^2}{2}\right). \tag{8.15}$$

This can be rewritten

$$-\mathbf{E}\cdot\mathbf{j} = \nabla\cdot\left(\frac{\mathbf{E}\times\mathbf{B}}{\mu_0}\right) + \frac{\partial}{\partial t}\left(\frac{\epsilon_0\,E^2}{2} + \frac{B^2}{2\mu_0}\right). \tag{8.16}$$

Thus, we obtain the desired conservation law,

$$\frac{\partial U}{\partial t} + \nabla\cdot\mathbf{u} = -\mathbf{E}\cdot\mathbf{j}, \tag{8.17}$$

where

$$U = \frac{\epsilon_0\,E^2}{2} + \frac{B^2}{2\mu_0} \tag{8.18}$$

is the electromagnetic energy density, and

$$\mathbf{u} = \frac{\mathbf{E}\times\mathbf{B}}{\mu_0} \tag{8.19}$$

is the electromagnetic energy flux. The latter quantity is usually called the *Poynting flux*, after its discoverer.

Let us see whether our expression for the electromagnetic energy flux makes sense. We all know that if we stand in the sun we get hot. This occurs because we absorb electromagnetic radiation emitted by the Sun.

So, radiation must transport energy. The electric and magnetic fields in electromagnetic radiation are mutually perpendicular, and are also perpendicular to the direction of propagation $\hat{\mathbf{k}}$ (this is a unit vector). Furthermore, $B = E/c$. Equation (4.90) can easily be transformed into the following relation between the electric and magnetic fields of an electromagnetic wave:

$$\mathbf{E} \times \mathbf{B} = \frac{E^2}{c}\,\hat{\mathbf{k}}. \tag{8.20}$$

Thus, the Poynting flux for electromagnetic radiation is

$$\mathbf{u} = \frac{E^2}{\mu_0\, c}\,\hat{\mathbf{k}} = \epsilon_0\, c\, E^2\,\hat{\mathbf{k}}. \tag{8.21}$$

This expression tells us that electromagnetic waves transport energy along their direction of propagation, which seems to make sense.

The energy density of electromagnetic radiation is

$$U = \frac{\epsilon_0\, E^2}{2} + \frac{B^2}{2\mu_0} = \frac{\epsilon_0\, E^2}{2} + \frac{E^2}{2\mu_0\, c^2} = \epsilon_0\, E^2, \tag{8.22}$$

using $B = E/c$. Note that the electric and magnetic fields in an electromagnetic wave have *equal* energy densities. Since electromagnetic waves travel at the speed of light, we would expect the energy flux through one square meter in one second to equal the energy contained in a volume of length c and unit cross-sectional area: *i.e.*, c times the energy density. Thus,

$$|\mathbf{u}| = c\, U = \epsilon_0\, c\, E^2, \tag{8.23}$$

which is in accordance with Equation (8.21).

As another example, consider a straight cylindrical wire of radius a, and uniform resistivity η. Suppose that the wire is coaxial with the z-axis. Let us adopt standard cylindrical polar coordinates (r, θ, z). If a current of uniform density $\mathbf{j} = j\,\mathbf{e}_z$ flows down the wire then Ohm's law tells us that there is a uniform longitudinal electric field $\mathbf{E} = E\,\mathbf{e}_z$ within the wire, where $E = \eta\, j$. According to Ampère's circuital law, the current also generates a circulating magnetic field, inside the wire, of the form $\mathbf{B} = (\mu_0\, r\, j/2)\,\mathbf{e}_\theta$. Hence, the Poynting flux, $\mathbf{u} = \mathbf{E} \times \mathbf{B}/\mu_0$, within the wire points *radially inward*, and is of magnitude $u = \eta\, j^2\, r/2$. The net energy flux into a cylindrical surface, coaxial with the wire, of radius r and length l is $U = u\, 2\pi\, r\, l = \eta\, j^2\, V(r)$, where $V(r) = \pi\, r^2\, l$ is the volume enclosed by the surface. However, $\eta\, j^2$ is the rate of electromagnetic energy loss per unit volume, due to ohmic heating. Hence, $\eta\, j^2\, V(r)$ represents the net rate of electromagnetic energy loss due to ohmic heating in the region lying within the cylindrical surface. Thus, we can see that this energy loss is balanced by the inward flux of electromagnetic energy across the surface. This flux represents energy which is ultimately derived from the battery which drives the current through the wire.

In the presence of diamagnetic and magnetic media, starting from Equation (6.117), we can derive an energy conservation law of the form

$$\frac{\partial U}{\partial t} + \nabla \cdot \mathbf{u} = -\mathbf{E} \cdot \mathbf{j}_t, \tag{8.24}$$

via analogous steps to those used to derive Equation (8.17). Here, the electromagnetic energy density is written

$$U = \frac{1}{2}\,\mathbf{E} \cdot \mathbf{D} + \frac{1}{2}\,\mathbf{B} \cdot \mathbf{H}, \tag{8.25}$$

which is consistent with Equation (8.18). The Poynting flux takes the form

$$\mathbf{u} = \mathbf{E} \times \mathbf{H}, \tag{8.26}$$

which is consistent with Equation (8.19). Of course, the above expressions are only valid for *linear* dielectric and magnetic media.

8.3 ELECTROMAGNETIC MOMENTUM

We have seen that electromagnetic waves carry energy. It turns out that they also carry momentum. Consider the following argument, due to Einstein. Suppose that we have a railroad car of mass M and length L which is free to move in one dimension—see Figure 8.1. Suppose that electromagnetic radiation of total energy E is emitted from one end of the car, propagates along the length of the car, and is then absorbed at the other end. The effective mass of this radiation is $m = E/c^2$ (from Einstein's famous relation $E = m\,c^2$). At first sight, the process described above appears to cause the center of mass of the system to spontaneously shift. This violates the law of momentum conservation (assuming the railway car is subject to no horizontal external forces). The only way in which the center of mass of the system can remain stationary is if the railway car *moves* in the opposite direction to the direction of propagation of the radiation. In fact, if the car moves by a distance x then the center of mass of the system is the same before and after the radiation pulse

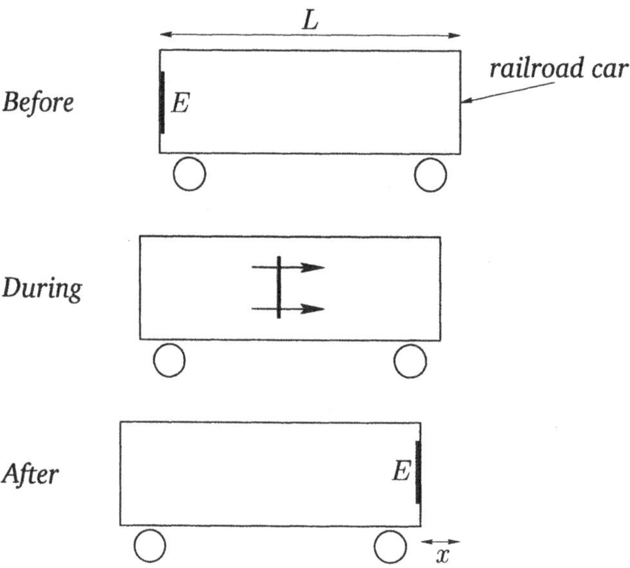

Figure 8.1: *Einstein's thought experiment regarding electromagnetic momentum.*

provided that

$$M\,x = m\,L = \frac{E}{c^2}\,L. \tag{8.27}$$

Incidentally, it is assumed that $m \ll M$ in this derivation.

But, what actually causes the car to move? If the radiation possesses momentum p then the car will recoil with the same momentum when the radiation is emitted. When the radiation hits the other end of the car then the car acquires momentum p in the opposite direction, which stops the motion. The time of flight of the radiation is L/c. So, the distance traveled by a mass M with momentum p in this time is

$$x = v\,t = \frac{p}{M}\frac{L}{c}, \tag{8.28}$$

giving

$$p = M\,x\,\frac{c}{L} = \frac{E}{c}. \tag{8.29}$$

Thus, the momentum carried by electromagnetic radiation equals its energy divided by the speed of light. The same result can be obtained from the well-known relativistic formula

$$E^2 = p^2c^2 + m^2c^4 \tag{8.30}$$

relating the energy E, momentum p, and mass m of a particle. According to quantum theory, electromagnetic radiation is made up of *massless* particles called *photons*. Thus,

$$p = \frac{E}{c} \tag{8.31}$$

for individual photons, so the same must be true of electromagnetic radiation as a whole. It follows from Equation (8.29) that the momentum density g of electromagnetic radiation equals its energy density over c, so

$$g = \frac{U}{c} = \frac{|\mathbf{u}|}{c^2} = \frac{\epsilon_0 E^2}{c}. \tag{8.32}$$

It is reasonable to suppose that the momentum points along the direction of the energy flow (this is obviously the case for photons), so the vector momentum density (which gives the direction, as well as the magnitude, of the momentum per unit volume) of electromagnetic radiation is

$$\mathbf{g} = \frac{\mathbf{u}}{c^2}. \tag{8.33}$$

Thus, the momentum density equals the energy flux over c^2.

Of course, the electric field associated with an electromagnetic wave oscillates rapidly, which implies that the previous expressions for the energy density, energy flux, and momentum density of electromagnetic radiation are also rapidly oscillating. It is convenient to average over many periods of the oscillation (this average is denoted $\langle\,\rangle$). Thus,

$$\langle U\rangle = \frac{\epsilon_0 E_0^2}{2}, \tag{8.34}$$

$$\langle \mathbf{u}\rangle = \frac{c\,\epsilon_0 E_0^2}{2}\,\hat{\mathbf{k}} = c\,\langle U\rangle\,\hat{\mathbf{k}}, \tag{8.35}$$

$$\langle \mathbf{g}\rangle = \frac{\epsilon_0 E_0^2}{2c}\,\hat{\mathbf{k}} = \frac{\langle U\rangle}{c}\,\hat{\mathbf{k}}, \tag{8.36}$$

where the factor $1/2$ comes from averaging $\cos^2(\omega t)$. Here, E_0 is the peak amplitude of the electric field associated with the wave.

If electromagnetic radiation possesses momentum then it must exert a force on bodies which absorb (or emit) radiation. Suppose that a body is placed in a beam of perfectly collimated radiation, which it absorbs completely. The amount of momentum absorbed per unit time, per unit cross-sectional area, is simply the amount of momentum contained in a volume of length c and unit cross-sectional area: *i.e.*, c times the momentum density, g. An absorbed momentum per unit time, per unit area, is equivalent to a pressure. In other words, the radiation exerts a pressure $c\,g$ on the body. Thus, the *radiation pressure* is given by

$$p = \frac{\epsilon_0\,E^2}{2} = \langle U \rangle. \tag{8.37}$$

So, the pressure exerted by collimated electromagnetic radiation is equal to its average energy density.

Consider a cavity filled with electromagnetic radiation. What is the radiation pressure exerted on the walls? In this situation, the radiation propagates in all directions with equal probability. Consider radiation propagating at an angle θ to the local normal to the wall. The amount of such radiation hitting the wall per unit time, per unit area, is proportional to $\cos\theta$. Moreover, the component of momentum normal to the wall which the radiation carries is also proportional to $\cos\theta$. Thus, the pressure exerted on the wall is the same as in Equation (8.37), except that it is weighted by the average of $\cos^2\theta$ over all solid angles, in order to take into account the fact that obliquely propagating radiation exerts a pressure which is $\cos^2\theta$ times that of normal radiation. The average of $\cos^2\theta$ over all solid angles is $1/3$, so for isotropic radiation

$$p = \frac{\langle U \rangle}{3}. \tag{8.38}$$

Clearly, the pressure exerted by isotropic radiation is one third of its average energy density.

The power incident on the surface of the Earth due to radiation emitted by the Sun is about $1300\ \mathrm{W\,m^{-2}}$. So, what is the radiation pressure? Since,

$$\langle |\mathbf{u}| \rangle = c\,\langle U \rangle = 1300\,\mathrm{W\,m^{-2}}, \tag{8.39}$$

then

$$p = \langle U \rangle \simeq 4 \times 10^{-6}\,\mathrm{N\,m^{-2}}. \tag{8.40}$$

Here, the radiation is assumed to be perfectly collimated. Thus, the radiation pressure exerted on the Earth is minuscule (for comparison, the pressure of the atmosphere is about $10^5\ \mathrm{N\,m^{-2}}$). Nevertheless, this small pressure due to radiation is important in outer space, since it is responsible for continuously sweeping dust particles out of the Solar System. It is quite common for comets to exhibit two separate tails. One (called the *gas tail*) consists of ionized gas, and is swept along by the Solar

Wind (a stream of charged particles and magnetic field-lines emitted by the Sun). The other (called the *dust tail*) consists of uncharged dust particles, and is swept radially outward (since light travels in straight-lines) from the Sun by radiation pressure. Two separate tails are observed if the local direction of the Solar Wind is not radially outward from the Sun (which is quite often the case).

The radiation pressure from sunlight is very weak. However, that produced by laser beams can be enormous (far higher than any conventional pressure which has ever been produced in a laboratory). For instance, the lasers used in Inertial Confinement Fusion (*e.g.*, the NOVA experiment in Lawrence Livermore National Laboratory) typically have energy fluxes of 10^{18} W m^{-2}. This translates to a radiation pressure of about 10^4 atmospheres!

8.4 MOMENTUM CONSERVATION

It follows from Equations (8.19) and (8.33) that the momentum density of electromagnetic fields can be written

$$\mathbf{g} = \epsilon_0\, \mathbf{E} \times \mathbf{B}. \tag{8.41}$$

Now, a momentum conservation equation for electromagnetic fields should take the integral form

$$-\frac{\partial}{\partial t}\int_V g_i\, dV = \int_S G_{ij}\, dS_j + \int_V [\rho\, \mathbf{E} + \mathbf{j} \times \mathbf{B}]_i\, dV. \tag{8.42}$$

Here, i and j run from 1 to 3 (1 corresponds to the x-direction, 2 to the y-direction, and 3 to the z-direction). Moreover, the Einstein summation convention is employed for repeated indices (*e.g.*, $a_j\, a_j \equiv \mathbf{a} \cdot \mathbf{a}$). Furthermore, the tensor G_{ij} represents the flux of the ith component of electromagnetic momentum in the j-direction. This tensor (a tensor is a direct generalization of a vector with two indices instead of one) is called the *momentum flux density tensor*. Hence, the above equation states that the rate of loss of electromagnetic momentum in some volume V is equal to the flux of electromagnetic momentum across the bounding surface S plus the rate at which momentum is transferred to matter inside V. The latter rate is, of course, just the net electromagnetic force acting on matter inside V: *i.e.*, the volume integral of the electromagnetic force density, $\rho\, \mathbf{E} + \mathbf{j} \times \mathbf{B}$. Now, a direct generalization of the divergence theorem states that

$$\int_S G_{ij}\, dS_j \equiv \int_V \frac{\partial G_{ij}}{\partial x_j}\, dV, \tag{8.43}$$

where $x_1 \equiv x$, $x_2 \equiv y$, *etc.* Hence, in differential form, our momentum conservation equation for electromagnetic fields is written

$$-\frac{\partial}{\partial t}\,[\epsilon_0\,\mathbf{E}\times\mathbf{B}]_i = \frac{\partial G_{ij}}{\partial x_j} + [\rho\,\mathbf{E} + \mathbf{j}\times\mathbf{B}]_i. \tag{8.44}$$

Let us now attempt to derive an equation of this form from Maxwell's equations.

Maxwell's equations are as follows:

$$\nabla\cdot\mathbf{E} = \frac{\rho}{\epsilon_0}, \tag{8.45}$$

$$\nabla\cdot\mathbf{B} = 0, \tag{8.46}$$

$$\nabla\times\mathbf{E} = -\frac{\partial\mathbf{B}}{\partial t}, \tag{8.47}$$

$$\nabla\times\mathbf{B} = \mu_0\,\mathbf{j} + \epsilon_0\mu_0\,\frac{\partial\mathbf{E}}{\partial t}. \tag{8.48}$$

We can cross Equation (8.48) divided by μ_0 with **B**, and rearrange, to give

$$-\epsilon_0\,\frac{\partial\mathbf{E}}{\partial t}\times\mathbf{B} = \frac{\mathbf{B}\times(\nabla\times\mathbf{B})}{\mu_0} + \mathbf{j}\times\mathbf{B}. \tag{8.49}$$

Next, let us cross **E** with Equation (8.47) times ϵ_0, rearrange, and add the result to the above equation. We obtain

$$-\epsilon_0\,\frac{\partial\mathbf{E}}{\partial t}\times\mathbf{B} - \epsilon_0\,\mathbf{E}\times\frac{\partial\mathbf{B}}{\partial t} = \epsilon_0\,\mathbf{E}\times(\nabla\times\mathbf{E}) + \frac{\mathbf{B}\times(\nabla\times\mathbf{B})}{\mu_0} + \mathbf{j}\times\mathbf{B}. \tag{8.50}$$

Next, making use of Equations (8.45) and (8.46), we get

$$\begin{aligned}-\frac{\partial}{\partial t}\,[\epsilon_0\,\mathbf{E}\times\mathbf{B}] &= \epsilon_0\,\mathbf{E}\times(\nabla\times\mathbf{E}) + \frac{\mathbf{B}\times(\nabla\times\mathbf{B})}{\mu_0}\\ &\quad - \epsilon_0\,(\nabla\cdot\mathbf{E})\,\mathbf{E} - \frac{1}{\mu_0}(\nabla\cdot\mathbf{B})\,\mathbf{B} + \rho\,\mathbf{E} + \mathbf{j}\times\mathbf{B}.\end{aligned} \tag{8.51}$$

Now, from vector field theory,

$$\nabla(E^2/2) \equiv \mathbf{E}\times(\nabla\times\mathbf{E}) + (\mathbf{E}\cdot\nabla)\mathbf{E}, \tag{8.52}$$

with a similar equation for **B**. Hence, Equation (8.51) takes the form

$$-\frac{\partial}{\partial t}\left[\epsilon_0\,\mathbf{E}\times\mathbf{B}\right] = \epsilon_0\left[\nabla(E^2/2) - (\nabla\cdot\mathbf{E})\,\mathbf{E} - (\mathbf{E}\cdot\nabla)\mathbf{E}\right] + \frac{1}{\mu_0}\left[\nabla(B^2/2) - (\nabla\cdot\mathbf{B})\,\mathbf{B} - (\mathbf{B}\cdot\nabla)\mathbf{B}\right] + \rho\,\mathbf{E} + \mathbf{j}\times\mathbf{B}. \tag{8.53}$$

Finally, when written in terms of components, the above equation becomes

$$-\frac{\partial}{\partial t}\left[\epsilon_0\,\mathbf{E}\times\mathbf{B}\right]_i = \frac{\partial}{\partial x_j}\left[\epsilon_0\,E^2\,\delta_{ij}/2 - \epsilon_0\,E_i\,E_j + B^2\,\delta_{ij}/2\,\mu_0 - B_i\,B_j/\mu_0\right] + \left[\rho\,\mathbf{E} + \mathbf{j}\times\mathbf{B}\right]_i, \tag{8.54}$$

since $[(\nabla\cdot\mathbf{E})\,\mathbf{E}]_i \equiv (\partial E_j/\partial x_j)\,E_i$, and $[(\mathbf{E}\cdot\nabla)\mathbf{E}]_i \equiv E_j\,(\partial E_i/\partial x_j)$. Here, δ_{ij} is a Kronecker delta symbol (*i.e.*, $\delta_{ij} = 1$ if $i = j$, and $\delta_{ij} = 0$ otherwise). Comparing the above equation with Equation (8.44), we conclude that the momentum flux density tensor of electromagnetic fields takes the form

$$G_{ij} = \epsilon_0\,(E^2\,\delta_{ij}/2 - E_i\,E_j) + (B^2\,\delta_{ij}/2 - B_i\,B_j)/\mu_0. \tag{8.55}$$

The momentum conservation equation (8.44) is sometimes written

$$[\rho\,\mathbf{E} + \mathbf{j}\times\mathbf{B}]_i = \frac{\partial T_{ij}}{\partial x_j} - \frac{\partial}{\partial t}\left[\epsilon_0\,\mathbf{E}\times\mathbf{B}\right]_i, \tag{8.56}$$

where

$$T_{ij} = -G_{ij} = \epsilon_0\,(E_i\,E_j - E^2\,\delta_{ij}/2) + (B_i\,B_j - B^2\,\delta_{ij}/2)/\mu_0 \tag{8.57}$$

is called the *Maxwell stress tensor*.

Consider a uniform electric field, $\mathbf{E} = E\,\mathbf{e}_z$. According to Equation (8.55), the momentum flux density tensor of such a field is

$$\mathbf{G} = \begin{pmatrix} \epsilon_0\,E^2/2 & 0 & 0 \\ 0 & \epsilon_0\,E^2/2 & 0 \\ 0 & 0 & -\epsilon_0\,E^2/2 \end{pmatrix}. \tag{8.58}$$

As is well-known, the momentum flux density tensor of a conventional gas of pressure p is written

$$\mathbf{G} = \begin{pmatrix} p & 0 & 0 \\ 0 & p & 0 \\ 0 & 0 & p \end{pmatrix}. \tag{8.59}$$

In other words, from any small volume element there is an equal *outward* momentum flux density p in all three Cartesian directions, which simply corresponds to an *isotropic* gas pressure, p. This suggests that a positive diagonal element in a momentum stress tensor corresponds to a *pressure* exerted in the direction of the corresponding Cartesian axis. Furthermore, a negative diagonal element corresponds to negative pressure, or *tension*, exerted in the direction of the corresponding Cartesian axis. Thus, we conclude, from Equation (8.58), that electric field-lines act rather like *mutually repulsive elastic bands*: *i.e.*, there is a pressure force acting perpendicular to the field-lines which tries to push them apart, whilst a tension force acting along the field-lines simultaneously tries to shorten them. As an example, we have seen that the normal electric field $E_\perp$ above the surface of a charged conductor exerts an outward pressure $\epsilon_0\, E_\perp^2/2$ on the surface. One way of interpreting this pressure is to say that it is due to the tension in the electric field-lines anchored in the surface. Likewise, the force of attraction between the two plates of a charged parallel plate capacitor can be attributed to the tension in the electric field-lines running between them.

It is easily demonstrated that the momentum flux density tensor of a uniform magnetic field, $\mathbf{B} = B\,\mathbf{e}_z$, is

$$\mathbf{G} = \begin{pmatrix} B^2/2\mu_0 & 0 & 0 \\ 0 & B^2/2\mu_0 & 0 \\ 0 & 0 & -B^2/2\mu_0 \end{pmatrix}. \tag{8.60}$$

Hence, magnetic field-lines also act like mutually repulsive elastic bands. For instance, the uniform field $\mathbf{B}$ inside a conventional solenoid exerts an outward pressure $B^2/2\mu_0$ on the windings which generate and confine it.

8.5 ANGULAR MOMENTUM CONSERVATION

An electromagnetic field which possesses a momentum density $\mathbf{g}$ must also possess an *angular momentum density*

$$\mathbf{h} = \mathbf{r} \times \mathbf{g}. \tag{8.61}$$

It follows that electromagnetic fields can exchange angular momentum, as well as linear momentum, with ordinary matter. As an illustration of this, consider the following famous example.

Suppose that we have two thin coaxial cylindrical conducting shells of radii a and c, where $a < c$. Let the length of both cylinders be l. Suppose, further, that the cylinders are free to rotate independently about their common axis. Finally, let the inner cylinder carry charge $-Q$, and the outer cylinder charge $+Q$ (where $Q > 0$). Now, suppose that a uniform coaxial cylindrical solenoid winding of radius b (where $a < b < c$), and number of turns per unit length N, is placed between the two cylinders, and energized with a current I. Note that the total angular momentum of this system is a conserved quantity, since the system is isolated.

Consider an initial state in which both cylinders are stationary. Ramping the solenoid current down to zero is observed to cause the two cylinders to start to *rotate*—the inner cylinder in the opposite sense to the sense of current circulation, and the outer cylinder in the same sense. In other words, the two cylinders acquire angular momentum when the current in the solenoid coil is ramped down. Where does this angular momentum come from? Clearly, it can only have come from the electric and magnetic fields in the region between the cylinders. Let us investigate further.

It is convenient to define cylindrical polar coordinates (r, θ, z) which are coaxial with the common axis of the two cylinders and the solenoid coil. As is easily demonstrated from Gauss' law, the electric field takes the form $\mathbf{E} = E_r\,\mathbf{e}_r$, where

$$E_r = \begin{cases} -Q/(2\pi\epsilon_0\, r\, l) & \text{for } a \leq r \leq c \\ 0 & \text{otherwise} \end{cases}. \tag{8.62}$$

Likewise, it is easily shown from Ampère's circuital law that the initial magnetic field is written $\mathbf{B} = B_z\,\mathbf{e}_z$, where

$$B_z = \begin{cases} \mu_0\, N\, I & \text{for } r \leq b \\ 0 & \text{otherwise} \end{cases}. \tag{8.63}$$

It follows from Equation (8.41) that the initial momentum density of the electromagnetic field is $\mathbf{g} = g_\theta\,\mathbf{e}_\theta$, where

$$g_\theta = \begin{cases} \mu_0\, N\, I\, Q/(2\pi\, r\, l) & \text{for } a \leq r \leq b \\ 0 & \text{otherwise} \end{cases}. \tag{8.64}$$

Hence, from Equation (8.61), the initial angular momentum density of the electromagnetic field is $\mathbf{h} = h_z\,\mathbf{e}_z$, where

$$h_z = \begin{cases} \mu_0\, N\, I\, Q/(2\pi\, l) & \text{for } a \leq r \leq b \\ 0 & \text{otherwise} \end{cases}. \tag{8.65}$$

In other words, the electromagnetic field possesses a *uniform* z-directed angular momentum density $\mathbf{h}$ in the region between the inner cylinder and the solenoid winding. It follows that the initial angular momentum content of the electromagnetic field, $\mathbf{L}$, is equal to $\mathbf{h}$ multiplied by the volume of this region. Hence, we obtain $\mathbf{L} = L_z\,\mathbf{e}_z$, where

$$L_z = \frac{(b^2 - a^2)\,\mu_0\,N\,I\,Q}{2}. \tag{8.66}$$

Of course, as the current in the solenoid winding is ramped down this electromagnetic angular momentum is lost, and, presumably, transferred to the two cylinders. Let us examine how this transfer is effected.

Any change in the current flowing in the solenoid winding generates an inductive electric field. From Faraday's law, this field takes the form $\mathbf{E} = E_\theta\,\mathbf{e}_\theta$, where

$$E_\theta = \begin{cases} -\mu_0\,N\,I\,r/2 & \text{for } r \leq b \\ -\mu_0\,N\,I\,b^2/(2\,r) & \text{otherwise} \end{cases}. \tag{8.67}$$

This electric field exerts a torque $\mathbf{T}_a = T_a\,\mathbf{e}_z$, where

$$T_a = -Q\,E_\theta(a)\,a = \frac{\mu_0\,N\,\dot{I}\,Q\,a^2}{2}, \tag{8.68}$$

on the inner cylinder, and a torque $\mathbf{T}_b = T_b\,\mathbf{e}_z$, where

$$T_b = Q\,E_\theta(b)\,b = -\frac{\mu_0\,N\,\dot{I}\,Q\,b^2}{2}, \tag{8.69}$$

on the outer cylinder. Thus, the net angular momentum acquired by the inner cylinder, as the current in the solenoid coil is ramped down, is $\mathbf{L}_a = L_a\,\mathbf{e}_z$, where

$$L_a = \int T_a\,dt = -\frac{\mu_0\,N\,I\,Q\,a^2}{2}. \tag{8.70}$$

Likewise, the net angular momentum acquired by the outer cylinder is $\mathbf{L}_b = L_b\,\mathbf{e}_z$, where

$$L_b = \int T_b\,dt = \frac{\mu_0\,N\,I\,Q\,b^2}{2}. \tag{8.71}$$

Hence, the net angular momentum acquired by the two cylinders, as a whole, is $\mathbf{L} = L_z\,\mathbf{e}_z$, where

$$L_z = \frac{(b^2 - a^2)\,\mu_0\,N\,I\,Q}{2}. \tag{8.72}$$

This, of course, is equal to the z-directed angular momentum lost by the electromagnetic field.

8.6 EXERCISES

8.1. A solenoid consists of a wire wound uniformly around a solid cylindrical core of radius a, length l, and permeability μ. Suppose that the wire has N turns per unit length, and carries a current I. What is the energy stored within the core? Suppose that the current in the wire is gradually ramped down. Calculate the integral of the Poynting flux (due to the induced electric field) over the surface of the core. Hence, demonstrate that the instantaneous rate of decrease of the energy within the core is always equal to the integral of the Poynting flux over its surface.

8.2. A coaxial cable consists of two thin coaxial cylindrical conducting shells of radii a and b (where $a < b$). Suppose that the inner conductor carries the longitudinal current I and the charge per unit length λ. Let the outer conductor carry equal and opposite current and charge per unit length. What is the flux of electromagnetic energy and momentum down the cable (in the direction of the inner current)? What is the electromagnetic pressure acting on the inner and outer conductors?

8.3. Consider a coaxial cable consisting of two thin coaxial cylindrical conducting shells of radii a and b (where $a < b$). Suppose that the inner conductor carries current per unit length I, circulating in the plane perpendicular to its axis, and charge per unit length λ. Let the outer conductor carry equal and opposite current per unit length and charge per unit length. What is the flux of electromagnetic angular momentum down the cable? What is the electromagnetic pressure acting on the inner and outer conductors?

8.4. Calculate the electrostatic force acting between two identical point electrical charges by finding the net electromagnetic momentum flux across a plane surface located half-way between the charges. Verify that the result is consistent with Coulomb's law.

8.5. Calculate the magnetic force per unit length acting between two long parallel straight wires carrying identical currents by finding the net electromagnetic momentum flux across a plane surface located half-way between the wires. Verify that the result is consistent with Ampère's law.

8.6. Calculate the mean force acting on a perfect mirror of area A which reflects normally incident electromagnetic radiation of peak electric field E_0. Suppose that the mirror only reflects a fraction f of the incident electromagnetic energy, and absorbs the remainder. What now is the force acting on the mirror?

8.7. A thin spherical conducting shell of radius a carries a charge Q. Use the concept of electric field-line tension to find the force of repulsion between any two halves of the shell.

8.8. A solid sphere of radius a carries a net charge Q which is uniformly distributed over its volume. What is the force of repulsion between any two halves of the sphere?

8.9. A U-shaped electromagnet of permeability $\mu \gg 1$, length l, pole separation d, and uniform cross-sectional area A is energized by a current I flowing in a winding with N turns. Find the force with which the magnet attracts a bar of the same material which is placed over both poles (and completely covers them).

8.10. An iron sphere of radius a and uniform magnetization $\mathbf{M} = M\,\mathbf{e}_z$ carries an electric charge Q. Find the net angular momentum of the electromagnetic field surrounding the sphere. Suppose that the magnetization of the sphere decays to zero. Demonstrate that the induced electric field exerts a torque on the sphere which is such as to impart to it a mechanical angular momentum equal to that lost by the electromagnetic field.

Chapter 9

ELECTROMAGNETIC RADIATION

9.1 INTRODUCTION

In this chapter, we shall employ Maxwell's equations to investigate the emission, scattering, propagation, absorption, reflection, and refraction of electromagnetic radiation.

9.2 THE HERTZIAN DIPOLE

Consider two small spherical conductors connected by a wire. Suppose that electric charge flows periodically back and forth between the spheres. Let $q(t)$ be the *instantaneous* charge on one of the conductors. The system is assumed to have zero net charge, so that the charge on the other conductor is $-q(t)$. Finally, let

$$q(t) = q_0 \sin(\omega t). \tag{9.1}$$

Now, we expect the oscillating current flowing in the wire connecting the two spheres to generate *electromagnetic radiation* (see Section 4.11). Let us consider the simple case in which the length of the wire is *small* compared to the wavelength of the emitted radiation. If this is the case then the current I flowing between the conductors has the same phase along the whole length of the wire. It follows that

$$I(t) = \frac{dq}{dt} = I_0 \cos(\omega t), \tag{9.2}$$

where $I_0 = \omega q_0$. This type of antenna is called a *Hertzian dipole*, after the German physicist Heinrich Hertz.

The magnetic vector potential generated by a current distribution $\mathbf{j}(\mathbf{r})$ is given by the well-known formula (see Section 4.12)

$$\mathbf{A}(\mathbf{r}, t) = \frac{\mu_0}{4\pi} \int \frac{[\mathbf{j}]}{|\mathbf{r} - \mathbf{r}'|} d^3\mathbf{r}', \tag{9.3}$$

where

$$[f] \equiv f(\mathbf{r}', t - |\mathbf{r} - \mathbf{r}'|/c). \tag{9.4}$$

Suppose that the wire is aligned along the z-axis, and extends from $z = -l/2$ to $z = l/2$. For a wire of negligible thickness, we can replace $\mathbf{j}(\mathbf{r}', t - |\mathbf{r} - \mathbf{r}'|/c)\, d^3\mathbf{r}'$ by $I(\mathbf{r}', t - |\mathbf{r} - \mathbf{r}'|/c)\, dz'\, \mathbf{e}_z$. Thus, $\mathbf{A}(\mathbf{r}, t) = A_z(\mathbf{r}, t)\, \mathbf{e}_z$, and

$$A_z(\mathbf{r}, t) = \frac{\mu_0}{4\pi} \int_{-l/2}^{l/2} \frac{I(z', t - |\mathbf{r} - z'\,\mathbf{e}_z|/c)}{|\mathbf{r} - z'\,\mathbf{e}_z|}\, dz'. \tag{9.5}$$

In the region $r \gg l$,

$$|\mathbf{r} - z'\,\mathbf{e}_z| \simeq r, \tag{9.6}$$

and

$$t - |\mathbf{r} - z'\,\mathbf{e}_z|/c \simeq t - r/c. \tag{9.7}$$

The maximum error involved in the latter approximation is $\Delta t \sim l/c$. This error (which is a time) must be much less than a period of oscillation of the emitted radiation, otherwise the phase of the radiation will be wrong. So we require that

$$\frac{l}{c} \ll \frac{2\pi}{\omega}, \tag{9.8}$$

which implies that $l \ll \lambda$, where $\lambda = 2\pi c/\omega$ is the wavelength of the emitted radiation. However, we have already assumed that the length of the wire l is much less than the wavelength of the radiation, and so the above inequality is automatically satisfied. Thus, in the *far field* region, $r \gg \lambda$, we can write

$$A_z(\mathbf{r}, t) \simeq \frac{\mu_0}{4\pi} \int_{-l/2}^{l/2} \frac{I(z', t - r/c)}{r}\, dz'. \tag{9.9}$$

This integral is easy to perform, since the current is uniform along the length of the wire. So, we get

$$A_z(\mathbf{r}, t) \simeq \frac{\mu_0\, l}{4\pi} \frac{I(t - r/c)}{r}. \tag{9.10}$$

The scalar potential is most conveniently evaluated using the Lorenz gauge condition (see Section 4.12)

$$\nabla \cdot \mathbf{A} = -\epsilon_0 \mu_0 \frac{\partial \phi}{\partial t}. \tag{9.11}$$

Now,

$$\nabla\cdot\mathbf{A} = \frac{\partial A_z}{\partial z} \simeq \frac{\mu_0\, l}{4\pi}\frac{\partial I(t-r/c)}{\partial t}\left(-\frac{z}{r^2\,c}\right) + O\left(\frac{1}{r^2}\right) \tag{9.12}$$

to leading order in r^{-1}. Thus, we obtain

$$\phi(\mathbf{r}, t) \simeq \frac{l}{4\pi\epsilon_0\, c}\,\frac{z}{r}\,\frac{I(t-r/c)}{r}. \tag{9.13}$$

Given the vector and scalar potentials, Equations (9.10) and (9.13), respectively, we can evaluate the associated electric and magnetic fields using (see Section 4.12)

$$\mathbf{E} = -\frac{\partial \mathbf{A}}{\partial t} - \nabla\phi, \tag{9.14}$$

$$\mathbf{B} = \nabla\times\mathbf{A}. \tag{9.15}$$

Note that we are only interested in *radiation fields*, which fall off like r^{-1} with increasing distance from the source. It is easily demonstrated that

$$\mathbf{E} \simeq -\frac{\omega\, l\, I_0}{4\pi\epsilon_0\, c^2}\,\sin\theta\,\frac{\sin[\omega\,(t-r/c)]}{r}\,\mathbf{e}_\theta, \tag{9.16}$$

and

$$\mathbf{B} \simeq -\frac{\omega\, l\, I_0}{4\pi\epsilon_0\, c^3}\,\sin\theta\,\frac{\sin[\omega\,(t-r/c)]}{r}\,\mathbf{e}_\varphi. \tag{9.17}$$

Here, (r, θ, φ) are standard spherical polar coordinates aligned along the z-axis. The above expressions for the far-field (*i.e.*, $r \gg \lambda$) electromagnetic fields generated by a localized oscillating current are also easily derived from Equations (4.190) and (4.191). Note that the fields are symmetric in the azimuthal angle φ. Moreover, there is no radiation along the axis of the oscillating dipole (*i.e.*, $\theta = 0$), and the maximum emission is in the plane perpendicular to this axis (*i.e.*, $\theta = \pi/2$)—see Figure 9.1.

The average power crossing a spherical surface S (whose radius is much greater than λ), centered on the dipole, is

$$P_{\rm rad} = \oint_S \langle\mathbf{u}\rangle\cdot d\mathbf{S}, \tag{9.18}$$

where the average is over a single period of oscillation of the wave, and the Poynting flux is given by (see Section 8.2)

$$\mathbf{u} = \frac{\mathbf{E}\times\mathbf{B}}{\mu_0} = \frac{\omega^2\, l^2\, I_0^2}{16\pi^2\epsilon_0\, c^3}\,\sin^2[\omega\,(t-r/c)]\,\frac{\sin^2\theta}{r^2}\,\mathbf{e}_r. \tag{9.19}$$

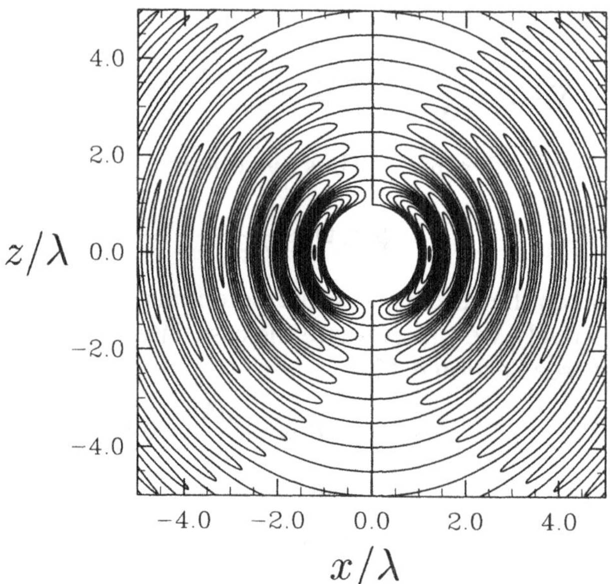

Figure 9.1: *Equally spaced contours of the normalized poloidal electric field* E_θ/E_0 *generated by a Hertzian dipole in the x-z plane at a fixed instant in time. Here,* $E_0 = \omega\, l\, I_0/4\pi\epsilon_0\, c^2$.

It follows that

$$\langle \mathbf{u} \rangle = \frac{\omega^2\, l^2\, I_0^2}{32\pi^2 \epsilon_0\, c^3} \frac{\sin^2\theta}{r^2}\, \mathbf{e}_r. \tag{9.20}$$

Note that the energy flux is *radially outward* from the source. The total power flux across S is given by

$$P_{\rm rad} = \frac{\omega^2\, l^2\, I_0^2}{32\pi^2 \epsilon_0\, c^3} \int_0^{2\pi} d\phi \int_0^{\pi} \frac{\sin^2\theta}{r^2}\, r^2 \sin\theta\, d\theta, \tag{9.21}$$

yielding

$$P_{\rm rad} = \frac{\omega^2\, l^2\, I_0^2}{12\pi \epsilon_0\, c^3}. \tag{9.22}$$

This total flux is independent of the radius of S, as is to be expected if energy is conserved.

Recall that for a resistor of resistance R the average ohmic heating power is

$$P_{\rm heat} = \langle I^2\,R\rangle = \frac{1}{2}\,I_0^2\,R, \tag{9.23}$$

assuming that $I = I_0\,\cos(\omega\,t)$. It is convenient to define the *radiation resistance* of a Hertzian dipole antenna:

$$R_{\rm rad} = \frac{P_{\rm rad}}{I_0^2/2}, \tag{9.24}$$

so that

$$R_{\rm rad} = \frac{2\pi}{3\epsilon_0\,c}\left(\frac{l}{\lambda}\right)^2, \tag{9.25}$$

where $\lambda = 2\pi\,c/\omega$ is the wavelength of the radiation. In fact,

$$R_{\rm rad} = 789\left(\frac{l}{\lambda}\right)^2 \text{ ohms.} \tag{9.26}$$

Now, in the theory of electrical circuits, an antenna is conventionally represented as a resistor whose resistance is equal to the characteristic radiation resistance of the antenna plus its real resistance. The power loss $I_0^2\,R_{\rm rad}/2$ associated with the radiation resistance is due to the emission of electromagnetic radiation, whereas the power loss $I_0^2\,R/2$ associated with the real resistance is due to ohmic heating of the antenna.

Note that the formula (9.26) is only valid for $l \ll \lambda$. This suggests that $R_{\rm rad} \ll R$ for most Hertzian dipole antennas: *i.e.*, the radiated power is swamped by the ohmic losses. Thus, antennas whose lengths are much less than that of the emitted radiation tend to be *extremely inefficient*. In fact, it is necessary to have $l \sim \lambda$ in order to obtain an efficient antenna. The simplest practical antenna is the *half-wave antenna*, for which $l = \lambda/2$. This can be analyzed as a series of Hertzian dipole antennas stacked on top of one another, each slightly out of phase with its neighbors. The characteristic radiation resistance of a half-wave antenna is

$$R_{\rm rad} = \frac{2.44}{4\pi\epsilon_0\,c} = 73 \text{ ohms.} \tag{9.27}$$

Antennas can also be used to receive electromagnetic radiation. The incoming wave induces a voltage in the antenna, which can be detected in an electrical circuit connected to the antenna. In fact, this process is equivalent to the emission of electromagnetic waves by the antenna viewed in reverse. It is easily demonstrated that antennas most readily detect electromagnetic radiation incident from those directions in which they preferentially emit radiation. Thus, a Hertzian dipole antenna is unable to detect radiation incident along its axis, and most efficiently detects radiation incident in the plane perpendicular to this axis. In the theory of electrical circuits, a receiving antenna is represented as a voltage source in series with a resistor. The voltage source, $V_0 \cos(\omega t)$, represents the voltage induced in the antenna by the incoming wave. The resistor, R_{rad}, represents the power re-radiated by the antenna (here, the real resistance of the antenna is neglected). Let us represent the detector circuit as a single load resistor, R_{load}, connected in series with the antenna. So, what value of R_{load} ensures that the maximum power is extracted from the wave and transmitted to the detector circuit?

According to Ohm's law,

$$V = V_0 \cos(\omega t) = I_0 \cos(\omega t) (R_{rad} + R_{load}), \tag{9.28}$$

where $I = I_0 \cos(\omega t)$ is the current induced in the circuit. The power input to the circuit is

$$P_{in} = \langle V I \rangle = \frac{V_0^2}{2 (R_{rad} + R_{load})}. \tag{9.29}$$

The power transferred to the load is

$$P_{load} = \langle I^2 R_{load} \rangle = \frac{R_{load} V_0^2}{2 (R_{rad} + R_{load})^2}. \tag{9.30}$$

Finally, the power re-radiated by the antenna is

$$P_{rad} = \langle I^2 R_{rad} \rangle = \frac{R_{rad} V_0^2}{2 (R_{rad} + R_{load})^2}. \tag{9.31}$$

Note that $P_{in} = P_{load} + P_{rad}$. The maximum power transfer to the load occurs when

$$\frac{\partial P_{load}}{\partial R_{load}} = \frac{V_0^2}{2} \left[\frac{R_{rad} - R_{load}}{(R_{rad} + R_{load})^3} \right] = 0. \tag{9.32}$$

Thus, the maximum transfer rate corresponds to

$$R_{load} = R_{rad}. \tag{9.33}$$

In other words, the resistance of the load circuit must match the radiation resistance of the antenna. For this optimum case,

$$P_{load} = P_{rad} = \frac{V_0^2}{8\,R_{rad}} = \frac{P_{in}}{2}. \tag{9.34}$$

So, in the optimum case *half* of the power absorbed by the antenna is immediately re-radiated. Clearly, an antenna which is receiving electromagnetic radiation is also emitting it. This (allegedly) is how the BBC catch people who do not pay their television license fee in the United Kingdom. They have vans which can detect the radiation emitted by a TV aerial whilst it is in use (they can even tell which channel you are watching!).

For a Hertzian dipole antenna interacting with an incoming wave whose electric field has an amplitude E_0, we expect

$$V_0 = E_0\, l. \tag{9.35}$$

Here, we have used the fact that the wavelength of the radiation is much longer than the length of the antenna. We have also assumed that the antenna is properly aligned (*i.e.*, the radiation is incident perpendicular to the axis of the antenna). The Poynting flux of the incoming wave is [see Equation (8.35)]

$$\langle u_{in} \rangle = \frac{\epsilon_0\, c\, E_0^2}{2}, \tag{9.36}$$

whereas the power transferred to a properly matched detector circuit is

$$P_{load} = \frac{E_0^2\, l^2}{8\,R_{rad}}. \tag{9.37}$$

Consider an idealized antenna in which all incoming radiation incident on some area A_{eff} is absorbed, and then magically transferred to the detector circuit, with no re-radiation. Suppose that the power absorbed from the idealized antenna matches that absorbed from the real antenna. This implies that

$$P_{load} = \langle u_{in} \rangle\, A_{eff}. \tag{9.38}$$

The quantity A_{eff} is called the *effective area* of the antenna: it is the area of the idealized antenna which absorbs as much net power from the incoming wave as the actual antenna. Thus,

$$P_{load} = \frac{E_0^2\, l^2}{8\,R_{rad}} = \frac{\epsilon_0\, c\, E_0^2}{2}\, A_{eff}, \tag{9.39}$$

giving

$$A_{\rm eff} = \frac{l^2}{4\epsilon_0\, c\, R_{\rm rad}} = \frac{3}{8\pi}\,\lambda^2. \tag{9.40}$$

So, it is clear that the effective area of a Hertzian dipole antenna is of order the *wavelength squared* of the incoming radiation.

For a properly aligned half-wave antenna,

$$A_{\rm eff} = 0.13\,\lambda^2. \tag{9.41}$$

Thus, the antenna, which is essentially one-dimensional with length $\lambda/2$, acts as if it is two-dimensional, with width $0.26\,\lambda$, as far as its absorption of incoming electromagnetic radiation is concerned.

9.3 ELECTRIC DIPOLE RADIATION

In the previous section, we examined the radiation emitted by a short electric dipole of oscillating dipole moment

$$\mathbf{p}(t) = p_0\,\sin(\omega\,t)\,\mathbf{e}_z, \tag{9.42}$$

where $p_0 = q_0\,l = I_0\,l/\omega$. We found that, in the far field region, the mean electromagnetic energy flux takes the form [see Equation (9.20)]

$$\langle\mathbf{u}\rangle = \frac{\omega^4\,p_0^2}{32\pi^2\epsilon_0\,c^3}\,\frac{\sin^2\theta}{r^2}\,\mathbf{e}_r, \tag{9.43}$$

assuming that the dipole is centered on the origin of our spherical polar coordinate system. The mean power radiated into the element of solid angle $d\Omega = \sin\theta\,d\theta\,d\varphi$, centered on the angular coordinates $(\theta,\,\varphi)$, is

$$dP = \langle\mathbf{u}(r,\theta,\varphi)\rangle\cdot\mathbf{e}_r\,r^2\,d\Omega. \tag{9.44}$$

Hence, the differential power radiated into this element of solid angle is simply

$$\frac{dP}{d\Omega} = \frac{\omega^4\,p_0^2}{32\pi^2\epsilon_0\,c^3}\,\sin^2\theta. \tag{9.45}$$

This formula completely specifies the radiation pattern of an oscillating electric dipole (provided that the dipole is much shorter in length than the wavelength of the emitted radiation). Of course, the power radiated into a given element of solid angle is independent of r, otherwise energy would not be conserved. Finally, the total radiated power is the integral of $dP/d\Omega$ over all solid angles.

9.4 THOMPSON SCATTERING

Consider a plane electromagnetic wave of angular frequency ω interacting with a *free* electron of mass m_e and charge $-e$. Suppose that the wave is polarized such that its associated electric field is parallel to the z-axis: *i.e.*,

$$\mathbf{E} = E_0 \sin(\omega t)\, \mathbf{e}_z. \tag{9.46}$$

Recall, from Section 4.7, that as long as the electron remains non-relativistic, the force exerted on it by the electromagnetic wave comes predominantly from the associated electric field. Hence, the electron's equation of motion can be written

$$m_e \frac{d^2 z}{dt^2} = -e\, E_0 \sin(\omega t), \tag{9.47}$$

which can be solved to give

$$z = \frac{e\, E_0}{m_e\, \omega^2} \sin(\omega t). \tag{9.48}$$

So, in response to the wave, the electron oscillates backward and forward in the direction of the wave electric field. It follows that the electron can be thought of as a sort of oscillating electric dipole, with dipole moment

$$\mathbf{p} = -e\, z\, \mathbf{e}_z = -p_0 \sin(\omega t)\, \mathbf{e}_z, \tag{9.49}$$

where $p_0 = e^2\, E_0/(m_e\, \omega^2)$. (For the moment, let us not worry about the positively charged component of the dipole.) Now, we know that an oscillating electric dipole emits electromagnetic radiation. Hence, it follows that a free electron placed in the path of a plane electromagnetic wave will radiate. To be more exact, the electron *scatters* electromagnetic radiation from the wave, since the radiation emitted by the electron is not necessarily in the same direction as the wave, and any energy radiated by the electron is ultimately extracted from the wave. This type of scattering is called *Thompson scattering*.

It follows from Equation (9.45) that the differential power scattered from a plane electromagnetic wave by a free electron into solid angle $d\,\Omega$ takes the form

$$\frac{dP}{d\Omega} = \frac{e^4\, E_0^2}{32\pi^2 \epsilon_0\, c^3\, m_e^2} \sin^2\theta. \tag{9.50}$$

Now, the mean energy flux of the incident electromagnetic wave is written

$$|\langle \mathbf{u} \rangle| = \frac{c\,\epsilon_0\,E_0^2}{2}. \tag{9.51}$$

It is helpful to introduce a quantity called the *differential scattering cross-section*. This is defined

$$\frac{d\sigma}{d\Omega} = \frac{dP/d\Omega}{|\langle \mathbf{u} \rangle|}, \tag{9.52}$$

and has units of area over solid angle. Somewhat figuratively, we can think of the electron as offering a target of area $d\sigma/d\Omega$ to the incident wave. Any wave energy which falls on this target is scattered into the solid angle $d\Omega$. Likewise, we can also define the *total scattering cross-section*,

$$\sigma = \oint \frac{d\sigma}{d\Omega}\,d\Omega, \tag{9.53}$$

which has units of area. Again, the electron effectively offers a target of area σ to the incident wave. Any wave energy which falls on this target is scattered in some direction or other. It follows from Equations (9.50) and (9.51) that the differential scattering cross-section for Thompson scattering is

$$\frac{d\sigma}{d\Omega} = r_e^2\,\sin^2\theta, \tag{9.54}$$

where the characteristic length

$$r_e = \frac{e^2}{4\pi\epsilon_0\,m_e\,c^2} = 2.82 \times 10^{-15}\,\text{m} \tag{9.55}$$

is called the *classical electron radius*. An electron effectively acts like it has a spatial extent r_e as far as its iteration with electromagnetic radiation is concerned. As is easily demonstrated, the total Thompson scattering cross-section is

$$\sigma_T = \frac{8\,\pi}{3}\,r_e^2 = 6.65 \times 10^{-29}\,\text{m}^2. \tag{9.56}$$

Note that both the differential and the total Thompson scattering cross-sections are completely *independent* of the frequency (or wavelength) of the incident radiation.

A scattering cross-section of $10^{-28}\,\text{m}^2$ does not sound like much. Nevertheless, Thompson scattering is one of the most important types of scattering in the Universe. Consider the Sun. It turns out that the mean mass density of the Sun is similar to that of water: *i.e.*, about $10^3\,\mathrm{kg\,m^{-3}}$. Hence, assuming that the Sun is predominantly made up of ionized Hydrogen, the mean number density of electrons in the Sun (which, of course, is the same as the number density of protons) is approximately $n_e \sim 10^3/m_p \sim 10^{30}\,\mathrm{m^{-3}}$, where $m_p \sim 10^{-27}$ kg is the mass of a proton. Let us consider how far, on average, a photon in the Sun travels before being scattered by a free electron. If we think of an individual photon as sweeping out a cylinder of cross-sectional area σ_T, then the photon will travel an average length l, such that a cylinder of area σ_T and length l contains about one free electron, before being scattered. Hence, $\sigma_T\, l\, n_e \sim 1$, or

$$l \sim \frac{1}{n_e\,\sigma_T} \sim 1\text{ cm}. \tag{9.57}$$

Given that the radius of the Sun is approximately 10^9 m, it is clear that solar photons are very strongly scattered by free electrons. In fact, it can easily be demonstrated that it takes a photon emitted in the solar core many thousands of years to fight its way to the surface because of Thompson scattering.

After the "Big Bang," when the Universe was very hot, it consisted predominately of ionized Hydrogen (and dark matter), and was consequently *opaque* to electromagnetic radiation, due to Thompson scattering. However, as the Universe expanded, it also cooled, and eventually became sufficiently cold (when the mean temperature was about 1000°C) for any free protons and electrons to combine to form molecular Hydrogen. It turns out that molecular Hydrogen does not scatter radiation anything like as effectively as free electrons (see the next section). Hence, as soon as the Universe became filled with molecular Hydrogen, it effectively became *transparent* to radiation. Indeed, the so-called *cosmic microwave background* is the remnant of radiation which was last scattered when the Universe was filled with ionized Hydrogen (*i.e.*, when it was about 1000°C). Astronomers can gain a great deal of information about the conditions in the early Universe by studying this radiation.

Incidentally, it is clear from Equations (9.55) and (9.56) that the scattering cross-section of a free particle of charge q and mass m is proportional to q^4/m^2. It follows that the scattering of electromagnetic radiation by free electrons is generally very much stronger than the scattering by free protons (assuming that the number densities of both species are similar).

9.5 RAYLEIGH SCATTERING

Let us now consider the scattering of electromagnetic radiation by neutral atoms. For instance, consider a Hydrogen atom. The atom consists of a light electron and a massive proton. As we have seen, the electron scatters radiation much more strongly than the proton, so let us concentrate on the response of the electron to an incident electromagnetic wave. Suppose that the wave electric field is again polarized in the z-direction, and is given by Equation (9.46). We can approximate the electron's equation of motion as

$$m_e \frac{d^2 z}{dt^2} = -m_e\, \omega_0^2\, z - e\, E_0 \sin(\omega\, t). \tag{9.58}$$

Here, the second term on the right-hand side represents the perturbing force due to the electromagnetic wave, whereas the first term represents the (linearized) force of electrostatic attraction between the electron and the proton. Here, we are very crudely modeling our Hydrogen atom as a *simple harmonic oscillator* of natural frequency ω_0. We can think of ω_0 as the typical frequency of electromagnetic radiation emitted by the atom after it is transiently disturbed. In other words, in our model, ω_0 should match the frequency of one of the spectral lines of Hydrogen. More generally, we can extend the above model to deal with just about any type of atom, provided that we set ω_0 to the frequency of a spectral line.

We can easily solve Equation (9.58) to give

$$z = \frac{e\, E_0}{m_e\,(\omega^2 - \omega_0^2)} \sin(\omega\, t). \tag{9.59}$$

Hence, the dipole moment of the electron takes the form $\mathbf{p} = -p_0 \sin(\omega\, t)\, \mathbf{e}_z$, where

$$p_0 = \frac{e^2\, E_0}{m_e\,(\omega^2 - \omega_0^2)}. \tag{9.60}$$

It follows, by analogy with the analysis in the previous section, that the differential and total scattering cross-sections of our model atom take the form

$$\frac{d\sigma}{d\Omega} = \frac{\omega^4}{(\omega^2 - \omega_0^2)^2}\, r_e^2 \sin^2\theta, \tag{9.61}$$

and

$$\sigma = \frac{\omega^4}{(\omega^2 - \omega_0^2)^2}\,\sigma_T, \tag{9.62}$$

respectively.

In the limit in which the frequency of the incident radiation is *much greater* than the natural frequency of the atom, Equations (9.61) and (9.62) reduce to the previously obtained expressions for scattering by a free electron. In other words, an electron in an atom acts very much like a free electron as far as high-frequency radiation is concerned. In the opposite limit, in which the frequency of the incident radiation is *much less* than the natural frequency of the atom, Equations (9.61) and (9.62) yield

$$\frac{d\sigma}{d\Omega} = \left(\frac{\omega}{\omega_0}\right)^4 r_e^2 \sin^2\theta, \tag{9.63}$$

and

$$\sigma = \left(\frac{\omega}{\omega_0}\right)^4 \sigma_T, \tag{9.64}$$

respectively. This type of scattering is called *Rayleigh scattering*. There are two features of Rayleigh scattering which are worth noting. First of all, it is much weaker than Thompson scattering (since $\omega \ll \omega_0$). Secondly, unlike Thompson scattering, it is highly *frequency dependent*. Indeed, it is clear, from the above formulae, that high-frequency (short wavelength) radiation is scattered far more effectively than low-frequency (long wavelength) radiation.

The most common example of Rayleigh scattering is the scattering of visible radiation from the Sun by neutral atoms (mostly Nitrogen and Oxygen) in the upper atmosphere. The frequency of visible radiation is much less than the typical emission frequencies of a Nitrogen or Oxygen atom (which lie in the ultraviolet band), so it is certainly the case that $\omega \ll \omega_0$. When the Sun is low in the sky, radiation from it has to traverse a comparatively long path through the atmosphere before reaching us. Under these circumstances, the scattering of direct solar light by neutral atoms in the atmosphere becomes noticeable (it is not noticeable when the Sun is high is the sky, and radiation from it consequently only has to traverse a relatively short path through the atmosphere before reaching us). According to Equation (9.64), blue light is scattered slightly more strongly than red light (since blue light has a slightly higher frequency than red light). Hence, when the Sun is low in the sky, it appears less bright, due to atmospheric scattering. However, it also appears *redder* than normal, because more blue light than red light is scattered out of the solar light-rays, leaving an excess of red light. Likewise, when we look up at the daytime sky, it does not appear black (like the sky on the Moon) because of light from solar radiation which grazes the atmosphere being scattered downward toward the surface of the Earth. Again, since blue light is scattered more effectively than red light, there is an excess of blue light scattered downward, and so the daytime sky appears *blue*.

Light from the Sun is unpolarized. However, when it is scattered it becomes polarized, because light is scattered preferentially in some directions rather than others. Consider a light-ray from the Sun which grazes the Earth's atmosphere. The light-ray contains light which is polarized such that the electric field is *vertical* to the ground, and light which is polarized such that the electric field is *horizontal* to the ground (and perpendicular to the path of the light-ray), in equal amounts. However, due to the $\sin^2\theta$ factor in the dipole emission formula (9.45) (where, in this case, θ is the angle between the direction of the wave electric field and the direction of scattering), very little light is scattered downward from the vertically polarized light compared to the horizontally polarized light. Moreover, the light scattered from the horizontally polarization is such that its electric field is preferentially perpendicular, rather than parallel, to the direction of propagation of the solar light-ray (*i.e.*, the direction to the Sun). Consequently, the blue light from the daytime sky is preferentially polarized in a direction *perpendicular* to the direction to the Sun.

9.6 PROPAGATION IN A DIELECTRIC MEDIUM

Consider the propagation of an electromagnetic wave through a uniform dielectric medium of dielectric constant ϵ. According to Equations (6.12) and (6.14), the dipole moment per unit volume induced in the medium by the wave electric field **E** is

$$\mathbf{P} = \epsilon_0\,(\epsilon - 1)\,\mathbf{E}. \tag{9.65}$$

There are no free charges or free currents in the medium. There is also no bound charge density (since the medium is uniform), and no magnetization current density (since the medium is non-magnetic). However, there is a *polarization current* due to the time-variation of the induced dipole moment per unit volume. According to Equation (6.109), this current is given by

$$\mathbf{j}_p = \frac{\partial \mathbf{P}}{\partial t}. \tag{9.66}$$

Hence, Maxwell's equations take the form

$$\nabla\cdot\mathbf{E} = 0, \tag{9.67}$$

$$\nabla\cdot\mathbf{B} = 0, \tag{9.68}$$

$$\nabla\times\mathbf{E} = -\frac{\partial \mathbf{B}}{\partial t}, \tag{9.69}$$

$$\nabla\times\mathbf{B} = \mu_0\,\mathbf{j}_p + \epsilon_0\,\mu_0\,\frac{\partial \mathbf{E}}{\partial t}. \tag{9.70}$$

According to Equations (9.65) and (9.66), the last of the above equations can be rewritten

$$\nabla\times\mathbf{B} = \epsilon_0\,\mu_0\,(\epsilon-1)\,\frac{\partial \mathbf{E}}{\partial t} + \epsilon_0\,\mu_0\,\frac{\partial \mathbf{E}}{\partial t} = \frac{\epsilon}{c^2}\,\frac{\partial \mathbf{E}}{\partial t}, \tag{9.71}$$

since $c = (\epsilon_0\,\mu_0)^{-1/2}$. Thus, Maxwell's equations for the propagation of electromagnetic waves through a dielectric medium are the same as Maxwell's equations for the propagation of waves through a vacuum (see Section 4.7), except that $c \to c/n$, where

$$n = \sqrt{\epsilon} \tag{9.72}$$

is called the *refractive index* of the medium in question. Hence, we conclude that electromagnetic waves propagate through a dielectric medium *slower* than through a vacuum by a factor n (assuming, of course, that $n > 1$). This conclusion (which was reached long before Maxwell's equations were invented) is the basis of all geometric optics involving refraction.

9.7 DIELECTRIC CONSTANT OF A GASEOUS MEDIUM

In Section 9.5, we discussed a rather crude model of an atom interacting with an electromagnetic wave. According to this model, the dipole moment $\mathbf{p}$ of the atom induced by the wave electric field $\mathbf{E}$ is given by

$$\mathbf{p} = \frac{e^2}{m_e\,(\omega_0^2 - \omega^2)}\,\mathbf{E}, \tag{9.73}$$

where ω_0 is the natural frequency of the atom (*i.e.*, the frequency of one of the atom's spectral lines), and ω the frequency of the incident radiation. Suppose that there are n atoms per unit volume. It follows that the induced dipole moment per unit volume of the assemblage of atoms takes the form

$$\mathbf{P} = \frac{n\,e^2}{m_e\,(\omega_0^2 - \omega^2)}\,\mathbf{E}. \tag{9.74}$$

Finally, a comparison with Equation (9.65) yields the following expression for the dielectric constant of the collection of atoms,

$$\epsilon = 1 + \frac{n\,e^2}{\epsilon_0\,m_e\,(\omega_0^2 - \omega^2)}. \tag{9.75}$$

The above formula works fairly well for dilute gases, although it is, of course, necessary to sum over all species and all important spectral lines.

Note that, in general, the dielectric "constant" of a gaseous medium (as far as electromagnetic radiation is concerned) is a function of the wave frequency, ω. Since the effective wave propagation speed through the medium is $c/\sqrt{\epsilon}$, it follows that waves of different frequencies travel through a gaesous medium at *different* speeds. This phenomenon is called *dispersion*, since it can be shown to cause short wave-pulses to spread out as they propagate through the medium. At low frequencies ($\omega \ll \omega_0$), however, our expression for ϵ becomes frequency independent, and so there is no dispersion of low-frequency waves by a gaseous medium.

9.8 DISPERSION RELATION OF A PLASMA

A plasma is very similar to a gaseous medium, expect that the electrons are *free*: *i.e.*, there is no restoring force due to nearby atomic nuclii. Hence, we can obtain an expression for the dielectric constant of a plasma from Equation (9.75) by setting ω_0 to zero, and n to the number density of electrons, n_e. We obtain

$$\epsilon = 1 - \frac{\omega_p^2}{\omega^2}, \tag{9.76}$$

where the characteristic frequency

$$\omega_p = \sqrt{\frac{n_e\,e^2}{\epsilon_0\,m_e}} \tag{9.77}$$

is called the *plasma frequency*. We can immediately see that formula (9.76) is problematic. For frequencies above the plasma frequency, the dielectric constant of a plasma is less than unity. Hence, the refractive index $n = \sqrt{\epsilon}$ is also less than unity. This would seem to imply that high-frequency electromagnetic waves can propagate through a plasma with a velocity c/n which is *greater* than the velocity of light in a vacuum. This appears to violate one of the principles of Relativity. On the other hand, for frequencies below the plasma frequency, the dielectric constant is *negative*, which would seem to imply that the refractive index $n = \sqrt{\epsilon}$ is *imaginary*. How should we interpret this?

Consider an infinite plane-wave of frequency ω, which is greater than the plasma frequency, propagating through a plasma. Suppose that the wave electric field takes the form

$$\mathbf{E} = E_0\, e^{i(kx-\omega t)}\, \mathbf{e}_z, \tag{9.78}$$

where it is understood that the physical electric field is the *real part* of the above expression. A peak or trough of the above wave travels at the so-called *phase-velocity*, which is given by

$$v_p = \frac{\omega}{k}. \tag{9.79}$$

Now, we have also seen that the phase-velocity of electromagnetic waves in a dielectric medium is $v_p = c/n = c/\sqrt{\epsilon}$, so

$$\omega^2 = \frac{k^2\, c^2}{\epsilon}. \tag{9.80}$$

It follows from Equation (9.76) that

$$\omega^2 = k^2\, c^2 + \omega_p^2 \tag{9.81}$$

in a plasma. The above type of expression, which effectively determines the wave frequency, ω, as a function of the wave-number, k, for the medium in question, is called a *dispersion relation* (since, amongst other things, it determines how fast wave-pulses disperse in the medium). According to the above dispersion relation, the phase-velocity of high-frequency waves propagating through a plasma is given by

$$v_p = \frac{c}{\sqrt{1 - \omega_p^2/\omega^2}}, \tag{9.82}$$

which is indeed greater than c. However, the Theory of Relativity does not forbid this. What the Theory of Relativity says is that *information* cannot travel at a velocity greater than c. However, the peaks and troughs of an infinite plane-wave, such as (9.78), *do not* carry any information.

We now need to consider how we could transmit information through a plasma (or any other dielectric medium) by means of electromagnetic waves. The easiest way would be to send a series of short discrete wave-pulses through the plasma, so that we could transmit information in a sort of Morse code. We can build up a wave-pulse from a suitable superposition of infinite plane-waves of different frequencies and wavelengths: *e.g.*,

$$E_z(x, t) = \int F(k)\, e^{i\,\phi(k)}\, dk, \tag{9.83}$$

where $\phi(k) = k\,x - \omega(k)\,t$, and $\omega(k)$ is determined from the dispersion relation (9.81). Now, it turns out that a relatively short wave-pulse can only be built up from a superposition of plane-waves with a relatively wide range of different k values. Hence, for a short wave-pulse, the integrand in the above formula consists of the product of a fairly slowly varying function, $F(k)$, and a rapidly oscillating function, $\exp[i\,\phi(k)]$. The latter function is rapidly oscillating because the phase $\phi(k)$ varies very rapidly with k, relative to $F(k)$. We expect the net result of integrating the product of a slowly varying function and rapidly oscillating function to be small, since the oscillations will generally average to zero. It follows that the integral (9.83) is dominated by those regions of k-space for which $\phi(k)$ varies *least rapidly* with k. Hence, the peak of the wave-pulse most likely corresponds to a maximum or minimum of $\phi(k)$: *i.e.*,

$$\frac{d\phi}{dk} = x - \frac{d\omega}{dk}\,t = 0. \tag{9.84}$$

Thus, we infer that the velocity of the wave-pulse (which corresponds to the velocity of the peak) is given by

$$v_g = \frac{d\omega}{dk}. \tag{9.85}$$

This velocity is called the *group-velocity*, and is different to the phase-velocity in dispersive media: *i.e.*, media for which ω is not *directly proportional* to k. (Of course, in a vacuum, $\omega = k\,c$, so the phase and group velocities are both equal to c.) The upshot of the above discussion is that information (*i.e.*, an individual wave-pulse) travels through a dispersive media at the group-velocity, rather than the phase-velocity. Hence, Relativity demands that the group-velocity, rather than the phase-velocity, must always be less than c.

What is the group-velocity for high-frequency waves propagating through a plasma? Well, differentiation of the dispersion relation (9.81) yields

$$\frac{\omega}{k}\,\frac{d\omega}{dk} = v_p\,v_g = c^2. \tag{9.86}$$

Hence, it follows from Equation (9.82) that

$$v_g = c\,\sqrt{1-\frac{\omega_p^{\,2}}{\omega^2}}, \tag{9.87}$$

which is less than c. We thus conclude that the dispersion relation (9.81) is indeed consistent with Relativity.

Let us now consider the propagation of low-frequency electromagnetic waves through a plasma. We can see, from Equations (9.82) and (9.87), that when the wave frequency, ω, falls below the plasma frequency, ω_p, both the phase and group velocities become imaginary. This indicates that the wave *attenuates* as it propagates. Consider, for instance, a plane-wave of frequency $\omega < \omega_p$. According to the dispersion relation (9.81), the associated wave-number is given by

$$k = i\,\sqrt{\omega_p^{\,2}-\omega^2}\,\Big/\,c = i\,|k|. \tag{9.88}$$

Hence, the wave electric field takes the form

$$E_z = E_0\,e^{\,i\,(i\,|k|\,x-\omega\,t)} = E_0\,e^{-|k|\,x}\,e^{-i\,\omega\,t}. \tag{9.89}$$

So, it can be seen that for $\omega < \omega_p$ electromagnetic waves in a plasma take the form of decaying *standing waves*, rather than traveling waves. We conclude that an electromagnetic wave, of frequency less than the plasma frequency, which is incident on a plasma will not propagate through the plasma. Instead, it will be *totally reflected*.

We can be sure that the incident wave is reflected by the plasma, rather than absorbed, by considering the energy flux of the wave in the plasma. It is easily demonstrated that the energy flux of an electromagnetic wave can be written

$$\mathbf{u} = \frac{\mathbf{E}\times\mathbf{B}}{\mu_0} = \frac{E^2}{\mu_0\,\omega}\,\mathbf{k}. \tag{9.90}$$

For a wave with a real frequency and a complex **k**-vector, the above formula generalizes to

$$\mathbf{u} = \frac{|E|^2}{\mu_0\,\omega}\,\mathrm{Re}(\mathbf{k}). \tag{9.91}$$

However, according to Equation (9.88), the **k**-vector for a low-frequency electromagnetic wave in a plasma is *purely imaginary*. It follows that the associated energy flux is *zero*. Hence, any low-frequency wave which is incident on the plasma must be totally reflected, since if there were any absorption of the wave energy then there would be a net energy flux into the plasma.

The outermost layer of the Earth's atmosphere consists of a partially ionized zone known as the *ionosphere*. The plasma frequency in the ionosphere is about 1 MHz, which lies at the upper end of the medium-wave band of radio frequencies. It follows that low-frequency radio signals (*i.e.*, all signals in the long-wave band, and most in the medium-wave band) are *reflected* off the ionosphere. For this reason, such signals can be detected over the horizon. Indeed, long-wave radio signals reflect multiple times off the ionosphere with very little loss (they also reflect multiple times off the Earth, which is enough of a conductor to act as a mirror for radio waves), and can consequently be detected all over the world. On the other hand, high-frequency radio signals (*i.e.*, all signals in the FM band) pass straight through the ionosphere. For this reason, such signals cannot be detected over the horizon, which accounts for the relatively local coverage of FM radio stations. Note, from Equation (9.77), that the plasma frequency is proportional to the square root of the number density of free electrons. Now, the level of ionization in the ionosphere is maintained by ultraviolet light from the Sun (which effectively knocks electrons out of neutral atoms). Of course, there is no such light at night, and the number density of free electrons in the ionosphere consequently drops as electrons and ions gradually recombine. It follows that the plasma frequency in the ionosphere also drops at night, giving rise to a marked deterioration in the reception of distant medium-wave radio stations.

9.9 FARADAY ROTATION

Consider a high-frequency electromagnetic wave propagating, along the z-axis, through a plasma with a longitudinal equilibrium magnetic field, $\mathbf{B} = B_0\,\mathbf{e}_z$. The equation of motion of an individual electron making up the plasma takes the form

$$m_e\,\frac{d\mathbf{v}}{dt} = -e\,(\mathbf{E} + B_0\,\mathbf{v}\times\mathbf{e}_z), \tag{9.92}$$

where the first term on the right-hand side is due to the wave electric field, and the second to the equilibrium magnetic field. (As usual, we can neglect the wave magnetic field, provided that the electron motion remains non-relativistic.) Of course, $\mathbf{v} = d\mathbf{r}/dt$, where $\mathbf{r}$ is the electron displacement from its equilibrium position. Suppose that all perturbed quantities vary with time like $\exp(-i\,\omega\,t)$, where ω is the wave frequency. It follows that

$$m_e\,\omega^2\,x = e\,(E_x - i\,\omega\,B_0\,y), \tag{9.93}$$

$$m_e\,\omega^2\,y = e\,(E_y + i\,\omega\,B_0\,x). \tag{9.94}$$

It is helpful to define

$$s_\pm = x \pm i\,y, \tag{9.95}$$

$$E_\pm = E_x \pm i\,E_y. \tag{9.96}$$

Using these new variables, Equations (9.93) and (9.94) can be rewritten

$$m_e\,\omega^2\,s_\pm = e\,(E_\pm \mp \omega\,B_0\,s_\pm), \tag{9.97}$$

which can be solved to give

$$s_\pm = \frac{e\,E_\pm}{m_e\,\omega\,(\omega \pm \Omega)}, \tag{9.98}$$

where $\Omega = e\,B_0/m_e$ is the so-called *cyclotron frequency* (*i.e.*, the characteristic gyration frequency of free electrons in the equilibrium magnetic field—see Section 3.7).

In terms of $s_\pm$, the electron displacement can be written

$$\mathbf{r} = s_+\,e^{\,i\,(k_+\,z-\omega\,t)}\,\mathbf{e}_+ + s_-\,e^{\,i\,(k_-\,z-\omega\,t)}\,\mathbf{e}_-, \tag{9.99}$$

where

$$\mathbf{e}_\pm = \frac{1}{2}\,(\mathbf{e}_x \mp i\,\mathbf{e}_y). \tag{9.100}$$

Likewise, in terms of $E_\pm$, the wave electric field takes the form

$$\mathbf{E} = E_+\,e^{\,i\,(k_+\,z-\omega\,t)}\,\mathbf{e}_+ + E_-\,e^{\,i\,(k_-\,z-\omega\,t)}\,\mathbf{e}_-. \tag{9.101}$$

Obviously, the actual displacement and electric field are the *real parts* of the above expressions. It follows from Equation (9.101) that E_+ corresponds to a constant amplitude electric field which rotates *clockwise* in the x-y plane (looking down the z-axis) as the wave propagates in the $+z$-direction, whereas E_- corresponds to a constant amplitude electric field which rotates *counter clockwise*. The former type of wave is termed *left-hand circularly polarized*, whereas the latter is termed *right-hand circularly polarized*. Note also that s_+ and s_- correspond to *circular* electron motion in opposite senses. With these insights, we conclude that Equation (9.98) indicates that individual electrons in the plasma have a slightly different response to left- and right-hand circularly polarized waves in the presence of a longitudinal magnetic field.

Following the analysis of Section 9.7, we can deduce from Equation (9.98) that the dielectric constant of the plasma for left- and right-hand circularly polarized waves is

$$\epsilon_\pm = 1 - \frac{\omega_p^2}{\omega\,(\omega \pm \Omega)}, \tag{9.102}$$

respectively. Hence, according to Equation (9.80), the dispersion relation for left- and right-hand circularly polarized waves becomes

$$k_\pm^2\,c^2 = \omega^2\left[1 - \frac{\omega_p^2}{\omega\,(\omega \pm \Omega)}\right], \tag{9.103}$$

respectively. In the limit $\omega \gg \omega_p, \Omega$, we obtain

$$k_\pm \simeq k \pm \Delta k, \tag{9.104}$$

where $k = \omega\,[1 - (1/2)\,\omega_p^2/\omega^2]/c$ and $\Delta k = (1/2)\,(\omega_p^2/\omega^2)\,\Omega/c$. In other words, in a magnetized plasma, left- and right-hand circularly polarized waves of the same frequency have slightly different wave-numbers.

Let us now consider the propagation of a *linearly polarized* electromagnetic wave through the plasma. Such a wave can be constructed via a superposition of left- and right-hand circularly polarized waves of *equal* amplitudes. So, the wave electric field can be written

$$\mathbf{E} = E_0\left[e^{\,i\,(k_+\,z-\omega\,t)}\,\mathbf{e}_+ + e^{\,i\,(k_-\,z-\omega\,t)}\,\mathbf{e}_-\right]. \tag{9.105}$$

It can easily be seen that at $z = 0$ the wave electric field is aligned along the x-axis. If left- and right-hand circularly polarized waves of the same frequency have the same wave-number (*i.e.*, if $k_+ = k_-$) then the wave electric field will continue to be aligned along the x-axis as the wave propagates in the $+z$-direction: *i.e.*, we will obtain a standard linearly polarized wave. However, we have just demonstrated that, in the presence of a longitudinal magnetic field, the wave-numbers k_+ and k_- are slightly different. What effect does this have on the polarization of the wave?

Taking the real part of Equation (9.105), and making use of Equation (9.104), and some standard trigonometrical identities, we obtain

$$\mathbf{E} = E_0\,[\cos(k\,z - \omega\,t)\,\cos(\Delta k\,z),\,\cos(k\,z - \omega\,t)\,\sin(\Delta k\,z),\,0]\,. \qquad (9.106)$$

The polarization angle of the wave (which is a convenient measure of its plane of polarization) is given by

$$\varphi = \tan^{-1}(E_y/E_x) = \Delta k\,z. \qquad (9.107)$$

Thus, we conclude that in the presence of a longitudinal magnetic field the polarization angle *rotates* as the wave propagates through the plasma. This effect is known as *Faraday rotation*. It is clear, from the above expression, that the rate of advance of the polarization angle with distance travelled by the wave is given by

$$\frac{d\varphi}{dz} = \Delta k = \frac{\omega_p^2\,\Omega}{2\,\omega^2\,c} = \frac{e^3}{2\epsilon_0\,m_e^2\,c}\,\frac{n_e\,B_0}{\omega^2}. \qquad (9.108)$$

Hence, a linearly polarized electromagnetic wave which propagates through a plasma with a (slowly) varying electron number density, $n_e(z)$, and longitudinal magnetic field, $B_0(z)$, has its plane of polarization rotated through a total angle

$$\Delta\varphi = \varphi - \varphi_0 = \frac{e^3}{2\epsilon_0\,m_e^2\,c}\,\frac{1}{\omega^2}\int n_e(z)\,B_0(z)\,dz. \qquad (9.109)$$

Note the very strong inverse variation of $\Delta\varphi$ with ω.

Pulsars are rapidly rotating neutron stars which emit regular blips of highly polarized radio waves. Hundreds of such objects have been found in our galaxy since the first was discovered in 1967. By measuring the variation of the angle of polarization, φ, of radio emission from a pulsar with frequency, ω, astronomers can effectively determine the line integral of $n_e\,B_0$ along the straight-line joining the pulsar to the Earth using formula (9.109). Here, n_e is the number density of free electrons in the interstellar medium, whereas B_0 is the parallel component of the galactic magnetic field. Obviously, in order to achieve this, astronomers must make the reasonable assumption that the radiation was emitted by the pulsar with a common angle of polarization, φ_0, over a wide range of different frequencies. By fitting Equation (9.109) to the data, and then extrapolating to large ω, it is then possible to determine φ_0, and, hence, the amount, $\Delta\varphi(\omega)$, through which the polarization angle of the radiation has rotated, at a given frequency, during its passage to Earth.

9.10 PROPAGATION IN A CONDUCTOR

Consider the propagation of an electromagnetic wave through a conducting medium which obeys Ohm's law:

$$\mathbf{j} = \sigma\,\mathbf{E}. \tag{9.110}$$

Here, σ is the *conductivity* of the medium in question. Maxwell's equations for the wave take the form:

$$\nabla\cdot\mathbf{E} = 0, \tag{9.111}$$

$$\nabla\cdot\mathbf{B} = 0, \tag{9.112}$$

$$\nabla\times\mathbf{E} = -\frac{\partial\mathbf{B}}{\partial t}, \tag{9.113}$$

$$\nabla\times\mathbf{B} = \mu_0\,\mathbf{j} + \epsilon\,\epsilon_0\mu_0\,\frac{\partial\mathbf{E}}{\partial t}, \tag{9.114}$$

where ϵ is the dielectric constant of the medium. It follows, from the above equations, that

$$\nabla\times\nabla\times\mathbf{E} = -\nabla^2\mathbf{E} = -\frac{\partial\nabla\times\mathbf{B}}{\partial t} = -\frac{\partial}{\partial t}\left[\mu_0\,\sigma\,\mathbf{E} + \epsilon\,\epsilon_0\mu_0\,\frac{\partial\mathbf{E}}{\partial t}\right]. \tag{9.115}$$

Looking for a wave-like solution of the form

$$\mathbf{E} = \mathbf{E}_0\,e^{\,i\,(k\,z-\omega\,t)}, \tag{9.116}$$

we obtain the dispersion relation

$$k^2 = \mu_0\,\omega\,(\epsilon\,\epsilon_0\,\omega + i\,\sigma). \tag{9.117}$$

Consider a "poor" conductor for which $\sigma \ll \epsilon\,\epsilon_0\,\omega$. In this limit, the dispersion relation (9.117) yields

$$k \simeq n\,\frac{\omega}{c} + i\,\frac{\sigma}{2}\sqrt{\frac{\mu_0}{\epsilon\,\epsilon_0}}, \tag{9.118}$$

where $n = \sqrt{\epsilon}$ is the refractive index. Substitution into Equation (9.116) gives

$$\mathbf{E} = \mathbf{E}_0\,e^{-z/d}\,e^{\,i\,(k_r\,z-\omega\,t)}, \tag{9.119}$$

where

$$d = \frac{2}{\sigma}\sqrt{\frac{\epsilon\,\epsilon_0}{\mu_0}}, \tag{9.120}$$

and $k_r = n\,\omega/c$. Thus, we conclude that the amplitude of an electromagnetic wave propagating through a conductor *decays exponentially* on some length-scale, d, which is termed the *skin-depth*. Note, from Equation (9.120), that the skin-depth for a poor conductor is *independent* of the frequency of the wave. Note, also, that $k_r\,d \gg 1$ for a poor conductor, indicating that the wave penetrates many wavelengths into the conductor before decaying away.

Consider a "good" conductor for which $\sigma \gg \epsilon\,\epsilon_0\,\omega$. In this limit, the dispersion relation (9.117) yields

$$k \simeq \sqrt{i\,\mu_0\,\sigma\,\omega}. \tag{9.121}$$

Substitution into Equation (9.116) again gives Equation (9.119), with

$$d = \frac{1}{k_r} = \sqrt{\frac{2}{\mu_0\,\sigma\,\omega}}. \tag{9.122}$$

It can be seen that the skin-depth for a good conductor *decreases* with increasing wave frequency. The fact that $k_r\,d = 1$ indicates that the wave only penetrates a few wavelengths into the conductor before decaying away.

Now the power per unit volume dissipated via ohmic heating in a conducting medium is

$$P = \mathbf{j}\cdot\mathbf{E} = \sigma\,E^2. \tag{9.123}$$

Consider an electromagnetic wave of the form (9.119). The mean power dissipated per unit area in the region $z > 0$ is written

$$\langle P\rangle = \frac{1}{2}\int_0^\infty \sigma\,E_0^2\,e^{-2\,z/d}\,dz = \frac{d\,\sigma}{4}\,E_0^2 = \sqrt{\frac{\sigma}{8\,\mu_0\,\omega}}\,E_0^2, \tag{9.124}$$

for a good conductor. Now, according to Equation (9.91), the mean electromagnetic power flux into the region $z > 0$ takes the form

$$\langle u\rangle = \left\langle\frac{\mathbf{E}\times\mathbf{B}\cdot\mathbf{e}_z}{\mu_0}\right\rangle_{z=0} = \frac{1}{2}\frac{E_0^2\,k_r}{\mu_0\,\omega} = \sqrt{\frac{\sigma}{8\,\mu_0\,\omega}}\,E_0^2. \tag{9.125}$$

It is clear, from a comparison of the previous two equations, that all of the wave energy which flows into the region $z > 0$ is dissipated via ohmic heating. We thus conclude that the attenuation of an electromagnetic wave propagating through a conductor is a direct consequence of ohmic power losses.

Consider a typical metallic conductor such as Copper, whose electrical conductivity at room temperature is about $6\times10^7\,(\Omega\,\mathrm{m})^{-1}$. Copper, therefore, acts as a good conductor for all electromagnetic waves of frequency below about 10^{18} Hz. The skin-depth in Copper for such waves is thus

$$d = \sqrt{\frac{2}{\mu_0\,\sigma\,\omega}} \simeq \frac{6}{\sqrt{f(\mathrm{Hz})}}\ \mathrm{cm}. \tag{9.126}$$

It follows that the skin-depth is about 6 cm at 1 Hz, but only about 2 mm at 1 kHz. This gives rise to the so-called *skin-effect* in copper wires, by which an oscillating electromagnetic signal of increasing frequency, transmitted along such a wire, is confined to an increasingly narrow layer (whose thickness is of order the skin-depth) on the surface of the wire.

The conductivity of seawater is only about $\sigma \simeq 5\,(\Omega\,\mathrm{m})^{-1}$. However, this is still sufficiently high for seawater to act as a good conductor for all radio frequency electromagnetic waves (*i.e.*, $f = \omega/2\pi < 10^9$ Hz). The skin-depth at 1 MHz ($\lambda \sim 2$ km) is about 0.2 m, whereas that at 1 kHz ($\lambda \sim 2000$ km) is still only about 7 m. This obviously poses quite severe restrictions for radio communication with submerged submarines. Either the submarines have to come quite close to the surface to communicate (which is dangerous), or the communication must be performed with extremely low-frequency (ELF) waves (*i.e.*, $f < 100$ Hz). Unfortunately, such waves have very large wavelengths ($\lambda > 20,000$ km), which means that they can only be efficiently generated by gigantic antennas.

9.11 DISPERSION RELATION OF A COLLISIONAL PLASMA

We have now investigated electromagnetic wave propagation through two different media possessing free electrons: *i.e.*, plasmas (see Section 9.8), and ohmic conductors (see Section 9.10). In the first case, we obtained the dispersion relation (9.81), whereas in the second we obtained the quite different dispersion relation (9.117). This leads us, quite naturally, to ask what the essential distinction is between the response of free electrons in a plasma to an electromagnetic wave, and that of free electrons in an ohmic conductor. It turns out that the main distinction is the relative strength of electron-ion *collisions*.

In the presence of electron-ion collisions, we can model the equation of motion of an individual electron in a plasma or a conductor as

$$m_e\,\frac{d\mathbf{v}}{dt} + m_e\,\nu\,\mathbf{v} = -e\,\mathbf{E}, \tag{9.127}$$

where **E** is the wave electric field. The collision term (*i.e.*, the second term on the left-hand side) takes the form of a drag force proportional to $-\mathbf{v}$. In the absence of the wave electric field, this force damps out any electron motion on the typical time-scale ν^{-1}. Since, in reality, an electron loses virtually all of its directed momentum during a collision with a much more massive ion, we can regard ν as the effective electron-ion collision frequency.

Assuming the usual $\exp(-i\,\omega\,t)$ time-dependence of perturbed quantities, we can solve Equation (9.127) to give

$$\mathbf{v} = -i\,\omega\,\mathbf{r} = -\frac{i\,\omega\,e\,\mathbf{E}}{m_e\,\omega\,(\omega + i\,\nu)}. \tag{9.128}$$

Hence, the perturbed current density can be written

$$\mathbf{j} = -e\,n_e\,\mathbf{v} = \frac{i\,n_e\,e^2\,\mathbf{E}}{m_e\,(\omega + i\,\nu)}, \tag{9.129}$$

where n_e is the number density of free electrons. It follows that the effective conductivity of the medium takes the form

$$\sigma = \frac{\mathbf{j}}{\mathbf{E}} = \frac{i\,n_e\,e^2}{m_e\,(\omega + i\,\nu)}. \tag{9.130}$$

Now, the mean rate of ohmic heating per unit volume in the medium is written

$$\langle P\rangle = \frac{1}{2}\,\mathrm{Re}(\sigma)\,E_0^{\,2}, \tag{9.131}$$

where E_0 is the amplitude of the wave electric field. Note that only the *real part* of σ contributes to ohmic heating, because the perturbed current must be *in phase* with the wave electric field in order for there to be a net heating effect. An imaginary σ gives a perturbed current which is in phase quadrature with the wave electric field. In this case, there is zero net transfer of power between the wave and the plasma over a wave period. We can see from Equation (9.130) that in the limit in which the wave frequency is much larger than the collision frequency (*i.e.*, $\omega \gg \nu$), the effective conductivity of the medium becomes purely imaginary:

$$\sigma \simeq \frac{i\,n_e\,e^2}{m_e\,\omega}. \tag{9.132}$$

In this limit, there is no loss of wave energy due to ohmic heating, and the medium acts like a conventional plasma. In the opposite limit, in which the wave frequency is much less than the collision frequency (*i.e.*, $\omega \ll \nu$), the effective conductivity becomes purely real:

$$\sigma \simeq \frac{n_e\, e^2}{m_e\, \nu}. \tag{9.133}$$

In this limit, ohmic heating losses are significant, and the medium acts like a conventional ohmic conductor.

Repeating the analysis of Section 9.7, we can derive the following dispersion relation from Equation (9.128):

$$k^2\, c^2 = \omega^2 - \frac{\omega_p^2\, \omega}{\omega + i\,\nu}. \tag{9.134}$$

It can be seen that, in the limit $\omega \gg \nu$, the above dispersion relation reduces to the dispersion relation (9.81) for a conventional (*i.e.*, collisionless) plasma. In the opposite limit, we obtain

$$k^2 = \frac{\omega^2}{c^2} + i\,\frac{\omega_p^2\, \omega}{\nu\, c^2} = \mu_0\, \omega\, (\epsilon_0\, \omega + i\, \sigma). \tag{9.135}$$

where use has been made of Eq (9.133). Of course, the above dispersion relation is identical to the dispersion relation (9.117) (with $\epsilon = 1$) which we previously derived for an ohmic conductor.

Our main conclusion from this section is that the dispersion relation (9.134) can be used to describe electromagnetic wave propagation through both a collisional plasma and an ohmic conductor. We can also deduce that in the low-frequency limit, $\omega \ll \nu$, a collisional plasma acts very much like an ohmic conductor, whereas in the high-frequency limit, $\omega \gg \nu$, an ohmic conductor acts very much like a collisionless plasma.

9.12 NORMAL REFLECTION AT A DIELECTRIC BOUNDARY

An electromagnetic wave of real (positive) frequency ω can be written

$$\mathbf{E}(\mathbf{r}, t) = \mathbf{E}_0\, e^{\,i\,(\mathbf{k}\cdot\mathbf{r} - \omega\, t)}, \tag{9.136}$$

$$\mathbf{B}(\mathbf{r}, t) = \mathbf{B}_0\, e^{\,i\,(\mathbf{k}\cdot\mathbf{r} - \omega\, t)}. \tag{9.137}$$

The wave-vector, **k**, indicates the direction of propagation of the wave, and also its phase-velocity, v, via

$$v = \frac{\omega}{k}. \tag{9.138}$$

Since the wave is transverse in nature, we must have $\mathbf{E}_0 \cdot \mathbf{k} = \mathbf{B}_0 \cdot \mathbf{k} = 0$. Finally, the familiar Maxwell equation

$$\nabla \times \mathbf{E} = -\frac{\partial \mathbf{B}}{\partial t} \tag{9.139}$$

leads us to the following relation between the constant vectors $\mathbf{E}_0$ and $\mathbf{B}_0$:

$$\mathbf{B}_0 = \frac{\hat{\mathbf{k}} \times \mathbf{E}_0}{v}. \tag{9.140}$$

Here, $\hat{\mathbf{k}} = \mathbf{k}/k$ is a unit vector pointing in the direction of wave propagation.

Suppose that the plane $z = 0$ forms the boundary between two different dielectric media. Let medium 1, of refractive index n_1, occupy the region $z < 0$, whilst medium 2, of refractive index n_2, occupies the region $z > 0$. Let us investigate what happens when an electromagnetic wave is incident on this boundary from medium 1.

Consider, first of all, the simple case of incidence *normal* to the boundary—see Figure 9.2. In this case, $\hat{\mathbf{k}} = +\mathbf{e}_z$ for the incident and transmitted waves, and $\hat{\mathbf{k}} = -\mathbf{e}_z$ for the reflected wave. Without loss of generality, we can assume that the incident wave is polarized in the x-direction. Hence, using Equation (9.140), the incident wave can

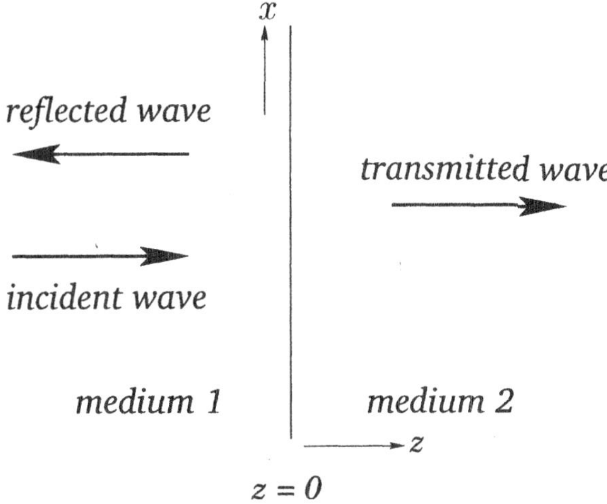

Figure 9.2: *Reflection at a dielectric boundary for the case of normal incidence.*

be written

$$\mathbf{E}(z,t) = E_i\, e^{i(k_1 z - \omega t)}\, \mathbf{e}_x, \tag{9.141}$$

$$\mathbf{B}(z,t) = \frac{E_i}{v_1}\, e^{i(k_1 z - \omega t)}\, \mathbf{e}_y, \tag{9.142}$$

where $v_1 = c/n_1$ is the phase-velocity in medium 1, and $k_1 = \omega/v_1$. Likewise, the reflected wave takes the form

$$\mathbf{E}(z,t) = E_r\, e^{i(-k_1 z - \omega t)}\, \mathbf{e}_x, \tag{9.143}$$

$$\mathbf{B}(z,t) = -\frac{E_r}{v_1}\, e^{i(-k_1 z - \omega t)}\, \mathbf{e}_y. \tag{9.144}$$

Finally, the transmitted wave can be written

$$\mathbf{E}(z,t) = E_t\, e^{i(k_2 z - \omega t)}\, \mathbf{e}_x, \tag{9.145}$$

$$\mathbf{B}(z,t) = \frac{E_t}{v_2}\, e^{i(k_2 z - \omega t)}\, \mathbf{e}_y, \tag{9.146}$$

where $v_2 = c/n_2$ is the phase-velocity in medium 2, and $k_2 = \omega/v_2$.

For the case of normal incidence, the electric and magnetic components of all three waves are *parallel* to the boundary between the two dielectric media. Hence, the appropriate boundary conditions to apply at $z = 0$ are

$$E_{\parallel 1} = E_{\parallel 2}, \tag{9.147}$$

$$B_{\parallel 1} = B_{\parallel 2}. \tag{9.148}$$

The latter condition derives from the general boundary condition $H_{\parallel 1} = H_{\parallel 2}$, and the fact that $\mathbf{B} = \mu_0\, \mathbf{H}$ in both media (which are assumed to be non-magnetic).

Application of the boundary condition (9.147) yields

$$E_i + E_r = E_t. \tag{9.149}$$

Likewise, application of the boundary condition (9.148) gives

$$\frac{E_i - E_r}{v_1} = \frac{E_t}{v_2}, \tag{9.150}$$

or

$$E_i - E_r = \frac{v_1}{v_2}\, E_t = \frac{n_2}{n_1}\, E_t, \tag{9.151}$$

since $v_1/v_2 = n_2/n_1$. Equations (9.149) and (9.151) can be solved to give

$$E_r = \left(\frac{n_1 - n_2}{n_1 + n_2}\right) E_i, \tag{9.152}$$

$$E_t = \left(\frac{2\,n_1}{n_1 + n_2}\right) E_t. \tag{9.153}$$

Thus, we have determined the amplitudes of the reflected and transmitted waves in terms of the amplitude of the incident wave.

It can be seen, first of all, that if $n_1 = n_2$ then $E_r = 0$ and $E_t = E_i$. In other words, if the two media have the same indices of refraction then there is no reflection at the boundary between them, and the transmitted wave is consequently equal in amplitude to the incident wave. On the other hand, if $n_1 \neq n_2$ then there is some reflection at the boundary. Indeed, the amplitude of the reflected wave is roughly proportional to the difference between n_1 and n_2. This has important practical consequences. We can only see a clean pane of glass in a window because some of the light incident on an air/glass boundary is reflected, due to the different refractive indicies of air and glass. As is well-known, it is a lot more difficult to see glass when it is submerged in water. This is because the refractive indices of glass and water are quite similar, and so there is very little reflection of light incident on a water/glass boundary.

According to Equation (9.152), $E_r/E_i < 0$ when $n_2 > n_1$. The negative sign indicates a 180° phase-shift of the reflected wave, with respect to the incident wave. We conclude that there is a 180° phase-shift of the reflected wave, relative to the incident wave, on reflection from a boundary with a medium of *greater* refractive index. Conversely, there is no phase-shift on reflection from a boundary with a medium of *lesser* refractive index.

The mean electromagnetic energy flux, or *intensity*, in the z-direction is simply

$$I = \frac{\langle \mathbf{E} \times \mathbf{B} \cdot \mathbf{e}_z \rangle}{\mu_0} = \frac{E_0\,B_0}{2\,\mu_0} = \frac{E_0^2}{2\,\mu_0\,v}. \tag{9.154}$$

The *coefficient of reflection*, R, is defined as the ratio of the intensities of the reflected and incident waves:

$$R = \frac{I_r}{I_i} = \left(\frac{E_r}{E_i}\right)^2. \tag{9.155}$$

Likewise, the *coefficient of transmission*, T, is the ratio of the intensities of the transmitted and incident waves:

$$T = \frac{I_t}{I_i} = \frac{v_1}{v_2}\left(\frac{E_t}{E_i}\right)^2 = \frac{n_2}{n_1}\left(\frac{E_t}{E_i}\right)^2. \tag{9.156}$$

Equations (9.152), (9.153), (9.155), and (9.156) yield

$$R = \left(\frac{n_1 - n_2}{n_1 + n_2}\right)^2, \tag{9.157}$$

$$T = \frac{n_2}{n_1}\left(\frac{2\,n_1}{n_1 + n_2}\right)^2. \tag{9.158}$$

Note that $R + T = 1$. In other words, any wave energy which is not reflected at the boundary is transmitted, and *vice versa*.

9.13 OBLIQUE REFLECTION AT A DIELECTRIC BOUNDARY

Let us now consider the case of incidence *oblique* to the boundary—see Figure 9.3. Suppose that the incident wave subtends an angle θ_i with the normal to the boundary, whereas the reflected and transmitted waves subtend angles θ_r and θ_t, respectively.

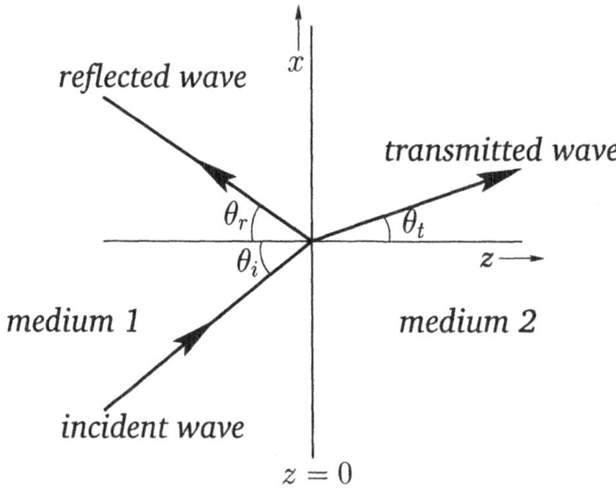

Figure 9.3: *Reflection at a dielectric boundary for the case of oblique incidence.*

The incident wave can be written

$$\mathbf{E}(\mathbf{r}, t) = \mathbf{E}_i\, e^{\,i\,(\mathbf{k}_i\cdot\mathbf{r}-\omega\,t)}, \tag{9.159}$$

$$\mathbf{B}(\mathbf{r}, t) = \mathbf{B}_i\, e^{\,i\,(\mathbf{k}_i\cdot\mathbf{r}-\omega\,t)}, \tag{9.160}$$

with analogous expressions for the reflected and transmitted waves. Since, in the case of oblique incidence, the electric and magnetic components of the wave fields are no longer necessarily parallel to the boundary, the boundary conditions (9.147) and (9.148) at $z = 0$ must be supplemented by the additional boundary conditions

$$\epsilon_1\, E_{\perp\,1} = \epsilon_2\, E_{\perp\,2}, \tag{9.161}$$

$$B_{\perp\,1} = B_{\perp\,2}. \tag{9.162}$$

Equation (9.161) derives from the general boundary condition $D_{\perp\,1} = D_{\perp\,2}$.

It follows from Equations (9.148) and (9.162) that both components of the magnetic field are continuous at the boundary. Hence, we can write

$$\mathbf{B}_i\, e^{\,i\,(\mathbf{k}_i\cdot\mathbf{r}-\omega\,t)} + \mathbf{B}_r\, e^{\,i\,(\mathbf{k}_r\cdot\mathbf{r}-\omega\,t)} = \mathbf{B}_t\, e^{\,i\,(\mathbf{k}_t\cdot\mathbf{r}-\omega\,t)} \tag{9.163}$$

at $z = 0$. Given that $\mathbf{B}_i$, $\mathbf{B}_r$, and $\mathbf{B}_t$ are constant vectors, the only way in which the above equation can be satisfied for all values of x and y is if

$$\mathbf{k}_i\cdot\mathbf{r} = \mathbf{k}_r\cdot\mathbf{r} = \mathbf{k}_t\cdot\mathbf{r} \tag{9.164}$$

throughout the $z = 0$ plane. This, in turn, implies that

$$k_{i\,x} = k_{r\,x} = k_{t\,x} \tag{9.165}$$

and

$$k_{i\,y} = k_{r\,y} = k_{t\,y}. \tag{9.166}$$

It immediately follows that if $k_{i\,y} = 0$ then $k_{r\,y} = k_{t\,y} = 0$. In other words, if the incident wave lies in the x-z plane then the reflected and transmitted waves also lie in the x-z plane. Another way of putting this is that the incident, reflected, and transmitted waves all lie in the *same* plane, known as the *plane of incidence*. This, of course, is one of the laws of geometric optics. From now on, we shall assume that the plane of incidence is the x-z plane.

Now, $k_i = k_r = \omega/v_1$ and $k_t = \omega/v_2$. Moreover,

$$\sin\theta_i = \frac{k_{x\,i}}{k_i}, \tag{9.167}$$

with similar expressions for θ_r and θ_t. Hence, according to Equation (9.165),

$$\sin\theta_r = \sin\theta_i, \tag{9.168}$$

which implies that $\theta_r = \theta_i$. Moreover,

$$\frac{\sin\theta_t}{\sin\theta_i} = \frac{v_2}{v_1} = \frac{n_1}{n_2}. \tag{9.169}$$

Of course, the above expressions correspond to the *law of reflection* and *Snell's law of refraction*, respectively.

For the case of oblique incidence, we need to consider *two* independent wave polarizations separately. The first polarization has all the wave electric fields parallel to the boundary whilst the second has all the wave magnetic fields parallel to the boundary.

Let us consider the first wave polarization. We can write unit vectors in the directions of propagation of the incident, reflected, and transmitted waves like so:

$$\hat{\mathbf{k}}_i = (\sin\theta_i,\ 0,\ \cos\theta_i)\,, \tag{9.170}$$

$$\hat{\mathbf{k}}_r = (\sin\theta_i,\ 0,\ -\cos\theta_i)\,, \tag{9.171}$$

$$\hat{\mathbf{k}}_t = (\sin\theta_t,\ 0,\ \cos\theta_t)\,. \tag{9.172}$$

The constant vectors associated with the incident wave are written

$$\mathbf{E}_i = E_i\,\mathbf{e}_y, \tag{9.173}$$

$$\mathbf{B}_i = \frac{E_i}{v_1}\,(-\cos\theta_i,\ 0,\ \sin\theta_i)\,, \tag{9.174}$$

where use has been made of Equation (9.140). Likewise, the constant vectors associated with the reflected and transmitted waves are

$$\mathbf{E}_r = E_r\,\mathbf{e}_y, \tag{9.175}$$

$$\mathbf{B}_r = \frac{E_r}{v_1}\,(\cos\theta_i,\ 0,\ \sin\theta_i)\,, \tag{9.176}$$

and

$$\mathbf{E}_t = E_t\,\mathbf{e}_y, \tag{9.177}$$

$$\mathbf{B}_t = \frac{E_t}{v_2}\,(-\cos\theta_t,\ 0,\ \sin\theta_t)\,, \tag{9.178}$$

respectively.

Now, the boundary condition (9.147) yields $E_{y\,1} = E_{y\,2}$, or

$$E_i + E_r = E_t. \tag{9.179}$$

Likewise, the boundary condition (9.162) gives $B_{z\,1} = B_{z\,2}$, or

$$(E_i + E_r)\,\frac{\sin\theta_i}{v_1} = E_t\,\frac{\sin\theta_t}{v_2}. \tag{9.180}$$

However, using Snell's law, (9.169), the above expression reduces to Equation (9.179). Finally, the boundary condition (9.148) yields $B_{x\,1} = B_{x\,2}$, or

$$(E_i - E_r)\,\frac{\cos\theta_i}{v_1} = E_t\,\frac{\cos\theta_t}{v_2}. \tag{9.181}$$

It is convenient to define the parameters

$$\alpha = \frac{\cos\theta_t}{\cos\theta_i}, \tag{9.182}$$

and

$$\beta = \frac{v_1}{v_2} = \frac{n_2}{n_1}. \tag{9.183}$$

Equations (9.179) and (9.181) can be solved in terms of these parameters to give

$$E_r = \left(\frac{1-\alpha\,\beta}{1+\alpha\,\beta}\right) E_i, \tag{9.184}$$

$$E_t = \left(\frac{2}{1+\alpha\,\beta}\right) E_i. \tag{9.185}$$

These relations are known as *Fresnel equations*.

The wave intensity in the z-direction is given by

$$I_z = \frac{\langle \mathbf{E}\times\mathbf{B}\cdot\mathbf{e}_z\rangle}{\mu_0} = \frac{E_0\,B_0\,\cos\theta}{2\,\mu_0} = \frac{E_0^2\,\cos\theta}{2\,\mu_0\,v}. \tag{9.186}$$

Hence, the coefficient of reflection is written

$$R = \left(\frac{E_r}{E_i}\right)^2 = \left(\frac{1-\alpha\,\beta}{1+\alpha\,\beta}\right)^2, \tag{9.187}$$

whereas the coefficient of transmission takes the form

$$T = \frac{\cos\theta_t}{\cos\theta_i}\,\frac{v_1}{v_2}\left(\frac{E_t}{E_i}\right)^2 = \alpha\,\beta\left(\frac{2}{1+\alpha\,\beta}\right)^2. \tag{9.188}$$

Note that it is again the case that $R + T = 1$.

Let us now consider the second wave polarization. In this case, the constant vectors associated with the incident, reflected, and transmitted waves are written

$$\mathbf{E}_i = E_i\,(\cos\theta_i,\,0,\,-\sin\theta_i), \tag{9.189}$$

$$\mathbf{B}_i = \frac{E_i}{v_1}\,\mathbf{e}_y, \tag{9.190}$$

and

$$\mathbf{E}_r = E_r\,(\cos\theta_i,\,0,\,\sin\theta_i), \tag{9.191}$$

$$\mathbf{B}_r = -\frac{E_r}{v_1}\,\mathbf{e}_y, \tag{9.192}$$

and

$$\mathbf{E}_t = E_t\,(\cos\theta_t,\,0,\,-\sin\theta_t), \tag{9.193}$$

$$\mathbf{B}_t = \frac{E_t}{v_2}\,\mathbf{e}_y, \tag{9.194}$$

respectively. The boundary condition (9.148) yields $B_{y\,1} = B_{y\,2}$, or

$$\frac{E_i - E_r}{v_1} = \frac{E_t}{v_2}. \tag{9.195}$$

Likewise, the boundary condition (9.147) gives $E_{x\,1} = E_{x\,2}$, or

$$(E_i + E_r)\,\cos\theta_i = E_t\,\cos\theta_t. \tag{9.196}$$

Finally, the boundary condition (9.161) yields $\epsilon_1\,E_{z\,1} = \epsilon_2\,E_{z\,2}$, or

$$\epsilon_1\,(E_i - E_r)\,\sin\theta_i = \epsilon_2\,E_i\,\sin\theta_t. \tag{9.197}$$

Making use of Snell's law, and the fact that $\epsilon = n^2$, the above expression reduces to Equation (9.195).

Solving Equations (9.165) and (9.196), we obtain

$$E_r = \left(\frac{\alpha-\beta}{\alpha+\beta}\right) E_i, \tag{9.198}$$

$$E_t = \left(\frac{2}{\alpha+\beta}\right) E_i. \tag{9.199}$$

The associated coefficients of reflection and transmission take the form

$$R = \left(\frac{\alpha-\beta}{\alpha+\beta}\right)^2, \tag{9.200}$$

$$T = \alpha\,\beta\left(\frac{2}{\alpha+\beta}\right)^2, \tag{9.201}$$

respectively. As usual, $R + T = 1$.

Note that at oblique incidence the Fresnel equations, (9.184) and (9.185), for the wave polarization in which the electric field is parallel to the boundary are *different* to the Fresnel equations, (9.198) and (9.199), for the wave polarization in which the magnetic field is parallel to the boundary. This implies that the coefficients of reflection and transmission for these two wave polarizations are, in general, *different*.

Figure 9.4 shows the coefficients of reflection and transmission for oblique incidence from air ($n_1 = 1.0$) to glass ($n_2 \simeq 1.5$). In general, it can be seen that the coefficient of reflection rises, and the coefficient of transmission falls, as the angle of incidence increases. Note, however, that for the wave polarization in which the magnetic field is parallel to the boundary there is a particular angle of incidence, known as the *Brewster angle*, at which the reflected intensity is *zero*. There is no similar behavior for the wave polarization in which the electric field is parallel to the boundary.

It follows from Equation (9.198) that the Brewster angle corresponds to the condition

$$\alpha = \beta, \tag{9.202}$$

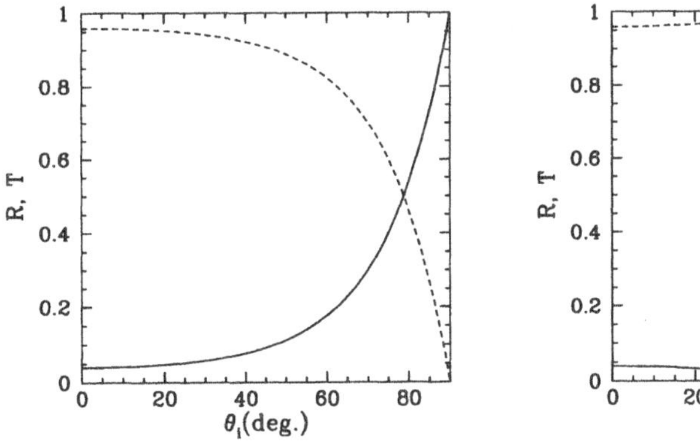

Figure 9.4: *Coefficients of reflection (solid curves) and transmission (dashed curves) for oblique incidence from air ($n = 1.0$) to glass ($n = 1.5$). The left-hand panel shows the wave polarization for which the electric field is parallel to the boundary, whereas the right-hand panel shows the wave polarization for which the magnetic field is parallel to the boundary.*

or

$$\beta^2 = \frac{\cos^2\theta_t}{\cos^2\theta_i} = \frac{1-\sin^2\theta_t}{1-\sin^2\theta_i} = \frac{1-\sin^2\theta_i/\beta^2}{1-\sin^2\theta_i}, \tag{9.203}$$

where use has been made of Snell's law. The above expression reduces to

$$\sin\theta_i = \frac{\beta}{\sqrt{1+\beta^2}}, \tag{9.204}$$

or $\tan\theta_i = \beta = n_2/n_1$. Hence, the Brewster angle satisfies

$$\theta_B = \tan^{-1}\left(\frac{n_2}{n_1}\right). \tag{9.205}$$

If unpolarized light is incident on an air/glass (say) boundary at the Brewster angle then the reflected light is 100% plane polarized.

9.14 TOTAL INTERNAL REFLECTION

Let us again consider an electromagnetic wave obliquely incident on a dielectric boundary. According to Equation (9.169), the angle of refraction θ_t is related to the angle of incidence θ_i via

$$\sin\theta_t = \frac{n_1}{n_2}\sin\theta_i. \tag{9.206}$$

This formula presents no problems when $n_1 < n_2$. However, if $n_1 > n_2$ then the formula predicts that $\sin\theta_t$ is greater than unity when the angle of incidence exceeds some critical angle given by

$$\theta_c = \sin^{-1}(n_2/n_1). \tag{9.207}$$

Obviously, in this situation, we can no longer interpret $\sin\theta_t$ as the sine of an angle. Moreover, $\cos\theta_t \equiv (1-\sin^2\theta_t)^{1/2}$ can no longer be interpreted as the cosine of an angle. However, these quantities still specify the wave-vector of the transmitted wave. In fact, from Equation (9.172),

$$\mathbf{k}_t = k_t\,(\sin\hat{\theta}_t,\, 0,\, i\,\cos\hat{\theta}_t), \tag{9.208}$$

where $k_t = n_2\,\omega/c$, $\cos\hat{\theta}_t = (\sin^2\hat{\theta}_t - 1)^{1/2}$, and

$$\sin\hat{\theta}_t = \frac{\sin\theta_i}{\sin\theta_c}. \tag{9.209}$$

Here, the hat on $\hat{\theta}_t$ is to remind us that this quantity is not a real angle. Now, the transmitted wave varies as

$$e^{i(\mathbf{k}_t\cdot\mathbf{r}-\omega t)} = e^{-k_t \cos\hat{\theta}_t}\, e^{i(k_t \sin\hat{\theta}_t-\omega t)}. \tag{9.210}$$

Hence, we conclude that when $\theta_i > \theta_c$ the transmitted wave is *evanescent*: *i.e.*, it decays exponentially, rather than propagating, in medium 2.

When $\theta_i > \theta_c$ the parameter α, defined in Equation (9.182), becomes *complex*. In fact, $\alpha \to i\,\hat{\alpha}$, where

$$\hat{\alpha} = \frac{\cos\hat{\theta}_t}{\cos\theta_i}. \tag{9.211}$$

Note that the parameter β, defined in Eq. (9.183), remains real. Hence, from Equations (9.184) and (9.198), the relationship between E_r and E_i for the two previously discussed wave polarizations, in which either the electric field or the magnetic field is parallel to the boundary, are

$$E_r = \left(\frac{1 - i\,\hat{\alpha}\,\beta}{1 + i\,\hat{\alpha}\,\beta}\right) E_i, \tag{9.212}$$

$$E_r = \left(\frac{i\,\hat{\alpha} - \beta}{i\,\hat{\alpha} + \beta}\right) E_i, \tag{9.213}$$

respectively. In both cases, the associated coefficients of reflection are *unity*: *i.e.*,

$$R = \left|\frac{E_r}{E_i}\right|^2 = 1. \tag{9.214}$$

In other words, the incident wave undergoes complete reflection at the boundary. This phenomenon is called *total internal reflection*, and occurs whenever a wave is incident on a boundary separating a medium of high refractive index from a medium of low refractive index, and the angle of incidence exceeds the critical angle, θ_c.

Figure 9.5 shows the coefficients of reflection and transmission for oblique incidence from water ($n_1 = 1.33$) to air ($n_2 = 1.0$). In this case, the critical angle is $\theta_c = 48.8^\circ$.

Note that when total internal reflection takes place the evanescent transmitted wave penetrates a few wavelengths into the lower refractive index medium, since (as is easily demonstrated) the amplitude of this wave is non-zero. The existence of the evanescent wave can be demonstrated using the apparatus pictured in Figure 9.6. Here, we have two right-angled glass prisms separated by a small air gap of width d. Light

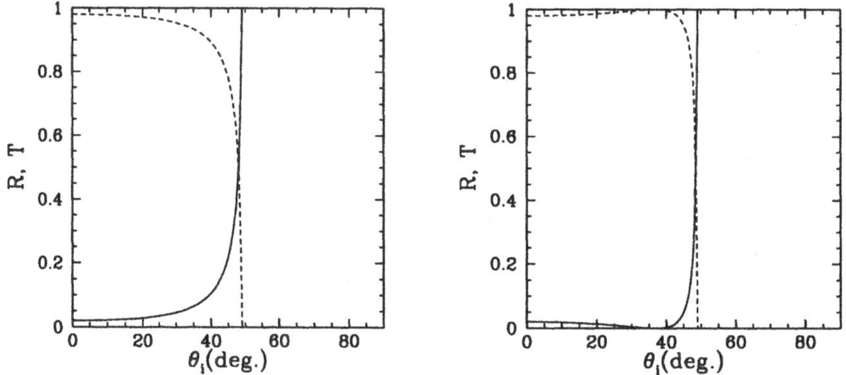

Figure 9.5: *Coefficients of reflection (solid curves) and transmission (dashed curves) for oblique incidence from water ($n = 1.33$) to air ($n = 1.0$). The left-hand panel shows the wave polarization for which the electric field is parallel to the boundary, whereas the right-hand panel shows the wave polarization for which the magnetic field is parallel to the boundary.*

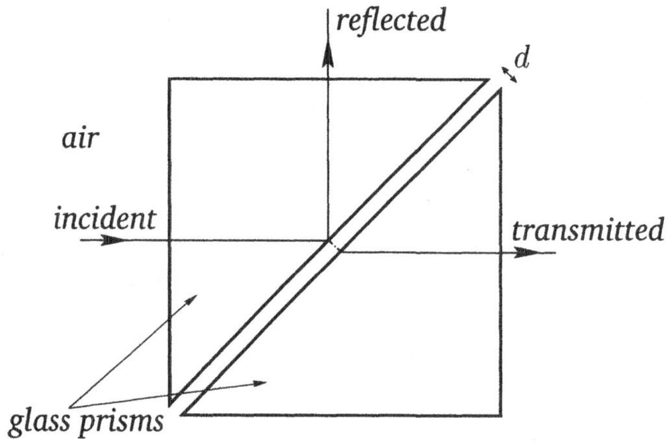

Figure 9.6: *Frustrated total internal reflection.*

incident on the internal surface of the first prism is internally reflected (assuming that $\theta_c < 45°$). However, if the spacing d is not too much larger than the wavelength of the light (in air) then the evanescent wave in the air gap still has a finite amplitude when it reaches the second prism. In this case, a detectable transmitted wave is excited in the second prism.

Obviously, the amplitude of this wave has an inverse exponential dependance on the width of the gap. This effect is called *frustrated total internal reflection* and is analogous to the *tunneling* of wave-functions through potential barriers in Quantum Mechanics.

9.15 OPTICAL COATINGS

Consider an optical instrument, such as a refracting telescope, which makes use of multiple glass lenses. Let us examine the light-ray running along the optical axis of the instrument. This ray is normally incident on all of the lenses. However, according to the analysis of Section 9.12, whenever the ray enters or leaves a lens it is partly reflected. In fact, for glass of refractive index 1.5 the transmission coefficient across an air/glass or a glass/air boundary is about 96%. Hence, approximately 8% of the light is lost each time the ray passes completely through a lens. Clearly, this level of attenuation is unacceptable in an instrument which contains many lens, especially if it is being used to view faint objects.

It turns out that the above-mentioned problem can be alleviated by coating all the lenses of the instrument in question with a thin layer of dielectric whose refractive index is intermediate between that of air and glass. Consider the situation shown in Figure 9.7. Here, a light-ray is normally incident on a boundary between medium 1, of refractive

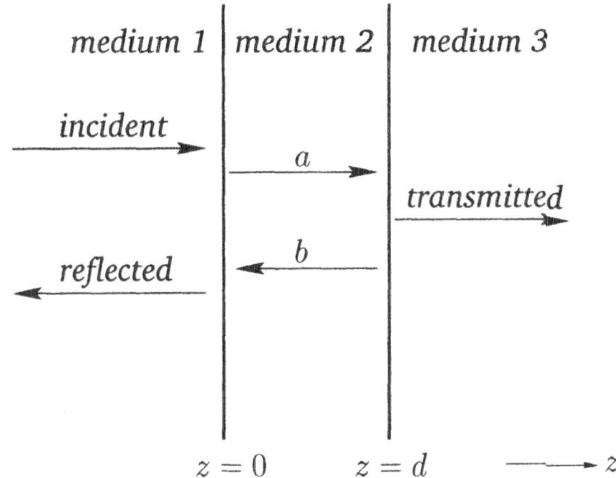

Figure 9.7: *An optical coating.*

index n_1, and medium 3, of refractive index n_3. Here, medium 1 represents air, and medium 3 represents glass, or *vice versa*. Suppose that the glass is covered with a thin optical coating, referred to as medium 2, of thickness d, and refractive index n_2.

In the notation of Section 9.12, the incident wave is written

$$\mathbf{E}(z,t) = E_i\, e^{i(k_1 z - \omega t)}\, \mathbf{e}_x, \tag{9.215}$$

$$\mathbf{B}(z,t) = \frac{E_i}{v_1}\, e^{i(k_1 z - \omega t)}\, \mathbf{e}_y, \tag{9.216}$$

where $v_1 = c/n_1$ is the phase-velocity in medium 1, and $k_1 = \omega/v_1$. Likewise, the reflected wave takes the form

$$\mathbf{E}(z,t) = E_r\, e^{i(-k_1 z - \omega t)}\, \mathbf{e}_x, \tag{9.217}$$

$$\mathbf{B}(z,t) = -\frac{E_r}{v_1}\, e^{i(-k_1 z - \omega t)}\, \mathbf{e}_y. \tag{9.218}$$

The wave traveling to the right in medium 2 is written

$$\mathbf{E}(z,t) = E_a\, e^{i(k_2 z - \omega t)}\, \mathbf{e}_x, \tag{9.219}$$

$$\mathbf{B}(z,t) = \frac{E_a}{v_2}\, e^{i(k_2 z - \omega t)}\, \mathbf{e}_y, \tag{9.220}$$

where $v_2 = c/n_2$ is the phase-velocity in medium 2, and $k_2 = \omega/v_2$. Likewise, the wave traveling to the left takes the form

$$\mathbf{E}(z,t) = E_b\, e^{i(-k_2 z - \omega t)}\, \mathbf{e}_x, \tag{9.221}$$

$$\mathbf{B}(z,t) = -\frac{E_b}{v_2}\, e^{i(-k_2 z - \omega t)}\, \mathbf{e}_y. \tag{9.222}$$

Finally, the transmitted wave is written

$$\mathbf{E}(z,t) = E_t\, e^{i[k_3 (z-d) - \omega t]}\, \mathbf{e}_x, \tag{9.223}$$

$$\mathbf{B}(z,t) = \frac{E_t}{v_3}\, e^{i[k_2 (z-d) - \omega t]}\, \mathbf{e}_y, \tag{9.224}$$

where $v_3 = c/n_3$ is the phase-velocity in medium 3, and $k_3 = \omega/v_3$.

Continuity of E_x and B_y at $z = 0$ yield

$$E_i + E_r = E_a + E_b, \tag{9.225}$$

$$\frac{E_i - E_r}{v_1} = = \frac{E_a - E_b}{v_2}, \tag{9.226}$$

respectively, whereas continuity of E_x and B_y at $z = d$ give

$$E_a\, e^{i k_2 d} + E_b\, e^{-i k_2 d} = E_t, \tag{9.227}$$

$$\frac{E_a\, e^{i k_2 d} - E_b\, e^{-i k_2 d}}{v_2} = \frac{E_t}{v_3}, \tag{9.228}$$

respectively. At this point, it is convenient to make the special choice $k_2\, d = \pi/2$. This corresponds to the optical coating being exactly *one-quarter* of a wavelength thick. It follows that

$$E_a - E_b = -i\, E_t, \tag{9.229}$$

$$E_a + E_b = -i\, \frac{n_3}{n_2}\, E_t. \tag{9.230}$$

The above equations can be solved to give

$$E_r = -\left(\frac{1-\alpha}{1+\alpha}\right) E_i, \tag{9.231}$$

$$E_t = \frac{n_1}{n_2}\, \frac{2\, i}{1+\alpha}\, E_i, \tag{9.232}$$

where

$$\alpha = \frac{n_1\, n_3}{n_2^2}. \tag{9.233}$$

Thus, the overall coefficient of reflection is

$$R = \left|\frac{E_r}{E_i}\right|^2 = \left(\frac{1-\alpha}{1+\alpha}\right)^2, \tag{9.234}$$

whereas the overall coefficient of transmission is

$$T = \frac{n_3}{n_1} \left|\frac{E_t}{E_i}\right|^2 = \frac{4\, \alpha}{(1+\alpha)^2}. \tag{9.235}$$

Suppose finally that

$$n_2 = \sqrt{n_1\, n_3}\,: \tag{9.236}$$

i.e., the refractive index of the coating is the *geometric mean* of that of air and glass. In this case, $\alpha = 1$, and it follows from Equations (9.234) and (9.235) that there is zero reflection, and 100% transmission, at the boundary. Hence, by coating lenses with a one-quarter wavelength thickness of a substance whose refractive index is (approximately) the geometric mean between those of air and glass (*e.g.*, Magnesium fluoride, whose refractive index is 1.38), we can completely eliminate unwanted reflections. This technique is widely used in high-quality optical instruments. Note that the physics of quarter-wavelength optical coatings is analogous to that of quarter-wave transformers in transmission lines—see Section 7.7.

9.16 REFLECTION AT A METALLIC BOUNDARY

Let us now consider the reflection of electromagnetic radiation by a metallic surface. This investigation is obviously relevant to optical instruments, such as reflecting telescopes, which make use of mirrors. For the sake of simplicity, we shall restrict our investigation to the case of normal incidence.

A metal is, by definition, a good electrical conductor. According to Equation (9.121), the wave-number of an electromagnetic wave of frequency ω in a good conductor of conductivity σ (and true dielectric constant unity) is

$$k \simeq \sqrt{i\,\mu_0\,\sigma\,\omega}. \tag{9.237}$$

Hence, it follows that the effective refractive index of the conductor is

$$n = \frac{k\,c}{\omega} \simeq \sqrt{\frac{i\,\sigma}{\epsilon_0\,\omega}}. \tag{9.238}$$

Note that the good conductor ordering $\sigma \gg \epsilon_0\,\omega$ ensures that $|n| \gg 1$.

For the case of a light-ray in air reflecting at normal incidence off a metal mirror, we can employ the previously derived formula (9.152) with $n_1 = 1$ and $n_2 = n$, where n is specified above. We obtain

$$\frac{E_r}{E_i} = \frac{1-n}{1+n} \simeq -1 + \frac{2}{n}, \tag{9.239}$$

where we have made use of the fact that $|n| \gg 1$. Hence, the coefficient of reflection of the mirror takes the form

$$R = \left|\frac{E_r}{E_i}\right|^2 \simeq 1 - \mathrm{Re}\left(\frac{4}{n}\right), \tag{9.240}$$

or

$$R \simeq 1 - \sqrt{\frac{8\,\epsilon_0\,\omega}{\sigma}}. \tag{9.241}$$

High-quality metallic mirrors are generally coated in Silver, whose conductivity is $6.3 \times 10^7\,(\Omega\,\mathrm{m})^{-1}$. It follows, from the above formula, that at optical frequencies ($\omega = 4 \times 10^{15}$ rad./s) the coefficient of reflection of a silvered mirror is $R \simeq 93.3\%$. This implies that about 7% of the light incident on a silvered mirror is absorbed, rather than being reflected. This rather severe light loss can be problematic in instruments, such as astronomical telescopes, which are used to view faint objects.

9.17 WAVE-GUIDES

A wave-guide is a hollow conducting pipe, of uniform cross-section, used to transport high-frequency electromagnetic waves (generally, in the microwave band) from one point to another. The main advantage of wave-guides is their relatively low level of radiation losses (since the electric and magnetic fields are completely enclosed by a conducting wall) compared to transmission lines.

Consider a vacuum-filled wave-guide which runs parallel to the z-axis. An electromagnetic wave trapped inside the wave-guide satisfies Maxwell's equations for free space:

$$\nabla \cdot \mathbf{E} = 0, \tag{9.242}$$

$$\nabla \cdot \mathbf{B} = 0, \tag{9.243}$$

$$\nabla \times \mathbf{E} = -\frac{\partial \mathbf{B}}{\partial t}, \tag{9.244}$$

$$\nabla \times \mathbf{B} = \frac{1}{c^2}\frac{\partial \mathbf{E}}{\partial t}. \tag{9.245}$$

Let $\partial/\partial t \equiv -i\,\omega$, and $\partial/\partial z \equiv i\,k$, where ω is the wave frequency, and k the wave-number parallel to the axis of the wave-guide. It follows that

$$\frac{\partial E_x}{\partial x} + \frac{\partial E_y}{\partial y} + i\,k\,E_z = 0, \tag{9.246}$$

$$\frac{\partial B_x}{\partial x} + \frac{\partial B_y}{\partial y} + i\,k\,B_z = 0, \tag{9.247}$$

$$i\,\omega\,B_x = \frac{\partial E_z}{\partial y} - i\,k\,E_y, \tag{9.248}$$

$$i\,\omega\,B_y = -\frac{\partial E_z}{\partial x} + i\,k\,E_x, \tag{9.249}$$

$$i\,\omega\,B_z = \frac{\partial E_y}{\partial x} - \frac{\partial E_x}{\partial y}, \tag{9.250}$$

$$i\,\frac{\omega}{c^2}\,E_x = -\frac{\partial B_z}{\partial y} + i\,k\,B_y, \tag{9.251}$$

$$i\,\frac{\omega}{c^2}\,E_y = \frac{\partial B_z}{\partial x} - i\,k\,B_x, \tag{9.252}$$

$$i\,\frac{\omega}{c^2}\,E_z = -\frac{\partial B_y}{\partial x} + \frac{\partial B_x}{\partial y}. \tag{9.253}$$

Equations (9.249) and (9.251) yield

$$E_x = i\left(\omega\,\frac{\partial B_z}{\partial y} + k\,\frac{\partial E_z}{\partial x}\right)\left(\frac{\omega^2}{c^2} - k^2\right)^{-1}, \tag{9.254}$$

and

$$B_y = i\left(\frac{\omega}{c^2}\frac{\partial E_z}{\partial x} + k\,\frac{\partial B_z}{\partial y}\right)\left(\frac{\omega^2}{c^2} - k^2\right)^{-1}. \tag{9.255}$$

Likewise, Equations (9.248) and (9.252) yield

$$E_y = i\left(-\omega\,\frac{\partial B_z}{\partial x} + k\,\frac{\partial E_z}{\partial y}\right)\left(\frac{\omega^2}{c^2} - k^2\right)^{-1}, \tag{9.256}$$

and

$$B_x = i\left(-\frac{\omega}{c^2}\frac{\partial E_z}{\partial y} + k\,\frac{\partial B_z}{\partial x}\right)\left(\frac{\omega^2}{c^2} - k^2\right)^{-1}. \tag{9.257}$$

These equations can be combined to give

$$\mathbf{E}_t = i\,(\omega\,\nabla B_z \times \mathbf{e}_z + k\,\nabla E_z)\left(\frac{\omega^2}{c^2} - k^2\right)^{-1}, \tag{9.258}$$

$$\mathbf{B}_t = i\left(-\frac{\omega}{c^2}\,\nabla E_z \times \mathbf{e}_z + k\,\nabla B_z\right)\left(\frac{\omega^2}{c^2} - k^2\right)^{-1}. \tag{9.259}$$

Here, $\mathbf{E}_t$ and $\mathbf{B}_t$ are the *transverse* electric and magnetic fields: *i.e.*, the electric and magnetic fields in the x-y plane. It is clear, from Equations (9.258) and (9.259), that the transverse fields are fully determined once the longitudinal fields, E_z and B_z, are known.

Substitution of Equations (9.258) and (9.259) into Equations (9.250) and (9.253) yields the equations satisfied by the longitudinal fields:

$$\left(\frac{\partial^2}{\partial x^2}+\frac{\partial^2}{\partial y^2}\right)E_z+\left(\frac{\omega^2}{c^2}-k^2\right)E_z=0, \tag{9.260}$$

$$\left(\frac{\partial^2}{\partial x^2}+\frac{\partial^2}{\partial y^2}\right)B_z+\left(\frac{\omega^2}{c^2}-k^2\right)B_z=0. \tag{9.261}$$

The remaining equations, (9.246) and (9.247), are automatically satisfied provided Equations (9.258)–(9.261) are satisfied.

We expect $\mathbf{E}=\mathbf{B}=\mathbf{0}$ inside the walls of the wave-guide, assuming that they are perfectly conducting. Hence, the appropriate boundary conditions at the walls are

$$E_{\parallel}=0, \tag{9.262}$$

$$B_{\perp}=0. \tag{9.263}$$

It follows, by inspection of Equations (9.258) and (9.259), that these boundary conditions are satisfied provided

$$E_z=0, \tag{9.264}$$

$$\hat{\mathbf{n}}\cdot\nabla B_z=0, \tag{9.265}$$

at the walls. Here, $\hat{\mathbf{n}}$ is a unit vector normal to the walls. Hence, the electromagnetic fields inside the wave-guide are fully specified by solving Equations (9.260) and (9.261), subject to the boundary conditions (9.264) and (9.265), respectively.

Equations (9.260) and (9.261) support two independent types of solution. The first type has $E_z=0$, and is consequently called a *transverse electric*, or TE, mode. Conversely, the second type has $B_z=0$, and is called a *transverse magnetic*, or TM, mode.

Consider the specific example of a *rectangular* wave-guide, with conducting walls at $x=0, a$, and $y=0, b$. For a TE mode, the longitudinal magnetic field can be written

$$B_z(x,y)=B_0\cos(k_x\,x)\cos(k_y\,y). \tag{9.266}$$

The boundary condition (9.265) requires that $\partial B_z/\partial x=0$ at $x=0,a$, and $\partial B_z/\partial y=0$ at $y=0,b$. It follows that

$$k_x=\frac{m\,\pi}{a}, \tag{9.267}$$

$$k_y=\frac{n\,\pi}{b}, \tag{9.268}$$

where $m = 0, 1, 2, \cdots$, and $n = 0, 1, 2, \cdots$. Clearly, there are many different kinds of TE mode, corresponding to the many different choices of m and n. Let us refer to a mode corresponding to a particular choice of m, n as a TE_{mn} mode. Note, however, that there is no TE_{00} mode, since $B_z(x, y)$ is uniform in this case. According to Equation (9.261), the dispersion relation for the TE_{mn} mode is given by

$$k^2 c^2 = \omega^2 - \omega_{mn}^2, \tag{9.269}$$

where

$$\omega_{mn} = c\,\pi \sqrt{\frac{m^2}{a^2} + \frac{n^2}{b^2}}. \tag{9.270}$$

According to the dispersion relation (9.269), k is imaginary for $\omega < \omega_{mn}$. In other words, for wave frequencies below ω_{mn}, the TE_{mn} mode fails to propagate down the wave-guide, and is instead attenuated. Hence, ω_{mn} is termed the *cut-off frequency* for the TE_{mn} mode. Assuming that $a > b$, the TE mode with the lowest cut-off frequency is the TE_{10} mode, where

$$\omega_{10} = \frac{c\,\pi}{a}. \tag{9.271}$$

For frequencies above the cut-off frequency, the phase-velocity of the TE_{mn} mode is given by

$$v_p = \frac{\omega}{k} = \frac{c}{\sqrt{1 - \omega_{mn}^2/\omega^2}}, \tag{9.272}$$

which is greater than c. However, the group-velocity takes the form

$$v_g = \frac{d\omega}{dk} = c\sqrt{1 - \omega_{mn}^2/\omega^2}, \tag{9.273}$$

which is always less than c. Of course, energy is transmitted down the wave-guide at the group-velocity, rather than the phase-velocity. Note that the group-velocity goes to zero as the wave frequency approaches the cut-off frequency.

For a TM mode, the longitudinal electric field can be written

$$E_z(x, y) = E_0 \sin(k_x\, x) \sin(k_y\, y). \tag{9.274}$$

The boundary condition (9.264) requires that $E_z = 0$ at $x = 0, a$, and $y = 0, b$. It follows that

$$k_x = \frac{m\,\pi}{a}, \tag{9.275}$$

$$k_y = \frac{n\,\pi}{b}, \tag{9.276}$$

where $m = 1, 2, \cdots$, and $n = 1, 2, \cdots$. The dispersion relation for the TM_{mn} mode is also given by Equation (9.269). Hence, Equations (9.272) and (9.273) also apply to TM modes. However, the TM mode with the lowest cut-off frequency is the TM_{11} mode, where

$$\omega_{11} = c\,\pi\sqrt{\frac{1}{a^2} + \frac{1}{b^2}} > \omega_{10}. \tag{9.277}$$

It follows that the mode with the lowest cut-off frequency is always a TE mode.

There is, in principle, a third type of mode which can propagate down a wave-guide. This third mode type is characterized by $E_z = B_z = 0$, and is consequently called a *transverse electromagnetic*, or TEM, mode. It is easily seen, from an inspection of Equations (9.248)–(9.253), that a TEM mode satisfies

$$\omega^2 = k^2\,c^2, \tag{9.278}$$

and

$$\mathbf{E}_t = -\nabla\phi, \tag{9.279}$$

$$\mathbf{B}_t = c^{-1}\,\nabla\phi \times \mathbf{e}_z, \tag{9.280}$$

where $\phi(x, y)$ satisfies

$$\nabla^2\phi = 0. \tag{9.281}$$

The boundary conditions (9.264) and (9.265) imply that

$$\phi = \text{constant} \tag{9.282}$$

at the walls. However, there is no non-trivial solution of Equations (9.281) and (9.282) for a conventional wave-guide. In other words, conventional wave-guides *do not* support TEM modes. It turns out that only wave-guides with *central conductors* support TEM modes. Consider, for instance, a coaxial wave-guide in which the electric and magnetic fields are trapped between two coaxial cylindrical conductors of radius a and b (with $b > a$). In this case, $\phi = \phi(r)$, and Equation (9.281) reduces to

$$\frac{1}{r}\frac{\partial}{\partial r}\left(r\,\frac{\partial\phi}{\partial r}\right) = 0, \tag{9.283}$$

where r is a standard cylindrical polar coordinate. The boundary condition (9.282) is automatically satisfied at $r = a$ and $r = b$. The above equation has the following non-trivial solution:

$$\phi(r) = \phi_b \ln(r/b). \tag{9.284}$$

Note, however, that the inner conductor *must* be present, otherwise $\phi \to \infty$ as $r \to 0$, which is unphysical. According to the dispersion relation (9.278), TEM modes have no cut-off frequency, and have the phase-velocity (and group-velocity) c. Indeed, this type of mode is the same as that supported by a transmission line (see Section 7.7).

9.18 EXERCISES

9.1. Consider an electromagnetic wave propagating through a non-dielectric, non-magnetic medium containing free charge density ρ and free current density $\mathbf{j}$. Demonstrate from Maxwell's equations that the associated wave equations take the form

$$\nabla^2\mathbf{E} - \frac{1}{c^2}\frac{\partial^2\mathbf{E}}{\partial t^2} = \frac{\nabla\rho}{\epsilon_0} + \mu_0\frac{\partial\mathbf{j}}{\partial t},$$

$$\nabla^2\mathbf{B} - \frac{1}{c^2}\frac{\partial^2\mathbf{B}}{\partial t^2} = -\mu_0\nabla\times\mathbf{j}.$$

9.2. A spherically symmetric charge distribution undergoes purely radial oscillations. Show that no electromagnetic waves are emitted. [Hint: Show that there is no magnetic field.]

9.3. A general electromagnetic wave-pulse propagating in the z-direction at velocity u is written

$$\mathbf{E} = P(z-ut)\,\mathbf{e}_x + Q(z-ut)\,\mathbf{e}_y + R(z-ut)\,\mathbf{e}_z,$$

$$\mathbf{B} = \frac{S(z-ut)}{u}\,\mathbf{e}_x + \frac{T(z-ut)}{u}\,\mathbf{e}_y + \frac{U(z-ut)}{u}\,\mathbf{e}_z,$$

where P, Q, R, S, T, and U are arbitrary functions. In order to exclude electrostatic and magnetostatic fields, these functions are subject to the constraint that $\langle P\rangle = \langle Q\rangle = \langle R\rangle = \langle S\rangle = \langle T\rangle = \langle U\rangle = 0$, where

$$\langle P\rangle = \int_{-\infty}^{\infty} P(x)\,dx.$$

Suppose that the pulse propagates through a uniform dielectric medium of dielectric constant ϵ. Demonstrate from Maxwell's equation that $u = c/\sqrt{\epsilon}$, $R = U = 0$, $S = -Q$, and $T = P$. Incidentally, this result implies that a general wave-pulse is characterized by *two* arbitrary functions, corresponding to the two possible independent polarizations of the pulse.

9.4. A medium is such that the product of the phase and group velocities of electromagnetic waves is equal to c^2 at all wave frequencies. Demonstrate that the dispersion relation for electromagnetic waves takes the form

$$\omega^2 = k^2\,c^2 + \omega_0^2,$$

where ω_0 is a constant.

9.5. Consider a uniform plasma of plasma frequency ω_p containing a uniform magnetic field $B_0\,\mathbf{e}_z$. Show that left-hand circularly polarized electromagnetic waves can only propagate parallel to the magnetic field provided that $\omega > -\Omega/2 + \sqrt{\Omega^2/4 + \omega_p^2}$, where $\Omega = e\,B_0/m_e$ is the electron cyclotron frequency. Demonstrate that right-hand circularly polarized electromagnetic waves can only propagate parallel to the magnetic field provided that their frequencies do not lie in the range $\Omega \leq \omega \leq \Omega/2 + \sqrt{\Omega^2/4 + \omega_p^2}$. You may neglect the finite mass of the ions.

9.6. Consider an electromagnetic wave propagating through a non-uniform dielectric medium whose dielectric constant ϵ is a function of $\mathbf{r}$. Demonstrate that the associated wave equations take the form

$$\nabla^2\mathbf{E} - \frac{\epsilon}{c^2}\frac{\partial^2\mathbf{E}}{\partial t^2} = -\nabla\left(\frac{\nabla\epsilon\cdot\mathbf{E}}{\epsilon}\right),$$

$$\nabla^2\mathbf{B} - \frac{\epsilon}{c^2}\frac{\partial^2\mathbf{B}}{\partial t^2} = -\frac{\nabla\epsilon\times(\nabla\times\mathbf{B})}{\epsilon}.$$

9.7. Consider a light-wave normally incident on a uniform pane of glass of thickness d and refractive index n. Show that the coefficient of transmission though the pane takes the form

$$T^{-1} = 1 + \left[\frac{n^2-1}{2n}\,\sin(k\,d)\right]^2,$$

where k is the wave-number within the glass.

9.8. Consider an electromagnetic wave obliquely incident on a plane boundary between two transparent magnetic media of relative permeabilities μ_1 and μ_2. Find the coefficients of reflection and transmission as functions of the angle of incidence for the wave polarizations in which all electric fields are parallel to the boundary and all magnetic fields are parallel to the boundary. Is there a Brewster angle? If so, what is it? Is it possible to obtain total reflection? If so, what is the critical angle of incidence required to obtain total reflection?

9.9. Suppose that a light-ray is incident on the front (air/glass) interface of a uniform pane of glass of refractive index n at the Brewster angle. Demonstrate that the refracted ray is also incident on the rear (glass/air) interface of the pane at the Brewster angle.

9.10. Consider an electromagnetic wave propagating through a good conductor. Demonstrate that the energy density of the wave's magnetic component dominates that of its electric component. In addition, show that the phase of the wave's magnetic component lags that of its electric component by 45°.

9.11. Demonstrate that the electric and magnetic fields inside a wave-guide are mutually orthogonal.

9.12. Consider a TE_{mn} mode in a rectangular wave-guide of dimensions a and b. Calculate the mean electromagnetic energy per unit length, as well as the mean electromagnetic energy flux down the wave-guide. Demonstrate that the ratio of the mean energy flux to the mean energy per unit length is equal to the group-velocity of the mode.

Chapter 10 RELATIVITY AND ELECTROMAGNETISM

10.1 INTRODUCTION

In this chapter, we shall discuss Maxwell's equations in the light of Einstein's Special Theory of Relativity.

10.2 THE RELATIVITY PRINCIPLE

Physical phenomena are conventionally described relative to some *frame of reference* which allows us to define fundamental quantities such as position and time. Of course, there are very many different ways of choosing a reference frame, but it is generally convenient to restrict our choice to the set of *rigid inertial frames*. A classical rigid reference frame is the imagined extension of a rigid body. For instance, the Earth determines a rigid frame throughout all space, consisting of all those points which remain rigidly at rest relative to the Earth, and to one another. We can associate an orthogonal Cartesian coordinate system S with such a frame, by choosing three mutually orthogonal planes within it, and measuring x, y, and z as perpendicular distances from these planes. A time coordinate must also be defined, in order that the system can be used to specify *events*. A rigid frame, endowed with such properties, is called a *Cartesian frame*. The description given above presupposes that the underlying geometry of space is *Euclidian*, which is reasonable provided that gravitational effects are negligible (we shall assume that this is the case). An *inertial frame* is a Cartesian frame in which free particles move without acceleration, in accordance with Newton's first law of motion. There are an infinite number of different inertial frames, moving with some constant velocity with respect to one another.

The key to understanding Special Relativity is Einstein's *Relativity Principle*, which states that:

All inertial frames are totally equivalent for the performance of all physical experiments.

In other words, it is impossible to perform a physical experiment which differentiates in any fundamental sense between different inertial frames. By definition, Newton's laws of motion take the same form in all inertial frames. Einstein generalized this result in his Special Theory of Relativity by asserting that *all* laws of Physics take the same form in all inertial frames.

Consider a wave-like disturbance. In general, such a disturbance propagates at a fixed velocity with respect to the medium in which the disturbance takes place. For instance, sound waves (at STP) propagate at 343 meters per second with respect to air. So, in the inertial frame in which air is stationary, sound waves appear to propagate at 343 meters per second. Sound waves appear to propagate at a different velocity in any inertial frame which is moving with respect to the air. However, this does not violate the Relativity Principle, since if the air were stationary in the second frame then sound waves would appear to propagate at 343 meters per second in that frame as well. In other words, exactly the same experiment (*e.g.*, the determination of the speed of sound relative to stationary air) performed in two different inertial frames of reference yields exactly the same result, in accordance with the relativity principle.

Consider, now, a wave-like disturbance which is self-regenerating, and does not require a medium through which to propagate. The most well-known example of such a disturbance is a light wave. Another example is a gravity wave. According to electromagnetic theory, the speed of propagation of a light wave through a vacuum is

$$c = \frac{1}{\sqrt{\epsilon_0\,\mu_0}} = 2.99729 \times 10^8 \text{ meters per second,} \tag{10.1}$$

where ϵ_0 and μ_0 are physical constants which can be evaluated by performing two simple experiments which involve measuring the force of attraction between two fixed changes and two fixed, parallel current-carrying wires. According to the Relativity Principle, these experiments must yield the same values for ϵ_0 and μ_0 in all inertial frames. Thus, the speed of light must be the same in all inertial frames. In fact, any disturbance which does not require a medium to propagate through must appear to travel at the *same velocity* in all inertial frames, otherwise we could differentiate inertial frames using the apparent propagation speed of the disturbance, which would violate the Relativity Principle.

10.3 THE LORENTZ TRANSFORMATION

Consider two Cartesian frames $S(x, y, z, t)$ and $S'(x', y', z', t')$ in the *standard configuration*, in which S' moves in the x-direction of S with uniform velocity v, and the corresponding axes of S and S′ remain parallel throughout the motion, having coincided at $t = t' = 0$. It is assumed that the same units of distance and time are adopted in both frames. Suppose that an *event* (*e.g.*, the flashing of a lightbulb, or the collision of two point particles) has coordinates (x, y, z, t) relative to S, and (x', y', z', t') relative to S′. The "common sense" relationship between these two sets of coordinates is given by the *Galilean transformation*:

$$x' = x - v\,t, \tag{10.2}$$

$$y' = y, \tag{10.3}$$

$$z' = z, \tag{10.4}$$

$$t' = t. \tag{10.5}$$

This transformation is tried and tested, and provides a very accurate description of our everyday experience. Nevertheless, it must be wrong! Consider a light wave which propagates along the x-axis in S with velocity c. According to the Galilean transformation, the apparent speed of propagation in S′ is $c - v$, which violates the Relativity Principle. Can we construct a new transformation which makes the velocity of light invariant between different inertial frames, in accordance with the Relativity Principle, but reduces to the Galilean transformation at low velocities, in accordance with our everyday experience?

Consider an event P, and a neighboring event Q, whose coordinates differ by dx, dy, dz, dt in S, and by dx', dy', dz', dt' in S′. Suppose that at the event P a flash of light is emitted, and that Q is an event in which some particle in space is illuminated by the flash. In accordance with the laws of light propagation, and the invariance of the velocity of light between different inertial frames, an observer in S will find that

$$dx^2 + dy^2 + dz^2 - c^2\,dt^2 = 0 \tag{10.6}$$

for $dt > 0$, and an observer in S′ will find that

$$dx'^2 + dy'^2 + dz'^2 - c^2\,dt'^2 = 0 \tag{10.7}$$

for $dt' > 0$. Any event near P whose coordinates satisfy *either* (10.6) *or* (10.7) is illuminated by the flash from P, and, therefore, its coordinates must satisfy *both* (10.6) *and* (10.7). Now, no matter what form the transformation between coordinates in the two inertial frames takes, the transformation between differentials at any fixed event P is linear and homogeneous. In other words, if

$$x' = F(x, y, z, t), \tag{10.8}$$

where F is a general function, then

$$dx' = \frac{\partial F}{\partial x}\,dx + \frac{\partial F}{\partial y}\,dy + \frac{\partial F}{\partial z}\,dz + \frac{\partial F}{\partial t}\,dt. \tag{10.9}$$

It follows that

$$\begin{aligned} dx'^2 + dy'^2 + dz'^2 - c^2\,dt'^2 = {} & a\,dx^2 + b\,dy^2 + c\,dz^2 + d\,dt^2 \\ & + g\,dx\,dt + h\,dy\,dt + k\,dz\,dt \\ & + l\,dy\,dz + m\,dx\,dz + n\,dx\,dy, \end{aligned} \tag{10.10}$$

where a, b, c, *etc.*, are functions of x, y, z, and t. We know that the right-hand side of the above expression vanishes for all real values of the differentials which satisfy Equation (10.6). It follows that the right-hand side is a multiple of the quadratic in Equation (10.6): *i.e.*,

$$dx'^2 + dy'^2 + dz'^2 - c^2\,dt'^2 = K\,(dx^2 + dy^2 + dz^2 - c^2\,dt^2), \tag{10.11}$$

where K is a function of x, y, z, and t. [We can prove this by substituting into Equation (10.10) the following obvious zeros of the quadratic in Equation (10.6): $(\pm 1, 0, 0, 1)$, $(0, \pm 1, 0, 1)$, $(0, 0, \pm 1, 1)$, $(0, 1/\sqrt{2}, 1/\sqrt{2}, 1)$, $(1/\sqrt{2}, 0, 1/\sqrt{2}, 1)$, $(1/\sqrt{2}, 1/\sqrt{2}, 0, 1)$: and solving the resulting conditions on the coefficients.] Note that K at P is also independent of the choice of standard coordinates in S and S'. Since the frames are Euclidian, the values of $dx^2 + dy^2 + dz^2$ and $dx'^2 + dy'^2 + dz'^2$ relevant to P and Q are independent of the choice of axes. Furthermore, the values of dt^2 and dt'^2 are independent of the choice of the origins of time. Thus, without affecting the value of K at P, we can choose coordinates such that $P = (0, 0, 0, 0)$ in both S and S'. Since the orientations of the axes in S and S' are, at present, arbitrary, and since inertial frames are isotropic, the relation of S and S' relative to each other, to the event P, and to the locus of possible events Q, is now completely symmetric. Thus, we can write

$$dx^2 + dy^2 + dz^2 - c^2\,dt^2 = K\,(dx'^2 + dy'^2 + dz'^2 - c^2\,dt'^2), \tag{10.12}$$

in addition to Equation (10.11). It follows that $K = \pm 1$. $K = -1$ can be dismissed immediately, since the intervals $dx^2 + dy^2 + dz^2 - c^2\,dt^2$

and $dx'^2 + dy'^2 + dz'^2 - c^2\, dt'^2$ must coincide exactly when there is no motion of S' relative to S. Thus,

$$dx'^2 + dy'^2 + dz'^2 - c^2\, dt'^2 = dx^2 + dy^2 + dz^2 - c^2\, dt^2. \qquad (10.13)$$

Equation (10.13) implies that the transformation equations between primed and unprimed coordinates must be *linear*. The proof of this statement is postponed until Section 10.7.

The linearity of the transformation allows the coordinate axes in the two frames to be orientated so as to give the *standard configuration* mentioned earlier. Consider a fixed plane in S with the equation $l\,x + m\,y + n\,z + p = 0$. In S', this becomes (say) $l\,(a_1\,x' + b_1\,y' + c_1\,z' + d_1\,t' + e_1) + m\,(a_2\,x' + \cdots) + n\,(a_3\,x' + \cdots) + p = 0$, which represents a moving plane unless $l\,d_1 + m\,d_2 + n\,d_3 = 0$. That is, unless the normal vector to the plane in S, (l, m, n), is perpendicular to the vector (d_1, d_2, d_3). All such planes intersect in lines which are fixed in both S and S', and which are parallel to the vector (d_1, d_2, d_3) in S. These lines must correspond to the direction of relative motion of the frames. By symmetry, two such planes which are orthogonal in S must also be orthogonal in S'. This allows the choice of two common coordinate planes.

Under a linear transformation, the finite coordinate differences satisfy the same transformation equations as the differentials. It follows from Equation (10.13), assuming that the events $(0, 0, 0, 0)$ coincide in both frames, that for any event with coordinates (x, y, z, t) in S and (x', y', z', t') in S', the following relation holds:

$$x^2 + y^2 + z^2 - c^2\, t^2 = x'^2 + y'^2 + z'^2 - c^2\, t'^2. \qquad (10.14)$$

By hypothesis, the coordinate planes $y = 0$ and $y' = 0$ coincide permanently. Thus, $y = 0$ must imply $y' = 0$, which suggests that

$$y' = A\,y, \qquad (10.15)$$

where A is a constant. We can reverse the directions of the x- and z-axes in S and S', which has the effect of interchanging the roles of these frames. This procedure does not affect Equation (10.15), but by symmetry we also have

$$y = A\,y'. \qquad (10.16)$$

It is clear that $A = \pm 1$. The negative sign can again be dismissed, since $y = y'$ when there is no motion between S and S'. The argument for z

is similar. Thus, we have

$$y' = y, \tag{10.17}$$

$$z' = z, \tag{10.18}$$

as in the Galilean transformation.

Equations (10.14), (10.17), and (10.18) yield

$$x^2 - c^2 t^2 = x'^2 - c^2 t'^2. \tag{10.19}$$

Since, $x' = 0$ must imply $x = v\,t$, we can write

$$x' = B\,(x - v\,t), \tag{10.20}$$

where B is a constant (possibly depending on v). It follows from the previous two equations that

$$t' = C\,x + D\,t, \tag{10.21}$$

where C and D are constants (possibly depending on v). Substituting Equations (10.20) and (10.21) into Equation (10.19), and comparing the coefficients of x^2, $x\,t$, and t^2, we obtain

$$B = D = \frac{1}{\pm(1 - v^2/c^2)^{1/2}}, \tag{10.22}$$

$$C = \frac{-v/c^2}{\pm(1 - v^2/c^2)^{1/2}}. \tag{10.23}$$

We must choose the positive sign in order to ensure that $x' \to x$ as $v/c \to 0$. Thus, collecting our results, the transformation between coordinates in S and S′ is given by

$$x' = \frac{x - v\,t}{(1 - v^2/c^2)^{1/2}}, \tag{10.24}$$

$$y' = y, \tag{10.25}$$

$$z' = z, \tag{10.26}$$

$$t' = \frac{t - v\,x/c^2}{(1 - v^2/c^2)^{1/2}}. \tag{10.27}$$

This is the famous *Lorentz transformation*. It ensures that the velocity of light is invariant between different inertial frames, and also reduces to the more familiar Galilean transform in the limit $v \ll c$. We can solve Equations (10.24)–(10.27) for x, y, z, and t, to obtain the *inverse Lorentz transformation*:

$$x = \frac{x' + v\,t'}{(1 - v^2/c^2)^{1/2}}, \tag{10.28}$$

$$y = y', \tag{10.29}$$

$$z = z', \tag{10.30}$$

$$t = \frac{t' + v\,x'/c^2}{(1 - v^2/c^2)^{1/2}}. \tag{10.31}$$

Not surprizingly, the inverse transformation is equivalent to a Lorentz transformation in which the velocity of the moving frame is $-v$ along the x-axis, instead of $+v$.

10.4 TRANSFORMATION OF VELOCITIES

Consider two frames, S and S$'$, in the standard configuration. Let **u** be the velocity of a particle in S. What is the particle's velocity in S$'$? The components of **u** are

$$u_1 = \frac{dx}{dt}, \tag{10.32}$$

$$u_2 = \frac{dy}{dt}, \tag{10.33}$$

$$u_3 = \frac{dz}{dt}. \tag{10.34}$$

Similarly, the components of $\mathbf{u}'$ are

$$u_1' = \frac{dx'}{dt'}, \tag{10.35}$$

$$u_2' = \frac{dy'}{dt'}, \tag{10.36}$$

$$u_3' = \frac{dz'}{dt'}. \tag{10.37}$$

Now we can write Equations (10.24)–(10.27) in the form $dx' = \gamma\,(dx - v\,dt)$, $dy' = dy$, $dz' = dz$, and $dt' = \gamma\,(dt - v\,dx/c^2)$, where

$$\gamma = (1 - v^2/c^2)^{-1/2} \tag{10.38}$$

is the well-known *Lorentz factor*. If we substitute these differentials into Equations (10.32)–(10.34), and make use of Equations (10.35)–(10.37), we obtain the transformation rules

$$u_1' = \frac{u_1 - v}{1 - u_1 v/c^2}, \tag{10.39}$$

$$u_2' = \frac{u_2}{\gamma\,(1 - u_1 v/c^2)}, \tag{10.40}$$

$$u_3' = \frac{u_3}{\gamma\,(1 - u_1 v/c^2)}. \tag{10.41}$$

As in the transformation of coordinates, we can obtain the inverse transform by interchanging primed and unprimed symbols, and replacing $+v$ with $-v$. Thus,

$$u_1 = \frac{u_1' + v}{1 + u_1' v/c^2}, \tag{10.42}$$

$$u_2 = \frac{u_2'}{\gamma\,(1 + u_1' v/c^2)}, \tag{10.43}$$

$$u_3 = \frac{u_3'}{\gamma\,(1 + u_1' v/c^2)}. \tag{10.44}$$

Equations (10.42)–(10.44) can be regarded as giving the resultant, $\mathbf{u} = (u_1, u_2, u_3)$, of two velocities, $\mathbf{v} = (v, 0, 0)$ and $\mathbf{u}' = (u_1', u_2', u_3')$, and are therefore usually referred to as the relativistic *velocity addition formulae*. The following relation between the magnitudes $u = (u_1^2 + u_2^2 + u_3^2)^{1/2}$ and $u' = (u_1'^2 + u_2'^2 + u_3'^2)^{1/2}$ of the velocities is easily demonstrated:

$$c^2 - u^2 = \frac{c^2\,(c^2 - u'^2)\,(c^2 - v^2)}{(c^2 + u_1'\,v)^2}. \tag{10.45}$$

If $u' < c$ and $v < c$ then the right-hand side is positive, implying that $u < c$. In other words, the resultant of two subluminal velocities is another subluminal velocity. It is evident that a particle can never attain the velocity of light relative to a given inertial frame, no matter how many subluminal velocity increments it is given. It follows that no inertial frame can ever appear to propagate with a superluminal velocity with respect to any other inertial frame (since we can track a given inertial frame using a particle which remains at rest at the origin of that frame).

According to Equation (10.45), if $u' = c$ then $u = c$, no matter what value v takes: *i.e.*, the velocity of light is invariant between different inertial frames. Note that the Lorentz transform only allows *one* such invariant velocity [*i.e.*, the velocity c which appears in Equations (10.24)–(10.27)]. Einstein's relativity principle tells us that any disturbance which propagates through a vacuum must appear to propagate at the same velocity in all inertial frames. It is now evident that *all* such disturbances must propagate at the velocity c. It follows immediately that all electromagnetic waves must propagate through the vacuum with this velocity, irrespective of their wavelength. In other words, it is impossible for there to be any dispersion of electromagnetic waves propagating through a vacuum. Furthermore, gravity waves must also propagate with the velocity c.

The Lorentz transformation implies that the velocities of propagation of all physical effects are limited by c in deterministic physics. Consider a general process by which an event P *causes* an event Q at a velocity $U > c$ in some frame S. In other words, *information* about the event P appears to propagate to the event Q with a superluminal velocity. Let us choose coordinates such that these two events occur on the x-axis with (finite) time and distance separations $\Delta t > 0$ and $\Delta x > 0$, respectively. The time separation in some other inertial frame S' is given by [see Equation (10.27)]

$$\Delta t' = \gamma\,(\Delta t - v\,\Delta x/c^2) = \gamma\,\Delta t\,(1 - v\,U/c^2). \tag{10.46}$$

Thus, for sufficiently large $v < c$ we obtain $\Delta t' < 0$: *i.e.*, there exist inertial frames in which cause and effect appear to be reversed. Of course, this is impossible in deterministic physics. It follows, therefore, that information can never appear to propagate with a superluminal velocity in any inertial frame, otherwise causality would be violated.

10.5 TENSORS

It is now convenient to briefly review the mathematics of *tensors*. Tensors are of primary importance in connection with *coordinate transforms*. They serve to isolate intrinsic geometric and physical properties from those that merely depend on coordinates.

A tensor of rank r in an n-dimensional space possesses n^r components which are, in general, functions of position in that space. A tensor of rank zero has one component, A, and is called a *scalar*. A tensor of rank one has n components, $(A_1, A_2, \cdots, A_n)$, and is called a *vector*. A tensor of rank two has n^2 components, which can be exhibited in matrix format. Unfortunately, there is no convenient way of exhibiting a higher rank tensor. Consequently, tensors are usually represented by a typical component: *e.g.*, the tensor A_{ijk} (rank 3), or the tensor A_{ijkl} (rank 4), *etc.* The suffixes $i, j, k, \cdots$ are always understood to range from 1 to n.

For reasons which will become apparent later on, we shall represent tensor components using both superscripts and subscripts. Thus, a typical tensor might look like A^{ij} (rank 2), or B^i_j (rank 2), *etc.* It is convenient to adopt the *Einstein summation convention*. Namely, if any suffix appears twice in a given term, once as a subscript and once as a superscript, a summation over that suffix (from 1 to n) is implied.

To distinguish between various different coordinate systems, we shall use primed and multiply primed suffixes. A first system of coordinates $(x^1, x^2, \cdots, x^n)$ can then be denoted by x^i, a second system $(x^{1'}, x^{2'}, \cdots, x^{n'})$ by $x^{i'}$, *etc.* Similarly, the general components of a tensor in various coordinate systems are distinguished by their suffixes. Thus, the components of some third-rank tensor are denoted A_{ijk} in the x^i system, by $A_{i'j'k'}$ in the $x^{i'}$ system, *etc.*

When making a coordinate transformation from one set of coordinates, x^i, to another, $x^{i'}$, it is assumed that the transformation in non-singular. In other words, the equations which express the $x^{i'}$ in terms of the x^i can be inverted to express the x^i in terms of the $x^{i'}$. It is also assumed that the functions specifying a transformation are differentiable. It is convenient to write

$$\frac{\partial x^{i'}}{\partial x^i} = p^{i'}_i, \tag{10.47}$$

$$\frac{\partial x^i}{\partial x^{i'}} = p^i_{i'}. \tag{10.48}$$

Note that

$$p^i_{i'} \, p^{i'}_j = \delta^i_j, \tag{10.49}$$

by the chain rule, where δ^i_j (the *Kronecker delta*) equals 1 or 0 when $i = j$ or $i \neq j$, respectively.

The formal definition of a tensor is as follows:

1. An entity having components $A_{ij\cdots k}$ in the x^i system and $A_{i'j'\cdots k'}$ in the $x^{i'}$ system is said to behave as a *covariant tensor* under the transformation $x^i \to x^{i'}$ if

$$A_{i'j'\cdots k'} = A_{ij\cdots k} \, p^i_{i'} \, p^j_{j'} \cdots p^k_{k'}. \tag{10.50}$$

2. Similarly, $A^{ij\cdots k}$ is said to behave as a *contravariant tensor* under $x^i \to x^{i'}$ if

$$A^{i'j'\cdots k'} = A^{ij\cdots k} p^{i'}_i \, p^{j'}_j \cdots p^{k'}_k. \tag{10.51}$$

3. Finally, $A^{i\cdots j}_{k\cdots l}$ is said to behave as a *mixed tensor* (contravariant in $i\cdots j$ and covariant in $k\cdots l$) under $x^i \to x^{i'}$ if

$$A^{i'\cdots j'}_{k'\cdots l'} = A^{i\cdots j}_{k\cdots l}\, p^{i'}_{i} \cdots p^{j'}_{j}\, p^{k}_{k'} \cdots p^{l}_{l'}. \tag{10.52}$$

When an entity is described as a tensor it is generally understood that it behaves as a tensor under *all* non-singular differentiable transformations of the relevant coordinates. An entity which only behaves as a tensor under a certain subgroup of non-singular differentiable coordinate transformations is called a *qualified tensor*, because its name is conventionally qualified by an adjective recalling the subgroup in question. For instance, an entity which only exhibits tensor behavior under Lorentz transformations is called a *Lorentz tensor*, or, more commonly, a *4-tensor*.

When applied to a tensor of rank zero (a scalar), the above definitions imply that $A' = A$. Thus, a scalar is a function of position only, and is independent of the coordinate system. A scalar is often termed an *invariant*.

The main theorem of tensor calculus is as follows:

If two tensors of the same type are equal in one coordinate system, then they are equal in all coordinate systems.

The simplest example of a contravariant vector (tensor of rank one) is provided by the differentials of the coordinates, dx^i, since

$$dx^{i'} = \frac{\partial x^{i'}}{\partial x^i}\, dx^i = dx^i\, p^{i'}_{i}. \tag{10.53}$$

The coordinates themselves do not behave as tensors under all coordinate transformations. However, since they transform like their differentials under linear homogeneous coordinate transformations, they do behave as tensors under such transformations.

The simplest example of a covariant vector is provided by the gradient of a function of position $\phi = \phi(x^1, \cdots, x^n)$, since if we write

$$\phi_i = \frac{\partial \phi}{\partial x^i}, \tag{10.54}$$

then we have

$$\phi_{i'} = \frac{\partial \phi}{\partial x^{i'}} = \frac{\partial \phi}{\partial x^i}\frac{\partial x^i}{\partial x^{i'}} = \phi_i\, p^{i}_{i'}. \tag{10.55}$$

An important example of a mixed second-rank tensor is provided by the Kronecker delta introduced previously, since

$$\delta^i_j\, p^{i'}_i\, p^j_{j'} = p^{i'}_j\, p^j_{j'} = \delta^{i'}_{j'}. \tag{10.56}$$

Tensors *of the same type* can be added or subtracted to form new tensors. Thus, if A_{ij} and B_{ij} are tensors, then $C_{ij} = A_{ij} \pm B_{ij}$ is a tensor of the same type. Note that the sum of tensors at different points in space is not a tensor if the p's are position dependent. However, under linear coordinate transformations the p's are constant, so the sum of tensors at different points behaves as a tensor under this particular type of coordinate transformation.

If A^{ij} and B_{ijk} are tensors, then $C^{ij}_{klm} = A^{ij}\, B_{klm}$ is a tensor of the type indicated by the suffixes. The process illustrated by this example is called *outer multiplication* of tensors.

Tensors can also be combined by *inner multiplication*, which implies at least one dummy suffix link. Thus, $C^j_{kl} = A^{ij}\, B_{ikl}$ and $C_k = A^{ij}\, B_{ijk}$ are tensors of the type indicated by the suffixes.

Finally, tensors can be formed by *contraction* from tensors of higher rank. Thus, if A^{ij}_{klm} is a tensor then $C^j_{kl} = A^{ij}_{ikl}$ and $C_k = A^{ij}_{kij}$ are tensors of the type indicated by the suffixes. The most important type of contraction occurs when no free suffixes remain: the result is a scalar. Thus, A^i_i is a scalar provided that A^j_i is a tensor.

Although we cannot usefully divide tensors, one by another, an entity like C^{ij} in the equation $A^j = C^{ij}\, B_i$, where A^i and B_i are tensors, can be formally regarded as the quotient of A^i and B_i. This gives the name to a particularly useful rule for recognizing tensors, the *quotient rule*. This rule states that *if a set of components, when combined by a given type of multiplication with all tensors of a given type yields a tensor, then the set is itself a tensor*. In other words, if the product $A^i = C^{ij}\, B_j$ transforms like a tensor for *all* tensors B_i then it follows that C^{ij} is a tensor.

Let

$$\frac{\partial A^{i\cdots j}_{k\cdots l}}{\partial x^m} = A^{i\cdots j}_{k\cdots l,m}. \tag{10.57}$$

Then if $A^{i\cdots j}_{k\cdots l}$ is a tensor, differentiation of the general tensor transformation (10.52) yields

$$A^{i'\cdots j'}_{k'\cdots l',m'} = A^{i\cdots j}_{k\cdots l,m}\, p^{i'}_i \cdots p^{j'}_j\, p^k_{k'} \cdots p^l_{l'}\, p^m_{m'} + P_1 + P_2 + \cdots, \tag{10.58}$$

where P_1, P_2, *etc.*, are terms involving derivatives of the p's. Clearly, $A^{i\cdots j}_{k\cdots l}$ is not a tensor under a general coordinate transformation. However, under a linear coordinate transformation (p's constant) $A^{i'\cdots j'}_{k'\cdots l',m'}$

behaves as a tensor of the type indicated by the suffixes, since the P_1, P_2, *etc.*, all vanish. Similarly, all higher partial derivatives,

$$A^{i\cdots j}_{k\cdots l,mn} = \frac{\partial A^{i\cdots j}_{k\cdots l}}{\partial x^m \partial x^n} \tag{10.59}$$

etc., also behave as tensors under linear transformations. Each partial differentiation has the effect of adding a new covariant suffix.

So far, the space to which the coordinates x^i refer has been without structure. We can impose a structure on it by defining the distance between all pairs of neighboring points by means of a *metric*,

$$ds^2 = g_{ij}\,dx^i\,dx^j, \tag{10.60}$$

where the g_{ij} are functions of position. We can assume that $g_{ij} = g_{ji}$ without loss of generality. The above metric is analogous to, but more general than, the metric of Euclidian n-space, $ds^2 = (dx^1)^2 + (dx^2)^2 + \cdots + (dx^n)^2$. A space whose structure is determined by a metric of the type (10.60) is called *Riemannian*. Since ds^2 is invariant, it follows from a simple extension of the quotient rule that g_{ij} must be a tensor. It is called the *metric tensor*.

The elements of the inverse of the matrix g_{ij} are denoted by g^{ij}. These elements are uniquely defined by the equations

$$g^{ij} g_{jk} = \delta^i_k. \tag{10.61}$$

It is easily seen that the g^{ij} constitute the elements of a contravariant tensor. This tensor is said to be *conjugate* to g_{ij}. The conjugate metric tensor is symmetric (*i.e.*, $g^{ij} = g^{ji}$) just like the metric tensor itself.

The tensors g_{ij} and g^{ij} allow us to introduce the important operations of *raising* and *lowering suffixes*. These operations consist of forming inner products of a given tensor with g_{ij} or g^{ij}. For example, given a contravariant vector A^i, we define its covariant components A_i by the equation

$$A_i = g_{ij}\,A^j. \tag{10.62}$$

Conversely, given a covariant vector B_i, we can define its contravariant components B^i by the equation

$$B^i = g^{ij}\,B_j. \tag{10.63}$$

More generally, we can raise or lower any or all of the free suffixes of any given tensor. Thus, if A_{ij} is a tensor we define $A^i{}_j$ by the equation

$$A^i{}_j = g^{ip} A_{pj}. \tag{10.64}$$

Note that once the operations of raising and lowering suffixes has been defined, the order of raised suffixes relative to lowered suffixes becomes significant.

By analogy with Euclidian space, we define the *squared magnitude* $(A)^2$ of a vector A^i with respect to the metric $g_{ij}\,dx^i\,dx^j$ by the equation

$$(A)^2 = g_{ij}\,A^i\,A^j = A_i\,A^i. \tag{10.65}$$

A vector A^i is termed a *null vector* if $(A)^2 = 0$. Two vectors A^i and B^i are said to be *orthogonal* if their inner product vanishes: *i.e.*, if

$$g_{ij}\,A^i\,B^j = A_i\,B^i = A^i\,B_i = 0. \tag{10.66}$$

Finally, let us consider differentiation with respect to an invariant distance, s. The vector dx^i/ds is a contravariant tensor, since

$$\frac{dx^{i'}}{ds} = \frac{\partial x^{i'}}{\partial x^i}\frac{dx^i}{ds} = \frac{dx^i}{ds}\,p_i^{i'}. \tag{10.67}$$

The derivative $d(A^{i\cdots j}{}_{k\cdots l})/ds$ of some tensor with respect to s is not, in general, a tensor, since

$$\frac{d(A^{i\cdots j}{}_{k\cdots l})}{ds} = A^{i\cdots j}{}_{k\cdots l,m}\frac{dx^m}{ds}, \tag{10.68}$$

and, as we have seen, the first factor on the right-hand side is not generally a tensor. However, under linear transformations it behaves as a tensor, so under linear transformations the derivative of a tensor with respect to an invariant distance behaves as a tensor of the same type.

10.6 PHYSICAL SIGNIFICANCE OF TENSORS

In this chapter, we shall only concern ourselves with coordinate transformations which transform an inertial frame into another inertial frame. This limits us to four classes of transformations: displacements of the coordinate axes, rotations of the coordinate axes, parity reversals (*i.e.*, $x, y, z \to -x, -y, -z$), and Lorentz transformations.

One of the central tenets of Physics is that experiments should be *reproducible*. In other words, if somebody performs a physical experiment today, and obtains a certain result, then somebody else performing the same experiment next week ought to obtain the same result, within the experimental errors. Presumably, in performing these hypothetical experiments, both experimentalists find it necessary to set up a coordinate frame. Usually, these two frames do not coincide. After all, the experiments are, in general, performed in different places and at different times. Also, the two experimentalists are likely to orientate their coordinate axes differently. Nevertheless, we still expect both experiments to yield the same result. What exactly do we mean by this statement? We do not mean that both experimentalists will obtain the same numbers when they measure something. For instance, the numbers used to denote the position of a point (*i.e.*, the coordinates of the point) are, in general, different in different coordinate frames. What we do expect is that any physically significant interrelation between physical quantities (*i.e.*, position, velocity, *etc.*) which appears to hold in the coordinate system of the first experimentalist will also appear to hold in the coordinate system of the second experimentalist. We usually refer to such interrelationships as *laws of Physics*. So, what we are really saying is that the laws of Physics do not depend on our choice of coordinate system. In particular, if a law of Physics is true in one coordinate system then it is automatically true in every other coordinate system, subject to the proviso that both coordinate systems are inertial.

Recall that tensors are geometric objects which possess the property that if a certain interrelationship holds between various tensors in one particular coordinate system, then the same interrelationship holds in any other coordinate system which is related to the first system by a certain class of transformations. It follows that *the laws of Physics are expressible as interrelationships between tensors*. In Special Relativity, the laws of Physics are only required to exhibit tensor behavior under transformations between different inertial frames: *i.e.*, translations, rotations, and Lorentz transformations. Parity inversion is a special type of transformation, and will be dealt with later on. In General Relativity, the laws of Physics are required to exhibit tensor behavior under *all* non-singular coordinate transformations.

10.7 SPACE-TIME

In Special Relativity, we are only allowed to use inertial frames to assign coordinates to events. There are many different types of inertial frames. However, it is convenient to adhere to those with *standard coordinates*. That is, spatial coordinates which are right-handed rectilinear Cartesians based on a standard unit of length, and time-scales based on a standard unit of time. We shall continue to assume that we are employing standard coordinates. However, from now on, we shall make no assumptions about the relative configuration of the two sets of spatial axes, and the origins of time, when dealing with two inertial frames. Thus, the most general transformation between two inertial frames consists of a Lorentz transformation in the standard configuration plus a translation (this includes a translation in time) and a rotation of the coordinate axes. The resulting transformation is called a *general Lorentz transformation*, as opposed to a Lorentz transformation in the standard configuration, which will henceforth be termed a *standard Lorentz transformation*.

In Section 10.3, we proved quite generally that corresponding differentials in two inertial frames S and S′ satisfy the relation

$$dx^2 + dy^2 + dz^2 - c^2\,dt^2 = dx'^2 + dy'^2 + dz'^2 - c^2\,dt'^2. \quad (10.69)$$

Thus, we expect this relation to remain invariant under a general Lorentz transformation. Since such a transformation is *linear*, it follows that

$$(x_2 - x_1)^2 + (y_2 - y_1)^2 + (z_2 - z_1)^2 - c^2\,(t_2 - t_1)^2 = \\ (x_2' - x_1')^2 + (y_2' - y_1')^2 + (z_2' - z_1')^2 - c^2\,(t_2' - t_1')^2, \quad (10.70)$$

where (x_1, y_1, z_1, t_1) and (x_2, y_2, z_2, t_2) are the coordinates of any two events in S, and the primed symbols denote the corresponding coordinates in S′. It is convenient to write

$$-dx^2 - dy^2 - dz^2 + c^2\,dt^2 = ds^2, \quad (10.71)$$

and

$$-(x_2 - x_1)^2 - (y_2 - y_1)^2 - (z_2 - z_1)^2 + c^2(t_2 - t_1)^2 = s^2. \quad (10.72)$$

The differential ds, or the finite number s, defined by these equations is called the *interval* between the corresponding events. Equations (10.71) and (10.72) express the fact that *the interval between two events is invariant*, in the sense that it has the same value in all inertial frames. In other words, the interval between two events is invariant under a general Lorentz transformation.

Let us consider entities defined in terms of four variables,

$$x^1 = x, \quad x^2 = y, \quad x^3 = z, \quad x^4 = c\,t, \quad (10.73)$$

and which transform as tensors under a general Lorentz transformation. From now on, such entities will be referred to as *4-tensors*.

Tensor analysis cannot proceed very far without the introduction of a non-singular tensor g_{ij}, the so-called *fundamental tensor*, which is used to define the operations of raising and lowering suffixes. The fundamental tensor is usually introduced using a metric $ds^2 = g_{ij}\, dx^i\, dx^j$, where ds^2 is a differential invariant. We have already come across such an invariant, namely

$$\begin{aligned} ds^2 &= -\,dx^2 - dy^2 - dz^2 + c^2\, dt^2 \\ &= -\,(dx^1)^2 - (dx^2)^2 - (dx^3)^2 + (dx^4)^2 \\ &= g_{\mu\nu}\, dx^\mu\, dx^\nu, \end{aligned} \tag{10.74}$$

where μ, ν run from 1 to 4. Note that the use of Greek suffixes is conventional in 4-tensor theory. Roman suffixes are reserved for tensors in three-dimensional Euclidian space, so-called *3-tensors*. The 4-tensor $g_{\mu\nu}$ has the components $g_{11} = g_{22} = g_{33} = -1$, $g_{44} = 1$, and $g_{\mu\nu} = 0$ when $\mu \neq \nu$, in all permissible coordinate frames. From now on, $g_{\mu\nu}$, as defined above, is adopted as the fundamental tensor for 4-tensors. $g_{\mu\nu}$ can be thought of as the *metric tensor* of the space whose points are the events (x^1, x^2, x^3, x^4). This space is usually referred to as *space-time*, for obvious reasons. Note that space-time cannot be regarded as a straightforward generalization of Euclidian 3-space to four dimensions, with time as the fourth dimension. The distribution of signs in the metric ensures that the time coordinate x^4 is not on the same footing as the three space coordinates. Thus, space-time has a non-isotropic nature which is quite unlike Euclidian space, with its positive definite metric. According to the Relativity Principle, all physical laws are expressible as interrelationships between 4-tensors in space-time.

A tensor of rank one is called a *4-vector*. We shall also have occasion to use ordinary vectors in three-dimensional Euclidian space. Such vectors are called *3-vectors*, and are conventionally represented by boldface symbols. We shall use the Latin suffixes i, j, k, *etc.*, to denote the components of a 3-vector: these suffixes are understood to range from 1 to 3. Thus, $\mathbf{u} = u^i = dx^i/dt$ denotes a velocity vector. For 3-vectors, we shall use the notation $u^i = u_i$ interchangeably: *i.e.*, the level of the suffix has no physical significance.

When tensor transformations from one frame to another actually have to be computed, we shall usually find it possible to choose coordinates in the standard configuration, so that the standard Lorentz transform applies. Under such a transformation, any contravariant 4-vector, T^μ, transforms according to the same scheme as the difference in coordinates $x_2^\mu - x_1^\mu$ between two points in space-time. It follows that

$$T^{1'} = \gamma\,(T^1 - \beta\,T^4), \tag{10.75}$$

$$T^{2'} = T^2, \tag{10.76}$$

$$T^{3'} = T^3, \tag{10.77}$$

$$T^{4'} = \gamma\,(T^4 - \beta\,T^1), \tag{10.78}$$

where $\beta = v/c$. Higher rank 4-tensors transform according to the rules (10.50)–(10.52). The transformation coefficients take the form

$$p_\mu^{\mu'} = \begin{pmatrix} +\gamma & 0 & 0 & -\gamma\,\beta \\ 0 & 1 & 0 & 0 \\ 0 & 0 & 1 & 0 \\ -\gamma\,\beta & 0 & 0 & +\gamma \end{pmatrix}, \tag{10.79}$$

$$p_{\mu'}^{\mu} = \begin{pmatrix} +\gamma & 0 & 0 & +\gamma\,\beta \\ 0 & 1 & 0 & 0 \\ 0 & 0 & 1 & 0 \\ +\gamma\,\beta & 0 & 0 & +\gamma \end{pmatrix}. \tag{10.80}$$

Often the first three components of a 4-vector coincide with the components of a 3-vector. For example, the x^1, x^2, x^3 in $R^\mu = (x^1, x^2, x^3, x^4)$ are the components of $\mathbf{r}$, the position 3-vector of the point at which the event occurs. In such cases, we adopt the notation exemplified by $R^\mu = (\mathbf{r}, c\,t)$. The covariant form of such a vector is simply $R_\mu = (-\mathbf{r}, c\,t)$. The squared magnitude of the vector is $(R)^2 = R_\mu R^\mu = -r^2 + c^2\,t^2$. The inner product $g_{\mu\nu}\,R^\mu\,Q^\nu = R_\mu\,Q^\mu$ of R^μ with a similar vector $Q^\mu = (\mathbf{q}, k)$ is given by $R_\mu\,Q^\mu = -\mathbf{r}\cdot\mathbf{q} + c\,t\,k$. The vectors R^μ and Q^μ are said to be *orthogonal* if $R_\mu\,Q^\mu = 0$.

Since a general Lorentz transformation is a *linear* transformation, the partial derivative of a 4-tensor is also a 4-tensor:

$$\frac{\partial A^{\nu\sigma}}{\partial x^\mu} = A^{\nu\sigma}{}_{,\mu}. \tag{10.81}$$

Clearly, a general 4-tensor acquires an extra covariant index after partial differentiation with respect to the contravariant coordinate x^μ. It is helpful to define a covariant derivative operator

$$\partial_\mu \equiv \frac{\partial}{\partial x^\mu} = \left(\nabla, \frac{1}{c}\frac{\partial}{\partial t}\right), \tag{10.82}$$

where

$$\partial_\mu A^{\nu\sigma} \equiv A^{\nu\sigma}{}_{,\mu}. \tag{10.83}$$

There is a corresponding contravariant derivative operator

$$\partial^\mu \equiv \frac{\partial}{\partial x_\mu} = \left(-\nabla, \frac{1}{c}\frac{\partial}{\partial t}\right), \tag{10.84}$$

where

$$\partial^\mu A^{\nu\sigma} \equiv g^{\mu\tau} A^{\nu\sigma}{}_{,\tau}. \tag{10.85}$$

The 4-divergence of a 4-vector, $A^\mu = (\mathbf{A}, A^0)$, is the invariant

$$\partial^\mu A_\mu = \partial_\mu A^\mu = \nabla\cdot\mathbf{A} + \frac{1}{c}\frac{\partial A^0}{\partial t}. \tag{10.86}$$

The four-dimensional Laplacian operator, or *d'Alembertian*, is equivalent to the invariant contraction

$$\Box \equiv \partial_\mu \partial^\mu = -\nabla^2 + \frac{1}{c^2}\frac{\partial^2}{\partial t^2}. \tag{10.87}$$

Recall that we still need to prove (from Section 10.3) that the invariance of the differential metric,

$$ds^2 = dx'^2 + dy'^2 + dz'^2 - c^2\,dt'^2 = dx^2 + dy^2 + dz^2 - c^2\,dt^2, \tag{10.88}$$

between two general inertial frames implies that the coordinate transformation between such frames is necessarily linear. To put it another way, we need to demonstrate that a transformation which transforms a metric $g_{\mu\nu}\,dx^\mu\,dx^\nu$ with constant coefficients into a metric $g_{\mu'\nu'}\,dx^{\mu'}\,dx^{\nu'}$ with constant coefficients must be linear. Now

$$g_{\mu\nu} = g_{\mu'\nu'}\,p^{\mu'}_{\mu}\,p^{\nu'}_{\nu}. \tag{10.89}$$

Differentiating with respect to x^σ, we get

$$g_{\mu'\nu'}\,p^{\mu'}_{\mu\sigma}\,p^{\nu'}_{\nu} + g_{\mu'\nu'}\,p^{\mu'}_{\mu}\,p^{\nu'}_{\nu\sigma} = 0, \tag{10.90}$$

where

$$p^{\mu'}_{\mu\sigma} = \frac{\partial p^{\mu'}_{\mu}}{\partial x^\sigma} = \frac{\partial^2 x^{\mu'}}{\partial x^\mu \partial x^\sigma} = p^{\mu'}_{\sigma\mu}, \tag{10.91}$$

etc. Interchanging the indices μ and σ yields

$$g_{\mu'\nu'}\,p^{\mu'}_{\mu\sigma}\,p^{\nu'}_{\nu} + g_{\mu'\nu'}\,p^{\mu'}_{\sigma}\,p^{\nu'}_{\nu\mu} = 0. \tag{10.92}$$

Interchanging the indices ν and σ gives

$$g_{\mu'\nu'}\, p^{\mu'}_{\sigma}\, p^{\nu'}_{\nu\mu} + g_{\mu'\nu'}\, p^{\mu'}_{\mu}\, p^{\nu'}_{\nu\sigma} = 0, \tag{10.93}$$

where the indices μ' and ν' have been interchanged in the first term. It follows from Equations (10.90), (10.92), and (10.93) that

$$g_{\mu'\nu'}\, p^{\mu'}_{\mu\sigma}\, p^{\nu'}_{\nu} = 0. \tag{10.94}$$

Multiplication by $p^{\nu}_{\sigma'}$ yields

$$g_{\mu'\nu'}\, p^{\mu'}_{\mu\sigma}\, p^{\nu'}_{\nu}\, p^{\nu}_{\sigma'} = g_{\mu'\sigma'}\, p^{\mu'}_{\mu\sigma} = 0. \tag{10.95}$$

Finally, multiplication by $g^{\nu'\sigma'}$ gives

$$g_{\mu'\sigma'}\, g^{\nu'\sigma'}\, p^{\mu'}_{\mu\sigma} = p^{\nu'}_{\mu\sigma} = 0. \tag{10.96}$$

This proves that the coefficients $p^{\nu'}_{\mu}$ are constants, and, hence, that the transformation is linear.

10.8 PROPER TIME

It is often helpful to write the invariant differential interval ds^2 in the form

$$ds^2 = c^2\, d\tau^2. \tag{10.97}$$

The quantity $d\tau$ is called the *proper time*. It follows that

$$d\tau^2 = -\frac{dx^2 + dy^2 + dz^2}{c^2} + dt^2. \tag{10.98}$$

Consider a series of events on the world-line of some material particle. If the particle has speed u then

$$d\tau^2 = dt^2\left[-\frac{dx^2 + dy^2 + dz^2}{c^2\, dt^2} + 1\right] = dt^2\left(1 - \frac{u^2}{c^2}\right), \tag{10.99}$$

implying that

$$\frac{dt}{d\tau} = \gamma(u). \tag{10.100}$$

It is clear that $dt = d\tau$ in the particle's rest frame. Thus, $d\tau$ corresponds to the time difference between two neighboring events on the particle's world-line, as measured by a clock attached to the particle (hence, the name *proper time*). According to Equation (10.100), the particle's clock appears to run slow, by a factor $\gamma(u)$, in an inertial frame in which the particle is moving with velocity u. This is the celebrated *time dilation* effect.

Let us consider how a small 4-dimensional volume element in space-time transforms under a general Lorentz transformation. We have

$$d^4x' = \mathcal{J}\, d^4x, \tag{10.101}$$

where

$$\mathcal{J} = \frac{\partial(x^{1'}, x^{2'}, x^{3'}, x^{4'})}{\partial(x^1, x^2, x^3, x^4)} \tag{10.102}$$

is the Jacobian of the transformation: *i.e.*, the determinant of the transformation matrix $p^{\mu'}_{\mu}$. A general Lorentz transformation is made up of a standard Lorentz transformation plus a displacement and a rotation. Thus, the transformation matrix is the *product* of that for a standard Lorentz transformation, a translation, and a rotation. It follows that the Jacobian of a general Lorentz transformation is the product of that for a standard Lorentz transformation, a translation, and a rotation. It is well-known that the Jacobian of the latter two transformations is unity, since they are both volume preserving transformations which do not affect time. Likewise, it is easily seen [*e.g.*, by taking the determinant of the transformation matrix (10.79)] that the Jacobian of a standard Lorentz transformation is also unity. It follows that

$$d^4x' = d^4x \tag{10.103}$$

for a general Lorentz transformation. In other words, a general Lorentz transformation preserves the volume of space-time. Since time is dilated by a factor γ in a moving frame, the volume of space-time can only be preserved if the volume of ordinary 3-space is reduced by the same factor. As is well-known, this is achieved by *length contraction* along the direction of motion by a factor γ.

10.9 4-VELOCITY AND 4-ACCELERATION

We have seen that the quantity dx^μ/ds transforms as a 4-vector under a general Lorentz transformation [see Equation (10.67)]. Since $ds \propto d\tau$

it follows that

$$U^\mu = \frac{dx^\mu}{d\tau} \tag{10.104}$$

also transforms as a 4-vector. This quantity is known as the *4-velocity*. Likewise, the quantity

$$A^\mu = \frac{d^2x^\mu}{d\tau^2} = \frac{dU^\mu}{d\tau} \tag{10.105}$$

is a 4-vector, and is called the *4-acceleration*.

For events along the world-line of a particle traveling with 3-velocity $\mathbf{u}$, we have

$$U^\mu = \frac{dx^\mu}{d\tau} = \frac{dx^\mu}{dt}\frac{dt}{d\tau} = \gamma(u)\,(\mathbf{u}, c), \tag{10.106}$$

where use has been made of Equation (10.100). This gives the relationship between a particle's 3-velocity and its 4-velocity. The relationship between the 3-acceleration and the 4-acceleration is less straightforward. We have

$$A^\mu = \frac{dU^\mu}{d\tau} = \gamma\frac{dU^\mu}{dt} = \gamma\frac{d}{dt}(\gamma\,\mathbf{u}, \gamma\,c) = \gamma\left(\frac{d\gamma}{dt}\mathbf{u} + \gamma\,\mathbf{a}, c\,\frac{d\gamma}{dt}\right), \tag{10.107}$$

where $\mathbf{a} = d\mathbf{u}/dt$ is the 3-acceleration. In the rest frame of the particle, $U^\mu = (\mathbf{0}, c)$ and $A^\mu = (\mathbf{a}, 0)$. It follows that

$$U_\mu\,A^\mu = 0 \tag{10.108}$$

(note that $U_\mu\,A^\mu$ is an invariant quantity). In other words, the 4-acceleration of a particle is always *orthogonal* to its 4-velocity.

10.10 THE CURRENT DENSITY 4-VECTOR

Let us now consider the laws of Electromagnetism. We wish to demonstrate that these laws are compatible with the Relativity Principle. In order to achieve this, it is necessary for us to make an *assumption* about the transformation properties of electric charge. The assumption we shall make, which is well substantiated experimentally, is that charge, unlike mass, is *invariant*. That is, the charge carried by a given particle has the same measure in all inertial frames. In particular, the charge carried by a particle does not vary with the particle's velocity.

Let us suppose, following Lorentz, that all charge is made up of elementary particles, each carrying the invariant amount e. Suppose that n is the number density of such charges at some given point and time, moving with velocity $\mathbf{u}$, as observed in a frame S. Let n_0 be the number density of charges in the frame S_0 in which the charges are momentarily at rest. As is well-known, a volume of measure V in S has measure $\gamma(u)\,V$ in S_0 (because of length contraction). Since observers in both frames must agree on how many particles are contained in the volume, and, hence, on how much charge it contains, it follows that $n = \gamma(u)\,n_0$. If $\rho = e\,n$ and $\rho_0 = e\,n_0$ are the charge densities in S and S_0, respectively, then

$$\rho = \gamma(u)\,\rho_0. \tag{10.109}$$

The quantity ρ_0 is called the *proper density*, and is obviously Lorentz invariant.

Suppose that x^μ are the coordinates of the moving charge in S. The *current density 4-vector* is constructed as follows:

$$J^\mu = \rho_0\,\frac{dx^\mu}{d\tau} = \rho_0\,U^\mu. \tag{10.110}$$

Thus,

$$J^\mu = \rho_0\,\gamma(u)\,(\mathbf{u}, c) = (\mathbf{j}, \rho\,c), \tag{10.111}$$

where $\mathbf{j} = \rho\,\mathbf{u}$ is the current density 3-vector. Clearly, charge density and current density transform as the time-like and space-like components of the same 4-vector.

Consider the invariant 4-divergence of J^μ:

$$\partial_\mu J^\mu = \nabla\cdot\mathbf{j} + \frac{\partial\rho}{\partial t}. \tag{10.112}$$

We know that one of the caveats of Maxwell's equations is the charge conservation law

$$\frac{\partial\rho}{\partial t} + \nabla\cdot\mathbf{j} = 0. \tag{10.113}$$

It is clear that this expression can be rewritten in the manifestly Lorentz invariant form

$$\partial_\mu J^\mu = 0. \tag{10.114}$$

This equation tells us that there are no net sources or sinks of electric charge in nature: *i.e.*, electric charge is neither created nor destroyed.

10.11 THE POTENTIAL 4-VECTOR

There are many ways of writing the laws of electromagnetism. However, the most obviously Lorentz invariant way is to write them in terms of the vector and scalar potentials (see Section 4.6). When written in this fashion, Maxwell's equations reduce to

$$\left(-\nabla^2 + \frac{1}{c^2}\frac{\partial^2}{\partial t^2}\right)\phi = \frac{\rho}{\epsilon_0}, \tag{10.115}$$

$$\left(-\nabla^2 + \frac{1}{c^2}\frac{\partial^2}{\partial t^2}\right)\mathbf{A} = \mu_0\,\mathbf{j}, \tag{10.116}$$

where ϕ is the scalar potential, and $\mathbf{A}$ the vector potential. Note that the differential operator appearing in these equations is the Lorentz invariant d'Alembertian, defined in Equation (10.87). Thus, the above pair of equations can be rewritten in the form

$$\Box\phi = \frac{\rho\,c}{c\,\epsilon_0}, \tag{10.117}$$

$$\Box(c\,\mathbf{A}) = \frac{\mathbf{j}}{c\,\epsilon_0}. \tag{10.118}$$

Maxwell's equations can be written in Lorentz invariant form provided that the entity

$$\Phi^{\mu} = (c\,\mathbf{A}, \phi) \tag{10.119}$$

transforms as a contravariant 4-vector. This entity is known as the *potential 4-vector*. It follows from Equations (10.111), (10.115), and (10.116) that

$$\Box\Phi^{\mu} = \frac{J^{\mu}}{c\,\epsilon_0}. \tag{10.120}$$

Thus, the field equations which govern classical electromagnetism can all be summed up in a single 4-vector equation.

10.12 GAUGE INVARIANCE

The electric and magnetic fields are obtained from the vector and scalar potentials according to the prescription (see Section 4.3)

$$\mathbf{E} = -\nabla\phi - \frac{\partial\mathbf{A}}{\partial t}, \tag{10.121}$$

$$\mathbf{B} = \nabla\times\mathbf{A}. \tag{10.122}$$

These fields are important, because they determine the electromagnetic forces exerted on charged particles. Note that the above prescription does not uniquely determine the two potentials. It is possible to make the following transformation, known as a *gauge transformation*, which leaves the fields unaltered (see Section 4.4):

$$\phi \to \phi + \frac{\partial \psi}{\partial t}, \tag{10.123}$$

$$\mathbf{A} \to \mathbf{A} - \nabla\psi, \tag{10.124}$$

where $\psi(\mathbf{r}, t)$ is a general scalar field. It is necessary to adopt some form of convention, generally known as a *gauge condition*, to fully specify the two potentials. In fact, there is only one gauge condition which is consistent with Equations (10.114). This is the *Lorenz gauge condition*,

$$\frac{1}{c^2}\frac{\partial \phi}{\partial t} + \nabla\cdot\mathbf{A} = 0. \tag{10.125}$$

Note that this condition can be written in the Lorentz invariant form

$$\partial_\mu \Phi^\mu = 0. \tag{10.126}$$

This implies that if the Lorenz gauge holds in one particular inertial frame then it automatically holds in all other inertial frames. A general gauge transformation can be written

$$\Phi^\mu \to \Phi^\mu + c\,\partial^\mu \psi. \tag{10.127}$$

Note that, even after the Lorentz gauge has been adopted, the potentials are undetermined to a gauge transformation using a scalar field, ψ, which satisfies the sourceless wave equation

$$\Box\psi = 0. \tag{10.128}$$

However, if we adopt sensible boundary conditions in both space and time then the only solution to the above equation is $\psi = 0$.

10.13 RETARDED POTENTIALS

We already know the solutions to Equations (10.117) and (10.118). They take the form (see Section 4.9)

$$\phi(\mathbf{r}, t) = \frac{1}{4\pi\epsilon_0}\int \frac{[\rho(\mathbf{r}')]}{|\mathbf{r} - \mathbf{r}'|}\, dV', \tag{10.129}$$

$$\mathbf{A}(\mathbf{r}, t) = \frac{\mu_0}{4\pi}\int \frac{[\mathbf{j}(\mathbf{r}')]}{|\mathbf{r} - \mathbf{r}'|}\, dV'. \tag{10.130}$$

The above equations can be combined to form the solution of the 4-vector wave Equation (10.120),

$$\Phi^{\mu} = \frac{1}{4\pi\epsilon_0\, c}\int \frac{[J^{\mu}]}{r}\, dV. \tag{10.131}$$

Here, the components of the 4-potential are evaluated at some event P in space-time, r is the distance of the volume element dV from P, and the square brackets indicate that the 4-current is to be evaluated at the retarded time: *i.e.*, at a time r/c before P.

But, does the right-hand side of Equation (10.131) really transform as a contravariant 4-vector? This is not a trivial question, since volume integrals in 3-space are not, in general, Lorentz invariant due to the length contraction effect. However, the integral in Equation (10.131) is not a straightforward volume integral, because the integrand is evaluated at the retarded time. In fact, the integral is best regarded as an integral over events in space-time. The events which enter the integral are those which intersect a spherical light wave launched from the event P and evolved backward in time. In other words, the events occur before the event P, and have zero interval with respect to P. It is clear that observers in all inertial frames will, at least, agree on which events are to be included in the integral, since both the interval between events, and the absolute order in which events occur, are invariant under a general Lorentz transformation.

We shall now demonstrate that all observers obtain the same value of dV/r for each elementary contribution to the integral. Suppose that S and S' are two inertial frames in the standard configuration. Let unprimed and primed symbols denote corresponding quantities in S and S', respectively. Let us assign coordinates $(0,0,0,0)$ to P, and $(x,y,z,c\,t)$ to the retarded event Q for which r and dV are evaluated. Using the standard Lorentz transformation, (10.24)–(10.27), the fact that the interval between events P and Q is zero, and the fact that both t and t' are negative, we obtain

$$r' = -c\,t' = -c\,\gamma\left(t - \frac{v\,x}{c^2}\right), \tag{10.132}$$

where v is the relative velocity between frames S' and S, γ is the Lorentz factor, and $r = \sqrt{x^2 + y^2 + z^2}$, *etc.* It follows that

$$r' = r\,\gamma\left(-\frac{c\,t}{r} + \frac{v\,x}{c\,r}\right) = r\,\gamma\left(1 + \frac{v}{c}\cos\theta\right), \tag{10.133}$$

where θ is the angle (in 3-space) subtended between the line PQ and the x-axis.

We now know the transformation for r. What about the transformation for dV? We might be tempted to set $dV' = \gamma\, dV$, according to the usual length contraction rule. However, this is incorrect. The contraction by a factor γ only applies if the whole of the volume is measured at the same time, which is not the case in the present problem. Now, the dimensions of dV along the y- and z-axes are the same in both S and S', according to Equations (10.24)–(10.27). For the x-dimension these equations give $dx' = \gamma\,(dx - v\, dt)$. The extremities of dx are measured at times differing by dt, where

$$dt = -\frac{dr}{c} = -\frac{dx}{c}\cos\theta. \tag{10.134}$$

Thus,

$$dx' = \gamma\left(1 + \frac{v}{c}\cos\theta\right) dx, \tag{10.135}$$

giving

$$dV' = \gamma\left(1 + \frac{v}{c}\cos\theta\right) dV. \tag{10.136}$$

It follows from Equations (10.133) and (10.136) that $dV'/r' = dV/r$. This result will clearly remain valid even when S and S' are not in the standard configuration.

Thus, dV/r is an invariant and, therefore, $[J^{\mu}]\, dV/r$ is a contravariant 4-vector. For linear transformations, such as a general Lorentz transformation, the result of adding 4-tensors evaluated at different 4-points is itself a 4-tensor. It follows that the right-hand side of Equation (10.131) is indeed a contravariant 4-vector. Thus, this 4-vector equation can be properly regarded as the solution to the 4-vector wave equation (10.120).

10.14 TENSORS AND PSEUDO-TENSORS

The totally antisymmetric fourth-rank tensor is defined

$$\epsilon^{\alpha\beta\gamma\delta} = \begin{cases} +1 & \text{for } \alpha, \beta, \gamma, \delta \text{ any even permutation of } 1,2,3,4 \\ -1 & \text{for } \alpha, \beta, \gamma, \delta \text{ any odd permutation of } 1,2,3,4 \\ 0 & \text{otherwise} \end{cases} . \tag{10.137}$$

The components of this tensor are invariant under a general Lorentz transformation, since

$$\epsilon^{\alpha\beta\gamma\delta}\,p_{\alpha}^{\alpha'}\,p_{\beta}^{\beta'}\,p_{\gamma}^{\gamma'}\,p_{\delta}^{\delta'} = \epsilon^{\alpha'\beta'\gamma'\delta'}\,|p_{\mu}^{\mu'}| = \epsilon^{\alpha'\beta'\gamma'\delta'}, \tag{10.138}$$

where $|p_{\mu}^{\mu'}|$ denotes the determinant of the transformation matrix, or the Jacobian of the transformation, which we have already established is unity for a general Lorentz transformation. We can also define a totally antisymmetric third-rank tensor ϵ^{ijk} which stands in the same relation to 3-space as $\epsilon^{\alpha\beta\gamma\delta}$ does to space-time. It is easily demonstrated that the elements of ϵ^{ijk} are invariant under a general translation or rotation of the coordinate axes. The totally antisymmetric third-rank tensor is used to define the cross product of two 3-vectors,

$$(\mathbf{a}\times\mathbf{b})^{i} = \epsilon^{ijk}\,a_j\,b_k, \tag{10.139}$$

and the curl of a 3-vector field,

$$(\nabla\times\mathbf{A})^{i} = \epsilon^{ijk}\,\frac{\partial A_k}{\partial x^j}. \tag{10.140}$$

The following two rules are often useful in deriving vector identities

$$\epsilon^{ijk}\,\epsilon_{iab} = \delta^{j}_{a}\,\delta^{k}_{b} - \delta^{j}_{b}\,\delta^{k}_{a}, \tag{10.141}$$

$$\epsilon^{ijk}\,\epsilon_{ijb} = 2\,\delta^{k}_{b}. \tag{10.142}$$

Up to now, we have restricted ourselves to three basic types of coordinate transformation: namely, translations, rotations, and standard Lorentz transformations. An arbitrary combination of these three transformations constitutes a general Lorentz transformation. Let us now extend our investigations to include a fourth type of transformation known as a *parity inversion*: *i.e.*, $x, y, z, \to -x, -y, -z$. A reflection is a combination of a parity inversion and a rotation. As is easily demonstrated, the Jacobian of a parity inversion is -1, unlike a translation, rotation, or standard Lorentz transformation, which all possess Jacobians of $+1$.

The prototype of all 3-vectors is the difference in coordinates between two points in space, $\mathbf{r}$. Likewise, the prototype of all 4-vectors is the difference in coordinates between two events in space-time, $R^{\mu} = (\mathbf{r}, c\,t)$. It is not difficult to appreciate that both of these objects are invariant under a parity transformation (in the sense that they correspond to the same geometric object before and after the transformation). It follows that any 3- or 4-tensor which is directly related to $\mathbf{r}$ and R^{μ}, respectively, is also invariant under a parity inversion. Such tensors include the distance between two points in 3-space, the interval between two points in space-time, 3-velocity, 3-acceleration, 4-velocity, 4-acceleration, and the metric tensor. Tensors which exhibit tensor behavior under translations, rotations, special Lorentz transformations, *and* are invariant under parity inversions, are termed *proper tensors*, or sometimes *polar tensors*. Since electric charge is clearly invariant under such transformations (*i.e.*, it is a proper scalar), it follows that 3-current and 4-current are proper vectors. It is also clear from Equation (10.120) that the scalar potential, the vector potential, and the potential 4-vector, are proper tensors.

It follows from Equation (10.137) that $\epsilon^{\alpha\beta\gamma\delta} \rightarrow -\epsilon^{\alpha\beta\gamma\delta}$ under a parity inversion. Tensors such as this, which exhibit tensor behavior under translations, rotations, and special Lorentz transformations, but are *not* invariant under parity inversions (in the sense that they correspond to different geometric objects before and after the transformation), are called *pseudo-tensors*, or sometimes *axial tensors*. Equations (10.139) and (10.140) imply that the cross product of two proper vectors is a pseudo-vector, and the curl of a proper vector field is a pseudo-vector field.

One particularly simple way of performing a parity transformation is to exchange positive and negative numbers on the three Cartesian axes. A proper vector is unaffected by such a procedure (*i.e.*, its magnitude and direction are the same before and after). On the other hand, a pseudo-vector ends up pointing in the opposite direction after the axes are renumbered.

What is the fundamental difference between proper tensors and pseudo-tensors? The answer is that all pseudo-tensors are defined according to a *handedness convention*. For instance, the cross product between two vectors is conventionally defined according to a right-hand rule. The only reason for this is that the majority of human beings are right-handed. Presumably, if the opposite were true then cross products, *etc.*, would be defined according to a left-hand rule, and would, therefore, take minus their conventional values. The totally antisymmetric tensor is the prototype pseudo-tensor, and is, of course, conventionally defined with respect to a right-handed spatial coordinate system. A parity inversion converts left into right, and *vice versa*, and, thereby, effectively swaps left- and right-handed conventions.

The use of conventions in Physics is perfectly acceptable provided that we recognize that they are conventions, and are *consistent* in our use of them. It follows that laws of Physics cannot contain mixtures of tensors and pseudo-tensors, otherwise they would depend on our choice of handedness convention.[1]

Let us now consider electric and magnetic fields. We know that

$$\mathbf{E} = -\nabla\phi - \frac{\partial \mathbf{A}}{\partial t}, \tag{10.143}$$

$$\mathbf{B} = \nabla \times \mathbf{A}. \tag{10.144}$$

We have already seen that the scalar and the vector potential are proper scalars and vectors, respectively. It follows that **E** is a proper vector, but that **B** is a pseudo-vector (since it is the curl of a proper vector). In order to fully appreciate the difference between electric and magnetic fields, let us consider a thought experiment first proposed by Richard Feynman. Suppose that we are in radio contact with a race of aliens, and are trying to explain to them our system of Physics. Suppose, further, that the aliens live sufficiently far away from us that there are no common objects which we can both see. The question is this: could we unambiguously explain to these aliens our concepts of electric and magnetic fields? We could certainly explain electric and magnetic lines of force. The former are the paths of charged particles (assuming that the particles are subject only to electric fields), and the latter can be mapped out using small test magnets. We could also explain how we put arrows on electric lines of force to convert them into electric field-lines: the arrows run from positive charges (*i.e.*, charges with the same sign as atomic nuclei) to negative charges. This explanation is unambiguous provided that our aliens live in a matter- (rather than an antimatter) dominated part of the Universe. But, could we explain how we put arrows on magnetic lines of force in order to convert them into magnetic field-lines? The answer is, no. By definition, magnetic field-lines emerge from the North poles of permanent magnets and converge on the corresponding South poles. The definition of the North pole of a magnet is simply that it possesses the same magnetic polarity as the South (geographic) pole of the Earth. This is obviously a convention. In fact, we could redefine magnetic field-lines to run from the South poles to the North poles of magnets without significantly altering our laws of Physics (we would just have to replace **B** by −**B** in all our equations). In a parity-inverted Universe, a North

[1] Here, we are assuming that the laws of Physics do not possess an intrinsic handedness. This is certainly the case for Mechanics and Electromagnetism. However, the weak interaction *does* possess an intrinsic handedness: *i.e.*, it is fundamentally different in a parity-inverted Universe. So, the equations governing the weak interaction do actually contain mixtures of tensors and pseudo-tensors.

pole becomes a South pole, and *vice versa*, so it is hardly surprising that $\mathbf{B} \to -\mathbf{B}$.

10.15 THE ELECTROMAGNETIC FIELD TENSOR

Let us now investigate whether we can write the components of the electric and magnetic fields as the components of some *proper* 4-tensor. There is an obvious problem here. How can we identify the components of the magnetic field, which is a pseudo-vector, with any of the components of a proper 4-tensor? The former components transform differently under parity inversion than the latter components. Consider a proper 3-tensor whose covariant components are written B_{ik}, and which is antisymmetric:

$$B_{ij} = -B_{ji}. \tag{10.145}$$

This immediately implies that all of the diagonal components of the tensor are zero. In fact, there are only three independent non-zero components of such a tensor. Could we, perhaps, use these components to represent the components of a pseudo-3-vector? Let us write

$$B^i = \frac{1}{2}\,\epsilon^{ijk}\,B_{jk}. \tag{10.146}$$

It is clear that B^i transforms as a contravariant pseudo-3-vector. It is easily seen that

$$B^{ij} = B_{ij} = \begin{pmatrix} 0 & B_z & -B_y \\ -B_z & 0 & B_x \\ B_y & -B_x & 0 \end{pmatrix}, \tag{10.147}$$

where $B^1 = B_1 \equiv B_x$, *etc.* In this manner, we can actually write the components of a pseudo-3-vector as the components of an antisymmetric proper 3-tensor. In particular, we can write the components of the magnetic field $\mathbf{B}$ in terms of an antisymmetric proper magnetic field 3-tensor which we shall denote B_{ij}.

Let us now examine Equations (10.143) and (10.144) more carefully. Recall that $\Phi_\mu = (-c\,\mathbf{A}, \phi)$ and $\partial_\mu = (\nabla, c^{-1}\partial/\partial t)$. It follows that we can write Equation (10.143) in the form

$$E_i = -\partial_i\Phi_4 + \partial_4\Phi_i. \tag{10.148}$$

Likewise, Equation (10.144) can be written

$$c\,B^i = \frac{1}{2}\,\epsilon^{ijk}\,c\,B_{jk} = -\epsilon^{ijk}\,\partial_j\Phi_k. \tag{10.149}$$

Let us multiply this expression by ϵ_{iab}, making use of the identity

$$\epsilon_{iab}\,\epsilon^{ijk} = \delta^j_a\,\delta^k_b - \delta^j_b\,\delta^k_a. \tag{10.150}$$

We obtain

$$\frac{c}{2}\,(B_{ab} - B_{ba}) = -\partial_a\Phi_b + \partial_b\Phi_a, \tag{10.151}$$

or

$$c\,B_{ij} = -\partial_i\Phi_j + \partial_j\Phi_i, \tag{10.152}$$

since $B_{ij} = -B_{ji}$.

Let us define a proper 4-tensor whose covariant components are given by

$$F_{\mu\nu} = \partial_\mu\Phi_\nu - \partial_\nu\Phi_\mu. \tag{10.153}$$

It is clear that this tensor is antisymmetric:

$$F_{\mu\nu} = -F_{\nu\mu}. \tag{10.154}$$

This implies that the tensor only possesses six independent non-zero components. Maybe it can be used to specify the components of **E** and **B**?

Equations (10.148) and (10.153) yield

$$F_{4i} = \partial_4\Phi_i - \partial_i\Phi_4 = E_i. \tag{10.155}$$

Likewise, Equations (10.152) and (10.153) imply that

$$F_{ij} = \partial_i\Phi_j - \partial_j\Phi_i = -c\,B_{ij}. \tag{10.156}$$

Thus,

$$F_{i4} = -F_{4i} = -E_i, \tag{10.157}$$

$$F_{ij} = -F_{ji} = -c\,B_{ij}. \tag{10.158}$$

In other words, the completely space-like components of the tensor specify the components of the magnetic field, whereas the hybrid space and time-like components specify the components of the electric field. The covariant components of the tensor can be written

$$F_{\mu\nu} = \begin{pmatrix} 0 & -c\,B_z & +c\,B_y & -E_x \\ +c\,B_z & 0 & -c\,B_x & -E_y \\ -c\,B_y & +c\,B_x & 0 & -E_z \\ +E_x & +E_y & +E_z & 0 \end{pmatrix}. \tag{10.159}$$

Not surprisingly, $F_{\mu\nu}$ is usually called the *electromagnetic field tensor*. The above expression, which appears in all standard textbooks, is very misleading. Taken at face value, it is simply wrong! We cannot form a proper 4-tensor from the components of a proper 3-vector and a pseudo-3-vector. The expression only makes sense if we interpret B_x (say) as representing the component B_{23} of the proper magnetic field 3-tensor B_{ij}

The contravariant components of the electromagnetic field tensor are given by

$$F^{i4} = -F^{4i} = +E^i, \tag{10.160}$$

$$F^{ij} = -F^{ji} = -c\,B^{ij}, \tag{10.161}$$

or

$$F^{\mu\nu} = \begin{pmatrix} 0 & -c\,B_z & +c\,B_y & +E_x \\ +c\,B_z & 0 & -c\,B_x & +E_y \\ -c\,B_y & +c\,B_x & 0 & +E_z \\ -E_x & -E_y & -E_z & 0 \end{pmatrix}. \tag{10.162}$$

Let us now consider two of Maxwell's equations:

$$\nabla\cdot\mathbf{E} = \frac{\rho}{\epsilon_0}, \tag{10.163}$$

$$\nabla\times\mathbf{B} = \mu_0\left(\mathbf{j} + \epsilon_0\,\frac{\partial\mathbf{E}}{\partial t}\right). \tag{10.164}$$

Recall that the 4-current is defined $J^\mu = (\mathbf{j}, \rho\,c)$. The first of these equations can be written

$$\partial_i E^i = \partial_i F^{i4} + \partial_4 F^{44} = \frac{J^4}{c\,\epsilon_0}. \tag{10.165}$$

since $F^{44} = 0$. The second of these equations takes the form

$$\epsilon^{ijk}\,\partial_j(c\,B_k) - \partial_4 E^i = \epsilon^{ijk}\,\partial_j(1/2\,\epsilon_{kab}\,c\,B^{ab}) + \partial_4 F^{4i} = \frac{J^i}{c\,\epsilon_0}. \tag{10.166}$$

Making use of Equation (10.150), the above expression reduces to

$$\frac{1}{2}\,\partial_j(c\,B^{ij} - c\,B^{ji}) + \partial_4 F^{4i} = \partial_j F^{ji} + \partial_4 F^{4i} = \frac{J^i}{c\,\epsilon_0}. \qquad (10.167)$$

Equations (10.165) and (10.167) can be combined to give

$$\partial_\mu F^{\mu\nu} = \frac{J^\nu}{c\,\epsilon_0}. \qquad (10.168)$$

This equation is consistent with the equation of charge continuity, $\partial_\mu J^\mu = 0$, because of the antisymmetry of the electromagnetic field tensor.

10.16 THE DUAL ELECTROMAGNETIC FIELD TENSOR

We have seen that it is possible to write the components of the electric and magnetic fields as the components of a proper 4-tensor. Is it also possible to write the components of these fields as the components of some *pseudo*-4-tensor? It is obvious that we cannot identify the components of the proper 3-vector **E** with any of the components of a pseudo-tensor. However, we can represent the components of **E** in terms of those of an antisymmetric pseudo-3-tensor E_{ij} by writing

$$E^i = \frac{1}{2}\,\epsilon^{ijk}\,E_{jk}. \qquad (10.169)$$

It is easily demonstrated that

$$E^{ij} = E_{ij} = \begin{pmatrix} 0 & E_z & -E_y \\ -E_z & 0 & E_x \\ E_y & -E_x & 0 \end{pmatrix}, \qquad (10.170)$$

in a right-handed coordinate system.

Consider the *dual electromagnetic field tensor*, $G^{\mu\nu}$, which is defined

$$G^{\mu\nu} = \frac{1}{2}\,\epsilon^{\mu\nu\alpha\beta}\,F_{\alpha\beta}. \qquad (10.171)$$

This tensor is clearly an antisymmetric pseudo-4-tensor. We have

$$G^{4i} = \frac{1}{2}\,\epsilon^{4ijk}\,F_{jk} = -\frac{1}{2}\,\epsilon^{ijk4}\,F_{jk} = \frac{1}{2}\,\epsilon^{ijk}\,c\,B_{jk} = c\,B^i, \qquad (10.172)$$

plus

$$G^{ij} = \frac{1}{2}\,(\epsilon^{ijk4}\,F_{k4} + \epsilon^{ij4k}\,F_{4k}) = \epsilon^{ijk}\,F_{k4}, \tag{10.173}$$

where use has been made of $F_{\mu\nu} = -F_{\nu\mu}$. The above expression yields

$$G^{ij} = -\epsilon^{ijk}\,E_k = -\frac{1}{2}\,\epsilon^{ijk}\epsilon_{kab}\,E^{ab} = -E^{ij}. \tag{10.174}$$

It follows that

$$G^{i4} = -G^{4i} = -c\,B^i, \tag{10.175}$$

$$G^{ij} = -G^{ji} = -E^{ij}, \tag{10.176}$$

or

$$G^{\mu\nu} = \begin{pmatrix} 0 & -E_z & +E_y & -c\,B_x \\ +E_z & 0 & -E_x & -c\,B_y \\ -E_y & +E_x & 0 & -c\,B_z \\ +c\,B_x & +c\,B_y & +c\,B_z & 0 \end{pmatrix}. \tag{10.177}$$

The above expression is, again, slightly misleading, since E_x stands for the component E^{23} of the pseudo-3-tensor E^{ij}, and not for an element of the proper 3-vector **E**. Of course, in this case, B_x really does represent the first element of the pseudo-3-vector **B**. Note that the elements of $G^{\mu\nu}$ are obtained from those of $F^{\mu\nu}$ by making the transformation $c\,B^{ij} \rightarrow E^{ij}$ and $E^i \rightarrow -c\,B^i$.

The covariant elements of the dual electromagnetic field tensor are given by

$$G_{i4} = -G_{4i} = +cB_i, \tag{10.178}$$

$$G_{ij} = -G_{ji} = -E_{ij}, \tag{10.179}$$

or

$$G_{\mu\nu} = \begin{pmatrix} 0 & -E_z & +E_y & +c\,B_x \\ +E_z & 0 & -E_x & +c\,B_y \\ -E_y & +E_x & 0 & +c\,B_z \\ -c\,B_x & -c\,B_y & -c\,B_z & 0 \end{pmatrix}. \tag{10.180}$$

The elements of $G_{\mu\nu}$ are obtained from those of $F_{\mu\nu}$ by making the transformation $c\,B_{ij} \rightarrow E_{ij}$ and $E_i \rightarrow -c\,B_i$.

Let us now consider the two Maxwell equations

$$\nabla\cdot\mathbf{B} = 0, \tag{10.181}$$

$$\nabla\times\mathbf{E} = -\frac{\partial\mathbf{B}}{\partial t}. \tag{10.182}$$

The first of these equations can be written

$$-\partial_i\,(c\,B^i) = \partial_i G^{i4} + \partial_4 G^{44} = 0, \tag{10.183}$$

since $G^{44} = 0$. The second equation takes the form

$$\epsilon^{ijk}\partial_j E_k = \epsilon^{ijk}\partial_j(1/2\,\epsilon_{kab}E^{ab}) = \partial_j E^{ij} = -\partial_4\,(c\,B^i), \tag{10.184}$$

or

$$\partial_j G^{ji} + \partial_4 G^{4i} = 0. \tag{10.185}$$

Equations (10.183) and (10.185) can be combined to give

$$\partial_\mu G^{\mu\nu} = 0. \tag{10.186}$$

Thus, we conclude that Maxwell's equations for the electromagnetic fields are equivalent to the following pair of 4-tensor equations:

$$\partial_\mu F^{\mu\nu} = \frac{J^\nu}{c\,\epsilon_0}, \tag{10.187}$$

$$\partial_\mu G^{\mu\nu} = 0. \tag{10.188}$$

It is obvious from the form of these equations that the laws of electromagnetism are invariant under translations, rotations, special Lorentz transformations, parity inversions, or any combination of these transformations.

10.17 TRANSFORMATION OF FIELDS

The electromagnetic field tensor transforms according to the standard rule

$$F^{\mu'\nu'} = F^{\mu\nu}\,p^{\mu'}_{\mu}\,p^{\nu'}_{\nu}. \tag{10.189}$$

This easily yields the celebrated rules for transforming electromagnetic fields:

$$E'_{\parallel} = E_{\parallel}, \tag{10.190}$$

$$B'_{\parallel} = B_{\parallel}, \tag{10.191}$$

$$\mathbf{E}'_{\perp} = \gamma\,(\mathbf{E}_{\perp} + \mathbf{v}\times\mathbf{B}), \tag{10.192}$$

$$\mathbf{B}'_{\perp} = \gamma\,(\mathbf{B}_{\perp} - \mathbf{v}\times\mathbf{E}/c^2), \tag{10.193}$$

where $\mathbf{v}$ is the relative velocity between the primed and unprimed frames, and the perpendicular and parallel directions are, respectively, perpendicular and parallel to $\mathbf{v}$.

At this stage, we may conveniently note two important invariants of the electromagnetic field. They are

$$\frac{1}{2}\,F_{\mu\nu}\,F^{\mu\nu} = c^2\,B^2 - E^2, \tag{10.194}$$

and

$$\frac{1}{4}\,G_{\mu\nu}\,F^{\mu\nu} = c\,\mathbf{E}\cdot\mathbf{B}. \tag{10.195}$$

The first of these quantities is a proper-scalar, and the second a pseudo-scalar.

10.18 POTENTIAL DUE TO A MOVING CHARGE

Suppose that a particle carrying a charge e moves with *uniform* velocity $\mathbf{u}$ through a frame S. Let us evaluate the vector potential, $\mathbf{A}$, and the scalar potential, ϕ, due to this charge at a given event P in S.

Let us choose coordinates in S so that $P = (0,0,0,0)$ and $\mathbf{u} = (u,0,0)$. Let S' be that frame in the standard configuration with respect to S in which the charge is (permanently) at rest at (say) the point (x', y', z'). In S', the potential at P is the usual potential due to a stationary charge,

$$\mathbf{A}' = \mathbf{0}, \tag{10.196}$$

$$\phi' = \frac{e}{4\pi\epsilon_0\,r'}, \tag{10.197}$$

where $r' = \sqrt{x'^2 + y'^2 + z'^2}$. Let us now transform these equations directly into the frame S. Since $A^\mu = (c\,\mathbf{A}, \phi)$ is a contravariant 4-vector, its components transform according to the standard rules

(10.75)–(10.78). Thus,

$$c\,A_1 = \gamma\left(c\,A_1' + \frac{u}{c}\,\phi'\right) = \frac{\gamma\,u\,e}{4\pi\,\epsilon_0\,c\,r'}, \tag{10.198}$$

$$c\,A_2 = c\,A_2' = 0, \tag{10.199}$$

$$c\,A_3 = c\,A_3' = 0, \tag{10.200}$$

$$\phi = \gamma\left(\phi' + \frac{u}{c}\,c\,A_1'\right) = \frac{\gamma\,e}{4\pi\epsilon_0\,r'}, \tag{10.201}$$

since $\beta = -u/c$ in this case. It remains to express the quantity r' in terms of quantities measured in S. The most physically meaningful way of doing this is to express r' in terms of *retarded* values in S. Consider the retarded event at the charge for which, by definition, $r' = -c\,t'$ and $r = -c\,t$. Using the standard Lorentz transformation, (10.24)–(10.27), we find that

$$r' = -c\,t' = -c\,\gamma\,(t - u\,x/c^2) = r\,\gamma\,(1 + u_r/c), \tag{10.202}$$

where $u_r = u\,x/r = \mathbf{r}\cdot\mathbf{u}/r$ denotes the radial velocity of the change in S. We can now rewrite Equations (10.198)–(10.201) in the form

$$\mathbf{A} = \frac{\mu_0\,e}{4\pi}\frac{[\mathbf{u}]}{[r + \mathbf{r}\cdot\mathbf{u}/c]}, \tag{10.203}$$

$$\phi = \frac{e}{4\pi\epsilon_0}\frac{1}{[r + \mathbf{r}\cdot\mathbf{u}/c]}, \tag{10.204}$$

where the square brackets, as usual, indicate that the enclosed quantities must be retarded. For a uniformly moving charge, the retardation of $\mathbf{u}$ is, of course, superfluous. However, since

$$\Phi^\mu = \frac{1}{4\pi\epsilon_0\,c}\int\frac{[J^\mu]}{r}\,dV, \tag{10.205}$$

it is clear that the potentials depend only on the (retarded) velocity of the charge, and not on its acceleration. Consequently, the expressions (10.203) and (10.204) give the correct potentials for an *arbitrarily* moving charge. They are known as the *Liénard-Wiechert potentials.*

10.19 FIELD DUE TO A MOVING CHARGE

Although the fields generated by a uniformly moving charge can be calculated from the expressions (10.203) and (10.204) for the potentials, it is simpler to calculate them from first principles.

Let a charge e, whose position vector at time $t = 0$ is $\mathbf{r}$, move with uniform velocity $\mathbf{u}$ in a frame S whose x-axis has been chosen in the direction of $\mathbf{u}$. We require to find the field strengths $\mathbf{E}$ and $\mathbf{B}$ at the event $P = (0, 0, 0, 0)$. Let S' be that frame in standard configuration with S in which the charge is permanently at rest. In S', the field is given by

$$\mathbf{B}' = \mathbf{0}, \tag{10.206}$$

$$\mathbf{E}' = -\frac{e}{4\pi\epsilon_0}\frac{\mathbf{r}'}{r'^3}. \tag{10.207}$$

This field must now be transformed into the frame S. The direct method, using Equations (10.190)–(10.193), is somewhat simpler here, but we shall use a somewhat indirect method because of its intrinsic interest.

In order to express Equations (10.206) and (10.207) in tensor form, we need the electromagnetic field tensor $F^{\mu\nu}$ on the left-hand side, and the position 4-vector $R^\mu = (\mathbf{r}, c\,t)$ and the scalar $e/(4\pi\epsilon_0\, r'^3)$ on the right-hand side. (We regard r' as an invariant for all observers.) To get a vanishing magnetic field in S', we multiply on the right by the 4-velocity $U^\mu = \gamma(u)\,(\mathbf{u}, c)$, thus tentatively arriving at the equation

$$F^{\mu\nu} = \frac{e}{4\pi\epsilon_0\, c\, r'^3}\, U^\mu\, R^\nu. \tag{10.208}$$

Recall that $F^{4i} = -E^i$ and $F^{ij} = -c\, B^{ij}$. However, this equation cannot be correct, because the antisymmetric tensor $F^{\mu\nu}$ can only be equated to another antisymmetric tensor. Consequently, let us try

$$F^{\mu\nu} = \frac{e}{4\pi\epsilon_0\, c\, r'^3}\,(U^\mu\, R^\nu - U^\nu\, R^\mu). \tag{10.209}$$

This is found to give the correct field at P in S', as long as R^μ refers to any event whatsoever at the charge. It only remains to interpret (10.209) in S. It is convenient to choose for R^μ that event at the charge at which $t = 0$ (not the retarded event). Thus,

$$F^{jk} = -c\, B^{jk} = \frac{e}{4\pi\epsilon_0\, c\, r'^3}\,\gamma(u)\,(u^j\, r^k - u^k\, r^j), \tag{10.210}$$

giving

$$B_i = \frac{1}{2}\,\epsilon_{ijk} B^{jk} = -\frac{\mu_0\, e}{4\pi\, r'^3}\,\gamma(u)\,\epsilon_{ijk}\, u^j\, r^k,. \tag{10.211}$$

or

$$\mathbf{B} = -\frac{\mu_0\, e\,\gamma}{4\pi\, r'^3}\,\mathbf{u}\times\mathbf{r}. \tag{10.212}$$

Likewise,

$$F^{4i} = -E^i = \frac{e\gamma}{4\pi\,\epsilon_0\, r'^3}\, r^i, \tag{10.213}$$

or

$$\mathbf{E} = -\frac{e\gamma}{4\pi\,\epsilon_0\, r'^3}\,\mathbf{r}. \tag{10.214}$$

Lastly, we must find an expression for r'^3 in terms of quantities measured in S at time $t = 0$. If t' is the corresponding time in S' at the charge, we have

$$r'^2 = r^2 + c^2 t'^2 = r^2 + \frac{\gamma^2 u^2 x^2}{c^2} = r^2\left(1 + \frac{\gamma^2 u_r^{\,2}}{c^2}\right). \tag{10.215}$$

Thus,

$$\mathbf{E} = -\frac{e}{4\pi\epsilon_0}\frac{\gamma}{r^3\,(1 + u_r^{\,2}\gamma^2/c^2)^{3/2}}\,\mathbf{r}, \tag{10.216}$$

$$\mathbf{B} = -\frac{\mu_0\, e}{4\pi}\frac{\gamma}{r^3\,(1 + u_r^{\,2}\gamma^2/c^2)^{3/2}}\,\mathbf{u}\times\mathbf{r} = \frac{1}{c^2}\,\mathbf{u}\times\mathbf{E}. \tag{10.217}$$

Note that **E** acts in line with the point which the charge occupies *at the instant of measurement*, despite the fact that, owing to the finite speed of propagation of all physical effects, the behavior of the charge during a finite period before that instant can no longer affect the measurement. Note also that, unlike Equations (10.203) and (10.204), the above expressions for the fields are not valid for an arbitrarily moving charge, nor can they be made valid by merely using retarded values. For whereas acceleration does not affect the potentials, it does affect the fields, which involve the derivatives of the potential.

For low velocities, $u/c \to 0$, Equations (10.216) and (10.217) reduce to the well-known Coulomb and Biot-Savart fields. However, at high velocities, $\gamma(u) \gg 1$, the fields exhibit some interesting behavior. The peak electric field, which occurs at the point of closest approach of the charge to the observation point, becomes equal to γ times its non-relativistic value. However, the duration of appreciable field strength at the point P is decreased. A measure of the time interval over which the field is appreciable is

$$\Delta t \sim \frac{b}{\gamma\, c}, \tag{10.218}$$

where b is the distance of closest approach (assuming $\gamma \gg 1$). As γ increases, the peak field increases in proportion, but its duration goes in the inverse proportion. The time integral of the field is independent of γ. As $\gamma \to \infty$, the observer at P sees electric and magnetic fields which are indistinguishable from the fields of a pulse of plane polarized radiation propagating in the x-direction. The direction of polarization is along the radius vector pointing toward the particle's actual position at the time of observation.

10.20 RELATIVISTIC PARTICLE DYNAMICS

Consider a particle which, in its instantaneous rest frame S_0, has mass m_0 and constant acceleration in the x-direction a_0. Let us transform to a frame S, in the standard configuration with respect to S_0, in which the particle's instantaneous velocity is u. What is the value of a, the particle's instantaneous x-acceleration, in S?

The easiest way in which to answer this question is to consider the acceleration 4-vector [see Equation (10.107)]

$$A^\mu = \gamma \left(\frac{d\gamma}{dt}\,\mathbf{u} + \gamma\,\mathbf{a}, c\,\frac{d\gamma}{dt} \right). \tag{10.219}$$

Using the standard transformation, (10.75)–(10.78), for 4-vectors, we obtain

$$a_0 = \gamma^3\, a, \tag{10.220}$$

$$\frac{d\gamma}{dt} = \frac{u\, a_0}{c^2}. \tag{10.221}$$

Equation (10.220) can be written

$$f = m_0\, \gamma^3\, \frac{du}{dt}, \tag{10.222}$$

where $f = m_0\, a_0$ is the constant force (in the x-direction) acting on the particle in S_0.

Equation (10.222) is equivalent to

$$f = \frac{d(m\, u)}{dt}, \tag{10.223}$$

where

$$m = \gamma\, m_0. \tag{10.224}$$

Thus, we can account for the ever-decreasing acceleration of a particle subject to a constant force [see Equation (10.220)] by supposing that the inertial mass of the particle increases with its velocity according to the rule (10.224). Henceforth, m_0 is termed the *rest mass*, and m the *inertial mass*.

The rate of increase of the particle's energy E satisfies

$$\frac{dE}{dt} = f\,u = m_0\,\gamma^3\,u\,\frac{du}{dt}. \tag{10.225}$$

This equation can be written

$$\frac{dE}{dt} = \frac{d(m\,c^2)}{dt}, \tag{10.226}$$

which can be integrated to yield Einstein's famous formula

$$E = m\,c^2. \tag{10.227}$$

The 3-momentum of a particle is defined

$$\mathbf{p} = m\,\mathbf{u}, \tag{10.228}$$

where $\mathbf{u}$ is its 3-velocity. Thus, by analogy with Equation (10.223), Newton's law of motion can be written

$$\mathbf{f} = \frac{d\mathbf{p}}{dt}, \tag{10.229}$$

where $\mathbf{f}$ is the 3-force acting on the particle.

The 4-momentum of a particle is defined

$$P^\mu = m_0\,U^\mu = \gamma\,m_0\,(\mathbf{u}, c) = (\mathbf{p}, E/c), \tag{10.230}$$

where U^μ is its 4-velocity. The 4-force acting on the particle obeys

$$\mathcal{F}^\mu = \frac{dP^\mu}{d\tau} = m_0\,A^\mu, \tag{10.231}$$

where A^μ is its 4-acceleration. It is easily demonstrated that

$$\mathcal{F}^\mu = \gamma\left(\mathbf{f}, c\,\frac{dm}{dt}\right) = \gamma\left(\mathbf{f}, \frac{\mathbf{f}\cdot\mathbf{u}}{c}\right), \tag{10.232}$$

since

$$\frac{dE}{dt} = \mathbf{f}\cdot\mathbf{u}. \tag{10.233}$$

10.21 FORCE ON A MOVING CHARGE

The electromagnetic 3-force acting on a charge e moving with 3-velocity **u** is given by the well-known formula

$$\mathbf{f} = e\,(\mathbf{E} + \mathbf{u} \times \mathbf{B}). \tag{10.234}$$

When written in component form this expression becomes

$$f_i = e\,(E_i + \epsilon_{ijk}\,u^j\,B^k), \tag{10.235}$$

or

$$f_i = e\,(E_i + B_{ij}\,u^j), \tag{10.236}$$

where use has been made of Equation (10.147).

Recall that the components of the **E** and **B** fields can be written in terms of an antisymmetric electromagnetic field tensor $F_{\mu\nu}$ via

$$F_{i4} = -F_{4i} = -E_i, \tag{10.237}$$

$$F_{ij} = -F_{ji} = -c\,B_{ij}. \tag{10.238}$$

Equation (10.236) can be written

$$f_i = -\frac{e}{\gamma\,c}\,(F_{i4}\,U^4 + F_{ij}\,U^j), \tag{10.239}$$

where $U^\mu = \gamma\,(\mathbf{u}, c)$ is the particle's 4-velocity. It is easily demonstrated that

$$\frac{\mathbf{f}\cdot\mathbf{u}}{c} = \frac{e}{c}\,\mathbf{E}\cdot\mathbf{u} = \frac{e}{c}\,E_i\,u^i = \frac{e}{\gamma\,c}(F_{4i}\,U^i + F_{44}\,U^4). \tag{10.240}$$

Thus, the 4-force acting on the particle,

$$\mathcal{F}_\mu = \gamma\left(-\mathbf{f}, \frac{\mathbf{f}\cdot\mathbf{u}}{c}\right), \tag{10.241}$$

can be written in the form

$$\mathcal{F}_\mu = \frac{e}{c}\,F_{\mu\nu}\,U^\nu. \tag{10.242}$$

The skew symmetry of the electromagnetic field tensor ensures that

$$\mathcal{F}_\mu\,U^\mu = \frac{e}{c}\,F_{\mu\nu}\,U^\mu\,U^\nu = 0. \tag{10.243}$$

This is an important result, since it ensures that electromagnetic fields do not change the rest mass of charged particles. In order to appreciate this, let us assume that the rest mass m_0 is not a constant. Since

$$\mathcal{F}_\mu = \frac{d(m_0\, U_\mu)}{d\tau} = m_0\, A_\mu + \frac{dm_0}{d\tau}\, U_\mu, \tag{10.244}$$

we can use the standard results $U_\mu\, U^\mu = c^2$ and $A_\mu\, U^\mu = 0$ to give

$$\mathcal{F}_\mu\, U^\mu = c^2\, \frac{dm_0}{d\tau}. \tag{10.245}$$

Thus, if rest mass is to remain an invariant, it is imperative that all laws of Physics predict 4-forces acting on particles which are orthogonal to the particles' 4-velocities. The laws of electromagnetism pass this test.

10.22 THE ELECTROMAGNETIC ENERGY TENSOR

Consider a continuous volume distribution of charged matter in the presence of an electromagnetic field. Let there be n_0 particles per unit proper volume (unit volume determined in the local rest frame), each carrying a charge e. Consider an inertial frame in which the 3-velocity field of the particles is $\mathbf{u}$. The number density of the particles in this frame is $n = \gamma(u)\, n_0$. The charge density and the 3-current due to the particles are $\rho = e\, n$ and $\mathbf{j} = e\, n\, \mathbf{u}$, respectively. Multiplying Equation (10.242) by the proper number density of particles, n_0, we obtain an expression

$$f_\mu = c^{-1}\, F_{\mu\nu}\, J^\nu \tag{10.246}$$

for the 4-force f_μ acting on unit proper volume of the distribution due to the ambient electromagnetic fields. Here, we have made use of the definition $J^\mu = e\, n_0\, U^\mu$. It is easily demonstrated, using some of the results obtained in the previous section, that

$$f^\mu = \left(\rho\, \mathbf{E} + \mathbf{j}\times\mathbf{B}, \frac{\mathbf{E}\cdot\mathbf{j}}{c}\right). \tag{10.247}$$

The above expression remains valid when there are many charge species (*e.g.*, electrons and ions) possessing different number density and 3-velocity fields. The 4-vector f^μ is usually called the *Lorentz force density*.

We know that Maxwell's equations reduce to

$$\partial_\mu F^{\mu\nu} = \frac{J^\nu}{c\, \epsilon_0}, \tag{10.248}$$

$$\partial_\mu G^{\mu\nu} = 0, \tag{10.249}$$

where $F^{\mu\nu}$ is the electromagnetic field tensor, and $G^{\mu\nu}$ is its dual. As is easily verified, Equation (10.249) can also be written in the form

$$\partial_\mu F_{\nu\sigma} + \partial_\nu F_{\sigma\mu} + \partial_\sigma F_{\mu\nu} = 0. \tag{10.250}$$

Equations (10.246) and (10.248) can be combined to give

$$f_\nu = \epsilon_0 \, F_{\nu\sigma} \, \partial_\mu F^{\mu\sigma}. \tag{10.251}$$

This expression can also be written

$$f_\nu = \epsilon_0 \left[\partial_\mu (F^{\mu\sigma} \, F_{\nu\sigma}) - F^{\mu\sigma} \, \partial_\mu F_{\nu\sigma}\right]. \tag{10.252}$$

Now,

$$F^{\mu\sigma} \, \partial_\mu F_{\nu\sigma} = \frac{1}{2} \, F^{\mu\sigma} (\partial_\mu F_{\nu\sigma} + \partial_\sigma F_{\mu\nu}), \tag{10.253}$$

where use has been made of the antisymmetry of the electromagnetic field tensor. It follows from Equation (10.250) that

$$F^{\mu\sigma} \, \partial_\mu F_{\nu\sigma} = -\frac{1}{2} \, F^{\mu\sigma} \, \partial_\nu F_{\sigma\mu} = \frac{1}{4} \, \partial_\nu (F^{\mu\sigma} \, F_{\mu\sigma}). \tag{10.254}$$

Thus,

$$f_\nu = \epsilon_0 \left[\partial_\mu (F^{\mu\sigma} \, F_{\nu\sigma}) - \frac{1}{4} \, \partial_\nu (F^{\mu\sigma} \, F_{\mu\sigma})\right]. \tag{10.255}$$

The above expression can also be written

$$f_\nu = -\partial_\mu T^\mu{}_\nu, \tag{10.256}$$

where

$$T^\mu{}_\nu = \epsilon_0 \left[F^{\mu\sigma} \, F_{\sigma\nu} + \frac{1}{4} \, \delta^\mu_\nu \, (F^{\rho\sigma} \, F_{\rho\sigma})\right] \tag{10.257}$$

is called the *electromagnetic energy tensor*. Note that $T^\mu{}_\nu$ is a proper 4-tensor. It follows from Equations (10.159), (10.162), and (10.194) that

$$T^i{}_j = \epsilon_0 \, E^i \, E_j + \frac{B^i \, B_j}{\mu_0} - \delta^i_j \, \frac{1}{2} \left(\epsilon_0 \, E^k \, E_k + \frac{B^k \, B_k}{\mu_0}\right), \tag{10.258}$$

$$T^{i}{}_{4} = -T^{4}{}_{i} = \frac{\epsilon^{ijk}\,E_j\,B_k}{\mu_0\,c}, \tag{10.259}$$

$$T^{4}_{4} = \frac{1}{2}\left(\epsilon_0\,E^k\,E_k + \frac{B^k\,B_k}{\mu_0}\right). \tag{10.260}$$

Equation (10.256) can also be written

$$f^{\nu} = -\partial_{\mu}T^{\mu\nu}, \tag{10.261}$$

where $T^{\mu\nu}$ is a symmetric tensor whose elements are

$$T^{ij} = -\,\epsilon_0\,E^i\,E^j - \frac{B^i\,B^j}{\mu_0} + \delta^{ij}\,\frac{1}{2}\left(\epsilon_0\,E^2 + \frac{B^2}{\mu_0}\right), \tag{10.262}$$

$$T^{i4} = T^{4i} = \frac{(\mathbf{E}\times\mathbf{B})^i}{\mu_0\,c}, \tag{10.263}$$

$$T^{44} = \frac{1}{2}\left(\epsilon_0\,E^2 + \frac{B^2}{\mu_0}\right). \tag{10.264}$$

Consider the time-like component of Equation (10.261). It follows from Equation (10.247) that

$$\frac{\mathbf{E}\cdot\mathbf{j}}{c} = -\partial_i T^{i4} - \partial_4 T^{44}. \tag{10.265}$$

This equation can be rearranged to give

$$\frac{\partial U}{\partial t} + \nabla\cdot\mathbf{u} = -\mathbf{E}\cdot\mathbf{j}, \tag{10.266}$$

where $U = T^{44}$ and $\mathbf{u}^i = c\,T^{i4}$, so that

$$U = \frac{\epsilon_0\,E^2}{2} + \frac{B^2}{2\,\mu_0}, \tag{10.267}$$

and

$$\mathbf{u} = \frac{\mathbf{E}\times\mathbf{B}}{\mu_0}. \tag{10.268}$$

The right-hand side of Equation (10.266) represents the rate per unit volume at which energy is transferred from the electromagnetic field to charged particles. It is clear, therefore, that Equation (10.266) is an *energy conservation equation* for the electromagnetic field (see Section 8.2). The proper 3-scalar U can be identified as the energy density of the electromagnetic field, whereas the proper 3-vector $\mathbf{u}$ is the energy flux due to the electromagnetic field: *i.e.*, the *Poynting flux*.

Consider the space-like components of Equation (10.261). It is easily demonstrated that these reduce to

$$\frac{\partial \mathbf{g}}{\partial t} + \nabla \cdot \mathbf{G} = -\rho\, \mathbf{E} - \mathbf{j} \times \mathbf{B}, \qquad (10.269)$$

where $G^{ij} = T^{ij}$ and $g^i = T^{4i}/c$, or

$$G^{ij} = -\epsilon_0\, E^i\, E^j - \frac{B^i\, B^j}{\mu_0} + \delta^{ij}\, \frac{1}{2}\left(\epsilon_0\, E^2 + \frac{B^2}{\mu_0}\right), \qquad (10.270)$$

and

$$\mathbf{g} = \frac{\mathbf{u}}{c^2} = \epsilon_0\, \mathbf{E} \times \mathbf{B}. \qquad (10.271)$$

Equation (10.269) is basically a *momentum conservation equation* for the electromagnetic field (see Section 8.4). The right-hand side represents the rate per unit volume at which momentum is transferred from the electromagnetic field to charged particles. The symmetric proper 3-tensor G^{ij} specifies the flux of electromagnetic momentum parallel to the ith axis crossing a surface normal to the jth axis. The proper 3-vector $\mathbf{g}$ represents the momentum density of the electromagnetic field. It is clear that the energy conservation law (10.266) and the momentum conservation law (10.269) can be combined together to give the relativistically invariant energy-momentum conservation law (10.261).

10.23 ACCELERATED CHARGES

Let us calculate the electric and magnetic fields observed at position x^i and time t due to a charge e whose *retarded* position and time are $x^{i'}$ and t', respectively. From now on (x^i, t) is termed the *field point* and $(x^{i'}, t')$ is termed the *source point*. It is assumed that we are given the retarded position of the charge as a function of its retarded time: *i.e.*, $x^{i'}(t')$. The retarded velocity and acceleration of the charge are

$$u^i = \frac{dx^{i'}}{dt'}, \qquad (10.272)$$

and

$$\dot{u}^i = \frac{du^i}{dt'}, \qquad (10.273)$$

respectively. The radius vector $\mathbf{r}$ is defined to extend *from* the retarded position of the charge *to* the field point, so that $r^i = x^i - x^{i\prime}$. (Note that this is the *opposite* convention to that adopted in Sections 10.18 and 10.19). It follows that

$$\frac{d\mathbf{r}}{dt'} = -\mathbf{u}. \tag{10.274}$$

The field and the source point variables are connected by the retardation condition

$$r(x^i, x^{i\prime}) = \left[(x^i - x^{i\prime})\,(x_i - x_i{}')\right]^{1/2} = c\,(t - t'). \tag{10.275}$$

The potentials generated by the charge are given by the Liénard-Wiechert formulae,

$$\mathbf{A}(x^i, t) = \frac{\mu_0\, e}{4\pi}\frac{\mathbf{u}}{s}, \tag{10.276}$$

$$\phi(x^i, t) = \frac{e}{4\pi\epsilon_0}\frac{1}{s}, \tag{10.277}$$

where $s = r - \mathbf{r}\cdot\mathbf{u}/c$ is a function both of the field point and the source point variables. Recall that the Liénard-Wiechert potentials are valid for accelerating, as well as uniformly moving, charges.

The fields $\mathbf{E}$ and $\mathbf{B}$ are derived from the potentials in the usual manner:

$$\mathbf{E} = -\nabla\phi - \frac{\partial \mathbf{A}}{\partial t}, \tag{10.278}$$

$$\mathbf{B} = \nabla \times \mathbf{A}. \tag{10.279}$$

However, the components of the gradient operator ∇ are partial derivatives at constant time, t, and *not* at constant time, t'. Partial differentiation with respect to the x^i compares the potentials at neighboring points at the same time, but these potential signals originate from the charge at different retarded times. Similarly, the partial derivative with respect to t implies constant x^i, and, hence, refers to the comparison of the potentials at a given field point over an interval of time during which the retarded coordinates of the source have changed. Since we only know the time variation of the particle's retarded position with respect to t' we must transform $\partial/\partial t|_{x^i}$ and $\partial/\partial x^i|_t$ to expressions involving $\partial/\partial t'|_{x^i}$ and $\partial/\partial x^i|_{t'}$.

Now, since $x^{i'}$ is assumed to be given as a function of t', we have

$$r(x^i, x^{i'}(t')) \equiv r(x^i, t') = c\,(t - t'), \tag{10.280}$$

which is a functional relationship between x^i, t, and t'. Note that

$$\left(\frac{\partial r}{\partial t'}\right)_{x^i} = -\frac{\mathbf{r}\cdot\mathbf{u}}{r}. \tag{10.281}$$

It follows that

$$\frac{\partial r}{\partial t} = c\left(1 - \frac{\partial t'}{\partial t}\right) = \frac{\partial r}{\partial t'}\frac{\partial t'}{\partial t} = -\frac{\mathbf{r}\cdot\mathbf{u}}{r}\frac{\partial t'}{\partial t}, \tag{10.282}$$

where all differentiation is at constant x^i. Thus,

$$\frac{\partial t'}{\partial t} = \frac{1}{1 - \mathbf{r}\cdot\mathbf{u}/r\,c} = \frac{r}{s}, \tag{10.283}$$

giving

$$\frac{\partial}{\partial t} = \frac{r}{s}\frac{\partial}{\partial t'}. \tag{10.284}$$

Similarly,

$$\nabla r = -c\,\nabla t' = \nabla' r + \frac{\partial r}{\partial t'}\nabla t' = \frac{\mathbf{r}}{r} - \frac{\mathbf{r}\cdot\mathbf{u}}{r}\nabla t', \tag{10.285}$$

where ∇' denotes differentiation with respect to x^i at constant t'. It follows that

$$\nabla t' = -\frac{\mathbf{r}}{s\,c}, \tag{10.286}$$

so that

$$\nabla = \nabla' - \frac{\mathbf{r}}{s\,c}\frac{\partial}{\partial t'}. \tag{10.287}$$

Equation (10.278) yields

$$\frac{4\pi\epsilon_0}{e}\,\mathbf{E} = \frac{\nabla s}{s^2} - \frac{\partial}{\partial t}\frac{\mathbf{u}}{s\,c^2}, \tag{10.288}$$

or

$$\frac{4\pi\epsilon_0}{e}\,\mathbf{E} = \frac{\nabla' s}{s^2} - \frac{\mathbf{r}}{s^3\,c}\frac{\partial s}{\partial t'} - \frac{r}{s^2\,c^2}\,\dot{\mathbf{u}} + \frac{r\,\mathbf{u}}{s^3\,c^2}\frac{\partial s}{\partial t'}. \tag{10.289}$$

However,

$$\nabla' s = \frac{\mathbf{r}}{r} - \frac{\mathbf{u}}{c}, \tag{10.290}$$

and

$$\frac{\partial s}{\partial t'} = \frac{\partial r}{\partial t'} - \frac{\mathbf{r}\cdot\dot{\mathbf{u}}}{c} + \frac{\mathbf{u}\cdot\mathbf{u}}{c} = -\frac{\mathbf{r}\cdot\mathbf{u}}{r} - \frac{\mathbf{r}\cdot\dot{\mathbf{u}}}{c} + \frac{u^2}{c}. \tag{10.291}$$

Thus,

$$\frac{4\pi\epsilon_0}{e}\mathbf{E} = \frac{1}{s^2\,r}\left(\mathbf{r} - \frac{r\,\mathbf{u}}{c}\right) + \frac{1}{s^3\,c}\left(\mathbf{r} - \frac{r\,\mathbf{u}}{c}\right)\left(\frac{\mathbf{r}\cdot\mathbf{u}}{r} - \frac{u^2}{c} + \frac{\mathbf{r}\cdot\dot{\mathbf{u}}}{c}\right) - \frac{r}{s^2\,c^2}\,\dot{\mathbf{u}}, \tag{10.292}$$

which reduces to

$$\frac{4\pi\epsilon_0}{e}\mathbf{E} = \frac{1}{s^3}\left(\mathbf{r} - \frac{r\,\mathbf{u}}{c}\right)\left(1 - \frac{u^2}{c^2}\right) + \frac{1}{s^3\,c^2}\left(\mathbf{r}\times\left[\left(\mathbf{r} - \frac{r\,\mathbf{u}}{c}\right)\times\dot{\mathbf{u}}\right]\right). \tag{10.293}$$

Similarly,

$$\frac{4\pi}{\mu_0\,e}\mathbf{B} = \nabla\times\frac{\mathbf{u}}{s} = -\frac{\nabla' s\times\mathbf{u}}{s^2} - \frac{\mathbf{r}}{s\,c}\times\left(\frac{\dot{\mathbf{u}}}{s} - \frac{\mathbf{u}}{s^2}\frac{\partial s}{\partial t'}\right), \tag{10.294}$$

or

$$\frac{4\pi}{\mu_0\,e}\mathbf{B} = -\frac{\mathbf{r}\times\mathbf{u}}{s^2\,r} - \frac{\mathbf{r}}{s\,c}\times\left[\frac{\dot{\mathbf{u}}}{s} + \frac{\mathbf{u}}{s^2}\left(\frac{\mathbf{r}\cdot\mathbf{u}}{r} + \frac{\mathbf{r}\cdot\dot{\mathbf{u}}}{c} - \frac{u^2}{c}\right)\right], \tag{10.295}$$

which reduces to

$$\frac{4\pi}{\mu_0\,e}\mathbf{B} = \frac{\mathbf{u}\times\mathbf{r}}{s^3}\left(1 - \frac{u^2}{c^2}\right) + \frac{1}{s^3\,c}\frac{\mathbf{r}}{r}\times\left(\mathbf{r}\times\left[\left(\mathbf{r} - \frac{r\,\mathbf{u}}{c}\right)\times\dot{\mathbf{u}}\right]\right). \tag{10.296}$$

A comparison of Equations (10.293) and (10.296) yields

$$\mathbf{B} = \frac{\mathbf{r}\times\mathbf{E}}{r\,c}. \tag{10.297}$$

Thus, the magnetic field is always perpendicular to **E** and the *retarded* radius vector **r**. Note that all terms appearing in the above formulae are retarded.

The electric field is composed of two separate parts. The first term in Equation (10.293) varies as $1/r^2$ for large distances from the charge. We can think of $\mathbf{r}_u = \mathbf{r} - r\,\mathbf{u}/c$ as the *virtual present radius vector*: *i.e.*, the radius vector directed from the position the charge would occupy at time t if it had continued with uniform velocity from its retarded position to the field point. In terms of $\mathbf{r}_u$, the $1/r^2$ field is simply

$$\mathbf{E}_{\text{induction}} = \frac{e}{4\pi\epsilon_0}\frac{1 - u^2/c^2}{s^3}\,\mathbf{r}_u. \tag{10.298}$$

We can rewrite the expression (10.216) for the electric field generated by a *uniformly* moving charge in the form

$$\mathbf{E} = \frac{e}{4\pi\epsilon_0}\frac{1 - u^2/c^2}{r_0^3\,(1 - u^2/c^2 + u_r{}^2/c^2)^{3/2}}\,\mathbf{r}_0, \tag{10.299}$$

where $\mathbf{r}_0$ is the radius vector directed from the *present* position of the charge at time t to the field point, and $u_r = \mathbf{u}\cdot\mathbf{r}_0/r_0$. For the case of uniform motion, the relationship between the retarded radius vector $\mathbf{r}$ and the actual radius vector $\mathbf{r}_0$ is simply

$$\mathbf{r}_0 = \mathbf{r} - \frac{r}{c}\,\mathbf{u}. \tag{10.300}$$

It is straightforward to demonstrate that

$$s = r_0\sqrt{1 - u^2/c^2 + u_r{}^2/c^2} \tag{10.301}$$

in this case. Thus, the electric field generated by a uniformly moving charge can be written

$$\mathbf{E} = \frac{e}{4\pi\epsilon_0}\frac{1 - u^2/c^2}{s^3}\,\mathbf{r}_0. \tag{10.302}$$

Since $\mathbf{r}_u = \mathbf{r}_0$ for the case of a uniformly moving charge, it is clear that Equation (10.298) is equivalent to the electric field generated by a uniformly moving charge located at the position the charge would occupy if it had continued with uniform velocity from its retarded position.

The second term in Equation (10.293),

$$\mathbf{E}_{\text{radiation}} = \frac{e}{4\pi\epsilon_0\,c^2}\frac{\mathbf{r}\times(\mathbf{r}_u\times\dot{\mathbf{u}})}{s^3}, \tag{10.303}$$

is of order $1/r$, and, therefore, represents a radiation field. Similar considerations hold for the two terms of Equation (10.296).

10.24 THE LARMOR FORMULA

Let us transform to the inertial frame in which the charge is instantaneously at rest at the origin at time $t = 0$. In this frame, $u \ll c$, so that $\mathbf{r}_u \simeq \mathbf{r}$ and $s \simeq r$ for events which are sufficiently close to the origin at $t = 0$ that the retarded charge still appears to travel with a velocity which is small compared to that of light. It follows from the previous section that

$$\mathbf{E}_{\rm rad} \simeq \frac{e}{4\pi\epsilon_0\,c^2}\frac{\mathbf{r}\times(\mathbf{r}\times\dot{\mathbf{u}})}{r^3}, \qquad (10.304)$$

$$\mathbf{B}_{\rm rad} \simeq \frac{e}{4\pi\epsilon_0\,c^3}\frac{\dot{\mathbf{u}}\times\mathbf{r}}{r^2}. \qquad (10.305)$$

Let us define spherical polar coordinates whose axis points along the direction of instantaneous acceleration of the charge. It is easily demonstrated that

$$E_\theta \simeq \frac{e}{4\pi\epsilon_0\,c^2}\frac{\sin\theta}{r}\,\dot{u}, \qquad (10.306)$$

$$B_\phi \simeq \frac{e}{4\pi\epsilon_0\,c^3}\frac{\sin\theta}{r}\,\dot{u}\,. \qquad (10.307)$$

These fields are identical to those of a radiating dipole whose axis is aligned along the direction of instantaneous acceleration (see Section 9.2). The radial Poynting flux is given by

$$\frac{E_\theta\,B_\phi}{\mu_0} = \frac{e^2}{16\pi^2\epsilon_0\,c^3}\frac{\sin^2\theta}{r^2}\,\dot{u}^2\,. \qquad (10.308)$$

We can integrate this expression to obtain the instantaneous power radiated by the charge

$$P = \frac{e^2}{6\pi\epsilon_0\,c^3}\,\dot{u}^2\,. \qquad (10.309)$$

This is known as *Larmor's formula*. Note that zero net momentum is carried off by the fields (10.306) and (10.307).

In order to proceed further, it is necessary to prove two useful theorems. The first theorem states that if a 4-vector field T^μ satisfies

$$\partial_\mu T^\mu = 0, \qquad (10.310)$$

and if the components of T^μ are non-zero only in a finite spatial region, then the integral over 3-space,

$$I = \int T^4 \, d^3x, \tag{10.311}$$

is an invariant. In order to prove this theorem, we need to use the 4-dimensional analog of Gauss's theorem, which states that

$$\int_V \partial_\mu T^\mu \, d^4x = \oint_S T^\mu \, dS_\mu, \tag{10.312}$$

where dS_μ is an element of the 3-dimensional surface S bounding the 4-dimensional volume V. The particular volume over which the integration is performed is indicated in Figure 10.1. The surfaces A and C are chosen so that the spatial components of T^μ vanish on A and C. This is always possible because it is assumed that the region over which the components of T^μ are non-zero is of finite extent. The surface B is chosen normal to the x^4-axis, whereas the surface D is chosen normal to the $x^{4'}$-axis. Here, the x^μ and the $x^{\mu'}$ are coordinates in two arbitrarily chosen inertial frames. It follows from Equation (10.312) that

$$\int T^4 \, dS_4 + \int T^{4'} \, dS_{4'} = 0. \tag{10.313}$$

Here, we have made use of the fact that $T^\mu \, dS_\mu$ is a scalar and, therefore, has the same value in all inertial frames. Since $dS_4 = -d^3x$ and

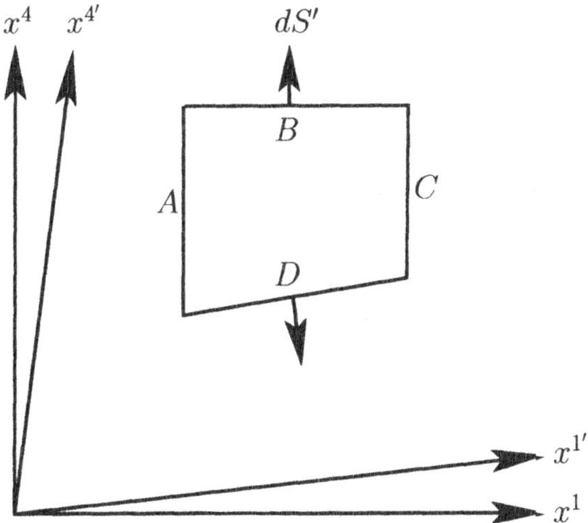

Figure 10.1: *An application of Gauss' theorem.*

$dS_{4'} = d^3x'$ it follows that $I = \int T^4\, d^3x$ is an invariant under a Lorentz transformation. Incidentally, the above argument also demonstrates that I is constant in time (just take the limit in which the two inertial frames are identical).

The second theorem is an extension of the first. Suppose that a 4-tensor field $Q^{\mu\nu}$ satisfies

$$\partial_\mu Q^{\mu\nu} = 0, \tag{10.314}$$

and has components which are only non-zero in a finite spatial region. Let A_μ be a 4-vector whose coefficients do not vary with position in space-time. It follows that $T^\mu = A_\nu\, Q^{\mu\nu}$ satisfies Equation (10.310). Therefore,

$$I = \int A_\nu\, Q^{4\nu}\, d^3x \tag{10.315}$$

is an invariant. However, we can write

$$I = A_\mu\, B^\mu, \tag{10.316}$$

where

$$B^\mu = \int Q^{4\mu}\, d^3x. \tag{10.317}$$

It follows from the quotient rule that if $A_\mu\, B^\mu$ is an invariant for arbitrary A_μ then B^μ must transform as a constant (in time) 4-vector.

These two theorems enable us to convert differential conservation laws into integral conservation laws. For instance, in differential form, the conservation of electrical charge is written

$$\partial_\mu J^\mu = 0. \tag{10.318}$$

However, from Equation (10.313) this immediately implies that

$$Q = \frac{1}{c}\int J^4\, d^3x = \int \rho\, d^3x \tag{10.319}$$

is an invariant. In other words, the total electrical charge contained in space is both constant in time, and the same in all inertial frames.

Suppose that S is the instantaneous rest frame of the charge. Let us consider the electromagnetic energy tensor $T^{\mu\nu}$ associated with all of the radiation emitted by the charge between times $t = 0$ and $t = dt$. According to Equation (10.261), this tensor field satisfies

$$\partial_\mu T^{\mu\nu} = 0, \tag{10.320}$$

apart from a region of space of measure zero in the vicinity of the charge. Furthermore, the region of space over which $T^{\mu\nu}$ is non-zero is clearly finite, since we are only considering the fields emitted by the charge in a small time interval, and these fields propagate at a finite velocity. Thus, according to the second theorem,

$$P^{\mu} = \frac{1}{c}\int T^{4\mu}\, d^3x \tag{10.321}$$

is a 4-vector. It follows from Section 10.22 that we can write $P^{\mu} = (d\mathbf{p}, dE/c)$, where $d\mathbf{p}$ and dE are the total momentum and energy carried off by the radiation emitted between times $t = 0$ and $t = dt$, respectively. As we have already mentioned, $d\mathbf{p} = 0$ in the instantaneous rest frame S. Transforming to an arbitrary inertial frame S', in which the instantaneous velocity of the charge is u, we obtain

$$dE' = \gamma(u)\left(dE + u\, dp^1\right) = \gamma\, dE. \tag{10.322}$$

However, the time interval over which the radiation is emitted in S' is $dt' = \gamma\, dt$. Thus, the instantaneous power radiated by the charge,

$$P' = \frac{dE'}{dt'} = \frac{dE}{dt} = P, \tag{10.323}$$

is the same in all inertial frames.

We can make use of the fact that the power radiated by an accelerating charge is Lorentz invariant to find a relativistic generalization of the Larmor formula, (10.309), which is valid in all inertial frames. We expect the power emitted by the charge to depend only on its 4-velocity and 4-acceleration. It follows that the Larmor formula can be written in Lorentz invariant form as

$$P = -\frac{e^2}{6\pi\epsilon_0\, c^3}\, A_{\mu}A^{\mu}, \tag{10.324}$$

since the 4-acceleration takes the form $A^{\mu} = (\dot{\mathbf{u}}, 0)$ in the instantaneous rest frame. In a general inertial frame,

$$-A_{\mu}A^{\mu} = \gamma^2\left(\frac{d\gamma}{dt}\,\mathbf{u} + \gamma\,\dot{\mathbf{u}}\right)^2 - \gamma^2 c^2\left(\frac{d\gamma}{dt}\right)^2, \tag{10.325}$$

where use has been made of Equation (10.107). Furthermore, it is easily demonstrated that

$$\frac{d\gamma}{dt} = \gamma^3\,\frac{\mathbf{u}\cdot\dot{\mathbf{u}}}{c^2}. \tag{10.326}$$

It follows, after a little algebra, that the relativistic generalization of Larmor's formula takes the form

$$P = \frac{e^2}{6\pi\epsilon_0\, c^3}\,\gamma^6\left[\dot{\mathbf{u}}^2 - \frac{(\mathbf{u}\times\dot{\mathbf{u}})^2}{c^2}\right]. \tag{10.327}$$

10.25 RADIATION LOSSES

Radiation losses often determine the maximum achievable energy in a charged particle accelerator. Let us investigate radiation losses in various different types of accelerator device using the relativistic Larmor formula.

For a linear accelerator, the motion is one-dimensional. In this case, it is easily demonstrated that

$$\frac{dp}{dt} = m_0\,\gamma^3\,\dot{u}, \tag{10.328}$$

where use has been made of Equation (10.326), and $p = \gamma\, m_0\, u$ is the particle momentum in the direction of acceleration (the x-direction, say). Here, m_0 is the particle rest mass. Thus, Equation (10.327) yields

$$P = \frac{e^2}{6\pi\epsilon_0\, m_0^2\, c^3}\left(\frac{dp}{dt}\right)^2. \tag{10.329}$$

The rate of change of momentum is equal to the force exerted on the particle in the x-direction, which, in turn, equals the change in the energy, E, of the particle per unit distance. Consequently,

$$P = \frac{e^2}{6\pi\epsilon_0\, m_0^2\, c^3}\left(\frac{dE}{dx}\right)^2. \tag{10.330}$$

Thus, in a linear accelerator, the radiated power depends on the external force acting on the particle, and not on the actual energy or momentum of the particle. It is obvious, from the above formula, that light particles, such as electrons, are going to radiate a lot more than heavier particles, such as protons. The ratio of the power radiated to the power supplied by the external sources is

$$\frac{P}{dE/dt} = \frac{e^2}{6\pi\epsilon_0\, m_0^2\, c^3}\,\frac{1}{u}\,\frac{dE}{dx} \simeq \frac{e^2}{6\pi\epsilon_0\, m_0\, c^2}\,\frac{1}{m_0\, c^2}\,\frac{dE}{dx}, \tag{10.331}$$

since $u \simeq c$ for a highly relativistic particle. It is clear, from the above expression, that the radiation losses in an electron linear accelerator are negligible unless the gain in energy is of order $m_e c^2 = 0.511$ MeV in a distance of $e^2/(6\pi\epsilon_0\, m_e\, c^2) = 1.28 \times 10^{-15}$ meters. That is 3×10^{14} MeV/meter. Typical energy gains are less than 10 MeV/meter. It follows, therefore, that radiation losses are completely negligible in linear accelerators, whether for electrons, or for other heavier particles.

The situation is quite different in circular accelerator devices, such as the synchrotron and the betatron. In such machines, the momentum **p** changes rapidly in direction as the particle rotates, but the change in energy per revolution is small. Furthermore, the direction of acceleration is always perpendicular to the direction of motion. It follows from Equation (10.327) that

$$P = \frac{e^2}{6\pi\epsilon_0\, c^3}\,\gamma^4\,\dot{u}^2 = \frac{e^2}{6\pi\epsilon_0\, c^3}\,\frac{\gamma^4\, u^4}{\rho^2}, \tag{10.332}$$

where ρ is the orbit radius. Here, use has been made of the standard result $\dot{u} = u^2/\rho$ for circular motion. The radiative energy loss per revolution is given by

$$\delta E = \frac{2\pi\,\rho}{u}\,P = \frac{e^2}{3\epsilon_0\, c^3}\,\frac{\gamma^4\, u^3}{\rho}. \tag{10.333}$$

For highly relativistic ($u \simeq c$) electrons, this expression yields

$$\delta E(\text{MeV}) = 8.85 \times 10^{-2}\,\frac{[E(\text{GeV})]^4}{\rho(\text{meters})}. \tag{10.334}$$

In the first electron synchrotrons, $\rho \sim 1$ meter, $E_{max} \sim 0.3$ GeV. Hence, $\delta E_{max} \sim 1$ keV per revolution. This was less than, but not negligible compared to, the energy gain of a few keV per turn. For modern electron synchrotrons, the limitation on the available radio-frequency power needed to overcome radiation losses becomes a major consideration, as is clear from the E^4 dependence of the radiated power per turn.

10.26 ANGULAR DISTRIBUTION OF RADIATION

In order to calculate the angular distribution of the energy radiated by an accelerated charge, we must think carefully about what is meant by the *rate of radiation* of the charge. This quantity is actually the amount of energy lost by the charge in a retarded time interval dt' during the emission of the signal. Thus,

$$P(t') = -\frac{dE}{dt'}, \tag{10.335}$$

where E is the energy of the charge. The Poynting vector

$$\frac{\mathbf{E}_{\rm rad}\times\mathbf{B}_{\rm rad}}{\mu_0} = \epsilon_0\,c\,E_{\rm rad}^{\,2}\,\frac{\mathbf{r}}{r}, \tag{10.336}$$

where use has been made of $\mathbf{B}_{\rm rad} = (\mathbf{r}\times\mathbf{E}_{\rm rad})/r\,c$ [see Equation (10.297)], represents the energy flux per unit actual time, t. Thus, the energy loss rate of the charge into a given element of solid angle $d\Omega$ is

$$\frac{dP(t')}{d\Omega}\,d\Omega = -\frac{dE(\theta,\varphi)}{dt'}\,d\Omega = \frac{dE(\theta,\varphi)}{dt}\,\frac{dt}{dt'}\,r^2\,d\Omega = \epsilon_0\,c\,E_{\rm rad}^{\,2}\,\frac{s}{r}\,r^2\,d\Omega, \tag{10.337}$$

where use has been made of Equation (10.283). Here, θ and φ are angular coordinates used to locate the element of solid angle. It follows from Equation (10.303) that

$$\frac{dP(t')}{d\Omega} = \frac{e^2\,r}{16\pi^2\,\epsilon_0\,c^3}\,\frac{[\mathbf{r}\times(\mathbf{r}_u\times\dot{\mathbf{u}})]^2}{s^5}. \tag{10.338}$$

Consider the special case in which the direction of acceleration coincides with the direction of motion. Let us define spherical polar coordinates whose axis points along this common direction. It is easily demonstrated that, in this case, the above expression reduces to

$$\frac{dP(t')}{d\Omega} = \frac{e^2\,\dot{u}^2}{16\pi^2\,\epsilon_0\,c^3}\,\frac{\sin^2\theta}{[1-(u/c)\,\cos\theta]^5}. \tag{10.339}$$

In the non-relativistic limit, $u/c \to 0$, the radiation pattern has the familiar $\sin^2\theta$ dependence of dipole radiation. In particular, the pattern is symmetric in the forward ($\theta < \pi/2$) and backward ($\theta > \pi/2$) directions. However, as $u/c \to 1$, the radiation pattern becomes more and more concentrated in the forward direction. The angle $\theta_{\rm max}$ for which the intensity is a maximum is

$$\theta_{\rm max} = \cos^{-1}\left[\frac{1}{3\,u/c}\left(\sqrt{1+15\,u^2/c^2}-1\right)\right]. \tag{10.340}$$

This expression yields $\theta_{\rm max} \to \pi/2$ as $u/c \to 0$, and $\theta_{\rm max} \to 1/(2\gamma)$ as $u/c \to 1$. Thus, for a highly relativistic charge, the radiation is emitted in a narrow cone whose axis is aligned along the direction of motion. In this case, the angular distribution (10.339) reduces to

$$\frac{dP(t')}{d\Omega} \simeq \frac{2\,e^2\,\dot{u}^2}{\pi^2\,\epsilon_0\,c^3}\,\gamma^8\,\frac{(\gamma\,\theta)^2}{[1+(\gamma\,\theta)^2]^5}. \tag{10.341}$$

The total power radiated by the charge is obtained by integrating Equation (10.339) over all solid angles. We obtain

$$P(t') = \frac{e^2\,\dot{u}^2}{8\pi\,\epsilon_0\,c^3}\int_0^{\pi}\frac{\sin^3\theta\,d\theta}{[1-(u/c)\,\cos\theta]^5} = \frac{e^2\,\dot{u}^2}{8\pi\,\epsilon_0\,c^3}\int_{-1}^{+1}\frac{(1-\mu^2)\,d\mu}{[1-(u/c)\,\mu]^5}. \tag{10.342}$$

It is easily verified that

$$\int_{-1}^{+1}\frac{(1-\mu^2)\,d\mu}{[1-(u/c)\,\mu]^5} = \frac{4}{3}\,\gamma^6. \tag{10.343}$$

Hence,

$$P(t') = \frac{e^2}{6\pi\,\epsilon_0\,c^3}\,\gamma^6\,\dot{u}^2, \tag{10.344}$$

which agrees with Equation (10.327), provided that $\mathbf{u}\times\dot{\mathbf{u}} = 0$.

10.27 SYNCHROTRON RADIATION

Synchrotron radiation (*i.e.*, radiation emitted by a charged particle constrained to follow a circular orbit by a magnetic field) is of particular importance in Astrophysics, since much of the observed radio frequency emission from supernova remnants and active galactic nuclei is thought to be of this type.

Consider a charged particle moving in a circle of radius a with constant angular velocity ω_0. Suppose that the orbit lies in the x-y plane. The radius vector pointing from the center of the orbit to the retarded position of the charge is defined

$$\rho = a\,(\cos\phi,\,\sin\phi,\,0), \tag{10.345}$$

where $\phi = \omega_0\,t'$ is the angle subtended between this vector and the x-axis. The retarded velocity and acceleration of the charge take the form

$$\mathbf{u} = \frac{d\rho}{dt'} = u\,(-\sin\phi,\,\cos\phi,\,0), \tag{10.346}$$

$$\dot{\mathbf{u}} = \frac{d\mathbf{u}}{dt'} = -\,\dot{u}\,(\cos\phi,\,\sin\phi,\,0), \tag{10.347}$$

where $u = a\,\omega_0$ and $\dot{u} = a\,\omega_0^2$. The observation point is chosen such that the radius vector $\mathbf{r}$, pointing from the retarded position of the charge to the observation point, is parallel to the y-z plane. Thus, we can write

$$\mathbf{r} = r\,(0,\, \sin\alpha,\, \cos\alpha), \tag{10.348}$$

where α is the angle subtended between this vector and the z-axis. As usual, we define θ as the angle subtended between the retarded radius vector $\mathbf{r}$ and the retarded direction of motion of the charge $\mathbf{u}$. It follows that

$$\cos\theta = \frac{\mathbf{u}\cdot\mathbf{r}}{u\,r} = \sin\alpha\,\cos\phi. \tag{10.349}$$

It is easily seen that

$$\dot{\mathbf{u}}\cdot\mathbf{r} = -\,\dot{u}\,r\,\sin\alpha\,\sin\phi. \tag{10.350}$$

A little vector algebra shows that

$$[\mathbf{r}\times(\mathbf{r}_u\times\dot{\mathbf{u}})]^2 = -(\mathbf{r}\cdot\dot{\mathbf{u}})^2\,r^2\,(1-u^2/c^2) + \dot{u}^2\,r^4\,(1-\mathbf{r}\cdot\mathbf{u}/r\,c)^2, \tag{10.351}$$

giving

$$[\mathbf{r}\times(\mathbf{r}_u\times\dot{\mathbf{u}})]^2 = \dot{u}^2\,r^4\left[\left(1-\frac{u}{c}\cos\theta\right)^2 - \left(1-\frac{u^2}{c^2}\right)\tan^2\phi\,\cos^2\theta\right]. \tag{10.352}$$

Making use of Equation (10.337), we obtain

$$\frac{dP(t')}{d\Omega} = \frac{e^2\,\dot{u}^2}{16\pi^2\epsilon_0\,c^3}\,\frac{[1-(u/c)\,\cos\theta)]^2 - (1-u^2/c^2)\tan^2\phi\,\cos^2\theta}{[1-(u/c)\,\cos\theta]^5}. \tag{10.353}$$

It is convenient to write this result in terms of the angles α and ϕ, instead of θ and ϕ. After a little algebra we obtain

$$\frac{dP(t')}{d\Omega} = \frac{e^2\,\dot{u}^2}{16\pi^2\epsilon_0\,c^3}\,\frac{[1-(u^2/c^2)]\cos^2\alpha + [(u/c)-\sin\alpha\,\cos\phi]^2}{[1-(u/c)\,\sin\alpha\,\cos\phi]^5}. \tag{10.354}$$

Let us consider the radiation pattern emitted in the plane of the orbit: *i.e.*, $\alpha = \pi/2$, with $\cos\phi = \cos\theta$. It is easily seen that

$$\frac{dP(t')}{d\Omega} = \frac{e^2\,\dot{u}^2}{16\pi^2\epsilon_0\,c^3}\,\frac{[(u/c)-\cos\theta]^2}{[1-(u/c)\,\cos\theta]^5}. \tag{10.355}$$

In the non-relativistic limit, the radiation pattern has a $\cos^2\theta$ dependence. Thus, the pattern is like that of dipole radiation where the axis is aligned along the instantaneous direction of acceleration. As the charge becomes more relativistic, the radiation lobe in the forward direction (*i.e.*, $0 < \theta < \pi/2$) becomes more focused and more intense. Likewise, the radiation lobe in the backward direction (*i.e.*, $\pi/2 < \theta < \pi$) becomes more diffuse. The radiation pattern has zero intensity at the angles

$$\theta_0 = \cos^{-1}(u/c). \tag{10.356}$$

These angles demark the boundaries between the two radiation lobes. In the non-relativistic limit, $\theta_0 = \pm\pi/2$, so the two lobes are of equal angular extents. In the highly relativistic limit, $\theta_0 \to \pm 1/\gamma$, so the forward lobe becomes highly concentrated about the forward direction ($\theta = 0$). In the latter limit, Equation (10.355) reduces to

$$\frac{dP(t')}{d\Omega} \simeq \frac{e^2\,\dot{u}^2}{2\pi^2\epsilon_0\,c^3}\,\gamma^6\,\frac{[1-(\gamma\,\theta)^2]^2}{[1+(\gamma\,\theta)^2]^5}. \tag{10.357}$$

Thus, the radiation emitted by a highly relativistic charge is focused into an intense beam, of angular extent $1/\gamma$, pointing in the instantaneous direction of motion. The maximum intensity of the beam scales like γ^6.

Integration of Equation (10.354) over all solid angle (making use of $d\Omega = \sin\alpha\,d\alpha\,d\phi$) yields

$$P(t') = \frac{e^2}{6\pi\epsilon_0\,c^3}\,\gamma^4\,\dot{u}^2, \tag{10.358}$$

which agrees with Equation (10.327), provided that $\mathbf{u}\cdot\dot{\mathbf{u}} = 0$. This expression can also be written

$$\frac{P}{m_0\,c^2} = \frac{2}{3}\frac{\omega_0^2\,r_0}{c}\,\beta^2\,\gamma^4, \tag{10.359}$$

where $r_0 = e^2/(4\pi\epsilon_0\,m_0\,c^2) = 2.82\times 10^{-15}$ meters is the *classical electron radius*, m_0 is the rest mass of the charge, and $\beta = u/c$. If the circular motion takes place in an orbit of radius a, perpendicular to a magnetic field $\mathbf{B}$, then ω_0 satisfies $\omega_0 = e\,B/m_0\,\gamma$. Thus, the radiated power is

$$\frac{P}{m_0\,c^2} = \frac{2}{3}\left(\frac{e\,B}{m_0}\right)^2\frac{r_0}{c}\,(\beta\,\gamma)^2, \tag{10.360}$$

and the radiated energy ΔE per revolution is

$$\frac{\Delta E}{m_0\,c^2} = \frac{4\pi\,r_0}{3\,a}\,\beta^3\,\gamma^4. \tag{10.361}$$

Let us consider the frequency distribution of the emitted radiation in the highly relativistic limit. Suppose, for the sake of simplicity, that the observation point lies in the plane of the orbit (*i.e.*, $\alpha = \pi/2$). Since the radiation emitted by the charge is beamed very strongly in the charge's instantaneous direction of motion, a fixed observer is only going to see radiation (at some later time) when this direction points almost directly toward the point of observation. This occurs once every rotation period, when $\phi \simeq 0$, assuming that $\omega_0 > 0$. Note that the point of observation is located many orbit radii away from the center of the orbit along the positive y-axis. Thus, our observer sees short periodic pulses of radiation from the charge. The repetition frequency of the pulses (in radians per second) is ω_0. Let us calculate the duration of each pulse. Since the radiation emitted by the charge is focused into a narrow beam of angular extent $\Delta\theta \sim 1/\gamma$, our observer only sees radiation from the charge when $\phi \lesssim \Delta\theta$. Thus, the observed pulse is emitted during a time interval $\Delta t' = \Delta\theta/\omega_0$. However, the pulse is received in a somewhat shorter time interval

$$\Delta t = \frac{\Delta\theta}{\omega_0}\left(1 - \frac{u}{c}\right), \tag{10.362}$$

because the charge is slightly closer to the point of observation at the end of the pulse than at the beginning. The above equation reduces to

$$\Delta t \simeq \frac{\Delta\theta}{2\,\omega_0\,\gamma^2} \sim \frac{1}{\omega_0\,\gamma^3}, \tag{10.363}$$

since $\gamma \gg 1$ and $\Delta\theta \sim 1/\gamma$. The width $\Delta\omega$ of the pulse in frequency space obeys $\Delta\omega\,\Delta t \sim 1$. Hence,

$$\Delta\omega = \gamma^3\,\omega_0. \tag{10.364}$$

In other words, the emitted frequency spectrum contains harmonics up to γ^3 times that of the cyclotron frequency, ω_0.

10.28 EXERCISES

10.1. Consider two Cartesian reference frames, S and S', in the standard configuration. Suppose that S' moves with constant velocity $v < c$ with respect to S along their common x-axis. Demonstrate that the Lorentz transformation between coordinates in the two frames can be written

$$x' = x \cosh\varphi - ct \sinh\varphi,$$

$$y' = y,$$

$$z' = z,$$

$$ct' = ct \cosh\varphi - x \sinh\varphi,$$

where $\tanh\varphi = v/c$. Show that the above transformation is equivalent to a rotation through an angle $i\,\varphi$, in the x–$i\,ct$ plane, in $(x, y, z, i\,ct)$ space.

10.2. Show that, in the standard configuration, two successive Lorentz transformations with velocities v_1 and v_2 are equivalent to a single Lorentz transformation with velocity

$$v = \frac{v_1 + v_2}{1 + v_1\, v_2/c^2}.$$

10.3. Let $\mathbf{r}$ and $\mathbf{r}'$ be the displacement vectors of some particle in the Cartesian reference frames S and S′, respectively. Suppose that frame S′ moves with velocity $\mathbf{v}$ with respect to frame S. Demonstrate that a general Lorentz transformation takes the form

$$\mathbf{r}' = \mathbf{r} + \left[\frac{(\gamma - 1)\,\mathbf{r}\cdot\mathbf{v}}{v^2} - \gamma\, t\right]\mathbf{v},$$

$$t' = \gamma\left[t - \frac{\mathbf{r}\cdot\mathbf{v}}{c^2}\right], \tag{10.365}$$

where $\gamma = (1 - v^2/c^2)^{-1/2}$. If $\mathbf{u} = d\mathbf{r}/dt$ and $\mathbf{u}' = d\mathbf{r}'/dt'$ are the particle's velocities in the two reference frames, respectively, demonstrate that a general velocity transformation is written

$$\mathbf{u}' = \frac{\mathbf{u} + \left[(\gamma - 1)\,\mathbf{u}\cdot\mathbf{v}/c^2 - \gamma\right]\mathbf{v}}{\gamma\,(1 - \mathbf{u}\cdot\mathbf{v}/c^2)}.$$

10.4. Let v be the Earth's approximately constant orbital speed. Demonstrate that the direction of starlight incident at right-angles to the Earth's instantaneous direction of motion appears slightly shifted in the Earth's instantaneous rest frame by an angle $\theta = \sin^{-1}(v/c)$. This effect is known as the *abberation* of starlight. Estimate the magnitude of θ (in arc seconds).

10.5. Let $\mathbf{E}$ and $\mathbf{B}$ be the electric and magnetic field, respectively, in some Cartesian reference frame S. Likewise, let $\mathbf{E}'$ and $\mathbf{B}'$ be the electric and magnetic field, respectively, in some other Cartesian frame S′, which moves with velocity $\mathbf{v}$ with respect to S. Demonstrate that the general transformation of fields takes the form

$$\mathbf{E}' = \gamma\,\mathbf{E} + \frac{1-\gamma}{v^2}\,(\mathbf{v}\cdot\mathbf{E})\,\mathbf{v} + \gamma\,(\mathbf{v}\times\mathbf{B}),$$

$$\mathbf{B}' = \gamma\,\mathbf{B} + \frac{1-\gamma}{v^2}\,(\mathbf{v}\cdot\mathbf{B})\,\mathbf{v} - \frac{\gamma}{c^2}\,(\mathbf{v}\times\mathbf{E}),$$

where $\gamma = (1 - v^2/c^2)^{-1/2}$.

10.6. A particle of rest mass m and charge e moves relativistically in a uniform magnetic field of strength B. Show that the particle's trajectory is a helix aligned along the direction of the field, and that the particle drifts parallel to the field at a uniform velocity, and gyrates in the plane perpendicular to the field with constant angular velocity

$$\Omega = \frac{e\,B}{\gamma\,m}.$$

Here, $\gamma = (1 - v^2/c^2)^{-1/2}$, and v is the particle's (constant) speed.

10.7. Let $P = \mathbf{E}\cdot\mathbf{B}$ and $Q = c^2B^2 - E^2$. Prove the following statements, assuming that E and B are not both zero.

(a) At any given event, $\mathbf{E}$ is perpendicular to $\mathbf{B}$ either in all frames of reference, or in none. Moreover, each of the three relations $E > cB$, $E = cB$, and $E < cB$ holds in all frames or in none.

(b) If $P = Q = 0$ then the field is said to be *null*. For a null field, $\mathbf{E}$ is perpendicular to $\mathbf{B}$, and $E = cB$, in all frames.

(c) If $P = 0$ and $Q \neq 0$ then there are infinitely many frames (with a common relative direction of motion) in which $E = 0$ or $B = 0$, according as $Q > 0$ or $Q < 0$, and none other. Precisely one of these frames moves in the direction $\mathbf{E}\times\mathbf{B}$, its velocity being E/B or c^2B/E, respectively.

(d) If $P \neq 0$ then there are infinitely many frames (with a common direction of motion) in which $\mathbf{E}$ is parallel to $\mathbf{B}$, and none other. Precisely one of these moves in the direction $\mathbf{E}\times\mathbf{B}$, its velocity being given by the smaller root of the quadratic equation $\beta^2 - R\,\beta + 1 = 0$, where $\beta = v/c$, and $R = (E^2 + c^2B^2)/|\mathbf{E}\times c\mathbf{B}|$. In order for β to be real we require $R > 2$. Demonstrate that this is always the case.

10.8. In the rest frame of a conducting medium, the current density satisfies Ohm's law $\mathbf{j}' = \sigma\,\mathbf{E}'$, where σ is the conductivity, and primes denote quantities in the rest frame.

(a) Taking into account the possibility of convection currents, as well as conduction currents, show that the covariant generalization of Ohm's law is

$$J^{\mu} - \frac{1}{c^2}(U_{\nu}J^{\nu})\,U^{\mu} = \frac{\sigma}{c}F^{\mu\nu}U_{\nu},$$

where U^{μ} is the 4-velocity of the medium, J^{μ} the 4-current, and $F^{\mu\nu}$ the electromagnetic field tensor.

(b) Show that if the medium has a velocity $\mathbf{v} = c\,\boldsymbol{\beta}$ with respect to some inertial frame then the 3-vector current in that frame is

$$\mathbf{j} = \gamma\,\sigma\,[\mathbf{E} + \boldsymbol{\beta}\times c\mathbf{B} - (\boldsymbol{\beta}\cdot\mathbf{E})\,\boldsymbol{\beta}] + \rho\,\mathbf{v}$$

where ρ is the charge density observed in the inertial frame.

10.9. Consider the relativistically covariant form of Maxwell's equations in the presence of magnetic monopoles. Demonstrate that it is possible to define a proper-4-current

$$J^{\mu} = (\mathbf{j},\, \rho\,c),$$

and a pseudo-4-current

$$J_m = (\mathbf{j}_m,\, \rho_m\,c),$$

where $\mathbf{j}$ and ρ are the flux and density of electric charges, respectively, whereas $\mathbf{j}_m$ and ρ_m are the flux and density of magnetic monopoles, respectively. Show that the conservation laws for electric charges and magnetic monopoles take the form

$$\partial_{\mu}J^{\mu} = 0,$$

$$\partial_{\mu}J^{\mu}_m = 0,$$

respectively. Finally, if $F^{\mu\nu}$ is the electromagnetic field tensor, and $G^{\mu\nu}$ its dual, show that Maxwell's equations are equivalent to

$$\partial_{\mu}F^{\mu\nu} = \frac{J^{\nu}}{\epsilon_0\,c},$$

$$\partial_{\mu}G^{\mu\nu} = \frac{J^{\nu}_m}{\epsilon_0\,c}.$$

10.10. Prove that the electromagnetic energy tensor satisfies the following two identities:

$$T^{\mu}{}_{\mu} = 0,$$

and

$$T^{\mu}{}_{\sigma}T^{\sigma}{}_{\nu} = \frac{I^2}{4}\,\delta^{\mu}_{\nu},$$

where

$$I^2 = \left(\frac{B^2}{\mu_0} - \epsilon_0 E^2\right)^2 + \frac{4\epsilon_0}{\mu_0}(\mathbf{E}\cdot\mathbf{B})^2.$$

10.11. A charge e moves in simple harmonic motion along the z-axis, such that its retarded position is $z(t') = a\cos(\omega_0 t')$.

(a) Show that the instantaneous power radiated per unit solid angle is

$$\frac{dP(t')}{d\Omega} = \frac{e^2 c\,\beta^4}{16\pi^2\epsilon_0\,a^2}\,\frac{\sin^2\theta\,\cos^2(\omega_0 t')}{[1+\beta\,\cos\theta\,\sin(\omega_0 t')]^5},$$

where $\beta = a\,\omega_0/c$, and θ is a standard spherical polar coordinate.

(b) By time averaging, show that the average power radiated per unit solid angle is

$$\frac{dP}{d\Omega} = \frac{e^2 c\,\beta^4}{128\pi^2\epsilon_0\,a^2}\left[\frac{4+\beta^2\cos^2\theta}{(1-\beta^2\cos^2\theta)^{7/2}}\right]\sin^2\theta.$$

(c) Sketch the angular distribution of the radiation for non-relativistic and ultra-relativistic motion.

10.12. The trajectory of a relativistic particle of charge e and rest mass m in a uniform magnetic field $\mathbf{B}$ is a helix aligned with the field. Let the pitch angle of the helix be α (so, $\alpha = 0$ corresponds to circular motion). By arguments similar to those used for synchrotron radiation, show that an observer far from the charge would detect radiation with a fundamental frequency

$$\omega_0 = \frac{\Omega}{\cos^2\alpha},$$

where $\Omega = e\,B/\gamma\,m$, and that the spectrum would extend up to frequencies of order

$$\omega_c = \gamma^3\,\Omega\,\cos\alpha.$$

Appendix A Physical Constants

Constant	Symbol	Value	Units
Electron Charge	e	-1.6022×10^{-19}	C
Electron Mass	m_e	9.1094×10^{-31}	kg
Proton Mass	m_p	1.6726×10^{-27}	kg
Speed of Light in Vacuum	c	2.9979×10^{8}	$\mathrm{m\,s^{-1}}$
Permittivity of Free Space	ϵ_0	8.8542×10^{-12}	$\mathrm{F\,m^{-1}}$
Permeability of Free Space	μ_0	$4\pi \times 10^{-7}$	$\mathrm{H\,m^{-1}}$
Gravitational Constant	G	6.6726×10^{-11}	$\mathrm{m^3\,s^{-1}\,kg^{-1}}$

Appendix B — Useful Vector Identities

Notation: **a**, **b**, **c**, **d** are general vectors; ϕ, ψ are general scalar fields; **A**, **B** are general vector fields.

$$\mathbf{a} \times (\mathbf{b} \times \mathbf{c}) = (\mathbf{a} \cdot \mathbf{c})\,\mathbf{b} - (\mathbf{a} \cdot \mathbf{b})\,\mathbf{c}$$

$$(\mathbf{a} \times \mathbf{b}) \times \mathbf{c} = (\mathbf{c} \cdot \mathbf{a})\,\mathbf{b} - (\mathbf{c} \cdot \mathbf{b})\,\mathbf{a}$$

$$(\mathbf{a} \times \mathbf{b}) \cdot (\mathbf{c} \times \mathbf{d}) = (\mathbf{a} \cdot \mathbf{c})\,(\mathbf{b} \cdot \mathbf{d}) - (\mathbf{a} \cdot \mathbf{d})\,(\mathbf{b} \cdot \mathbf{c})$$

$$(\mathbf{a} \times \mathbf{b}) \times (\mathbf{c} \times \mathbf{d}) = (\mathbf{a} \times \mathbf{b} \cdot \mathbf{d})\,\mathbf{c} - (\mathbf{a} \times \mathbf{b} \cdot \mathbf{c})\,\mathbf{d}$$

$$\nabla \cdot \nabla \times \mathbf{A} = 0$$

$$\nabla \times \nabla\phi = 0$$

$$\nabla^2\mathbf{A} = \nabla\,(\nabla \cdot \mathbf{A}) - \nabla \times \nabla \times \mathbf{A}$$

$$\nabla(\phi\,\psi) = \phi\,\nabla\psi + \psi\,\nabla\phi$$

$$\nabla(\mathbf{A} \cdot \mathbf{B}) = \mathbf{A} \times (\nabla \times \mathbf{B}) + \mathbf{B} \times (\nabla \times \mathbf{A}) + (\mathbf{A} \cdot \nabla)\mathbf{B} + (\mathbf{B} \cdot \nabla)\mathbf{A}$$

$$\nabla \cdot (\phi\,\mathbf{A}) = \phi\,\nabla \cdot \mathbf{A} + \mathbf{A} \cdot \nabla\phi$$

$$\nabla \cdot (\mathbf{A} \times \mathbf{B}) = \mathbf{B} \cdot \nabla \times \mathbf{A} - \mathbf{A} \cdot \nabla \times \mathbf{B}$$

$$\nabla \times (\phi\,\mathbf{A}) = \phi\,\nabla \times \mathbf{A} + \nabla\phi \times \mathbf{A}$$

$$\nabla \times (\mathbf{A} \times \mathbf{B}) = \mathbf{A}\,(\nabla \cdot \mathbf{B}) - \mathbf{B}\,(\nabla \cdot \mathbf{A}) + (\mathbf{B} \cdot \nabla)\mathbf{A} - (\mathbf{A} \cdot \nabla)\mathbf{B}$$

Appendix C: Gaussian Units

In 1960 physicists throughout the world adopted the so-called SI (short for the *Système International*) system of units , whose standard measures of length, mass, time, and electric charge are the meter, kilogram, second, and coloumb, respectively. Nowadays, the SI system is employed, almost exclusively, in most areas of Physics. In fact, only one area of Physics has proved at all resistant to the adoption of SI units, and that, unfortunately, is Electromagnetism, where the previous system of units, the so-called *Gaussian* system, is still widely used. Incidentally, the standard units of length, mass, time, and electric charge in the Gaussian system are the centimeter, gram, second, and statcoloumb, respectively.

Why would anyone wish to adopt a different set units in Electromagnetism to that used in most other branches of Physics? The answer is that the laws of Electromagnetism look a lot prettier in the Gaussian system than in the SI system. In particular, there are no ϵ_0 s and μ_0 s in any of the formulae. In fact, within the Gaussian system, the only normalizing constant appearing in Maxwell's equations is the velocity of light in vacuum, c. However, there is a severe price to pay for the aesthetic advantages of the Gaussian system.

Electromagnetic formulae can be converted from the SI to the Gaussian system via the following transformations

$$\epsilon_0 \rightarrow \frac{1}{4\pi}, \tag{C.1}$$

$$\mu_0 \rightarrow \frac{4\pi}{c^2}, \tag{C.2}$$

$$B \rightarrow \frac{B}{c}. \tag{C.3}$$

Transformation (C.3) also applies to quantities which are directly related to magnetic field-strength, such as vector potential. Converting electromagnetic formulae from the Gaussian to the SI system (or any other system) is far less straightforward. As an example of this, consider Coulomb's law in SI units:

$$\mathbf{f}_2 = \frac{q_1\, q_2}{4\pi\epsilon_0} \frac{\mathbf{r}_2 - \mathbf{r}_1}{|\mathbf{r}_2 - \mathbf{r}_1|^3}. \tag{C.4}$$

Employing the above transformation, this formula converts to

$$\mathbf{f}_2 = q_1\, q_2 \frac{\mathbf{r}_2 - \mathbf{r}_1}{|\mathbf{r}_2 - \mathbf{r}_1|^3} \tag{C.5}$$

in Gaussian units. However, applying the inverse transformation is problematic. In Equation (C.5), the geometric 4π in the SI formula has canceled with the $1/4\pi$ obtained from transforming ϵ_0 to give unity. Unfortunately, it is not at all obvious that the inverse transformation should generate a factor $4\pi\epsilon_0$ in the denominator. In order to understand the origin of this difficulty, it is necessary to consider *dimensionality*.

There are four fundamental quantities in Electrodynamics: mass, length, time, and charge, denoted M, L, T, and Q, respectively. Each of these quantities has its own particular units, since mass, length, time, and charge are fundamentally different (in a physical sense) from one another. The units of a general physical quantity, such as force or capacitance, can always be expressed as some appropriate power law combination of the four fundamental units, M, L, T, and Q. Now, all laws of Physics, and all equations derived from such laws, must be *dimensionally consistent* (*i.e.*, all terms on the left-hand and right-hand sides must possess the same power law combination of the four fundamental units) in order to ensure that the laws of Physics are *independent of the choice of units* (which is, after all, completely arbitrary). Equation (C.4) makes dimensional sense because the constant ϵ_0 possesses the units $\mathrm{M}^{-1}\,\mathrm{L}^{-3}\,\mathrm{T}^2\,\mathrm{Q}^2$. Likewise, the Biot-Savart law only makes dimensional sense because the constant μ_0 possesses the units $\mathrm{M\,L\,Q}^{-2}$. On the other hand, Eq. (C.5) does not make dimensional sense: *i.e.*, the right-hand side and the left-hand side appear to possess different units. In fact, we can only reconcile Eqs. (C.4) and (C.5) if we divide the right-hand side of (C.5) by some constant, $4\pi\epsilon_0$, say, with dimensions $\mathrm{M}^{-1}\,\mathrm{L}^{-3}\,\mathrm{T}^2\,\mathrm{Q}^2$, which happens to have the numerical value *unity* for the particular choice of units in the Gaussian scheme. Likewise, the Gaussian version of the Biot-Savart law contains a hidden constant with the numerical value unity which nevertheless possesses *dimensions*. So, it can be seen that the apparent simplicity of the equations of Electrodynamics in the Gaussian scheme is only achieved at the expense of wrecking their dimensionality. It is difficult to transform out of the Gaussian scheme because the aforementioned hidden constants, which lurk in virtually all of the equations of Electromagnetism, do not necessarily have the value unity in other schemes.

Appendix D FURTHER READING

Foundations of Electromagnetic Theory, 3rd Edition, J.R. Reitz, F.J. Milford, R.W. Christy (Addison-Wesley, Reading MA, 1960).

Classical Electricity and Magnetism, 2nd Edition, W.K.H. Panofsky, M. Philips (Addison-Wesley, Reading MA, 1962).

Special Relativity, 2nd Edition, W. Rindler (Interscience, New York NY, 1966).

Electromagnetic Fields and Waves, 2nd Edition, P. Lorrain, D.R. Corson (W.H. Freeman & Co., San Francisco CA, 1970).

Classical Electrodynamics, 2nd Edition, J.D. Jackson (John Wiley & Sons, New York NY, 1975).

Electromagnetism, I.S. Grant, W.R. Phillips (John Wiley & Sons, Chichester UK, 1975).

Essential Relativity: Special, General, and Cosmological, 2nd Rev. Edition, W. Rindler (Springer-Verlag, New York NY, 1980).

Introduction to Electrodynamics, 2nd Edition, D.J. Griffiths (Prentice-Hall, Englewood Cliffs NJ, 1989).

Classical Electromagnetic Radiation, 3rd Edition, M.A. Heald, J.B. Marion (Saunders College Publishing, Fort Worth TX, 1995).

Classical Electrodynamics, W. Greiner (Springer-Verlag, New York NY, 1998).

Index

U

V

CPSIA information can be obtained
at www.ICGtesting.com
Printed in the USA

9 781934 015209